ABYSSAL ENVIRONMENT
AND ECOLOGY OF
THE WORLD OCEANS

ABYSSAL ENVIRONMENT AND ECOLOGY OF THE WORLD OCEANS

ROBERT J. MENZIES

Professor of Oceanography
Florida State University

ROBERT Y. GEORGE

Adjunct Professor of Oceanography
Florida State University

GILBERT T. ROWE

Assistant Scientist
Woods Hole Oceanographic Institution

A WILEY-INTERSCIENCE PUBLICATION

JOHN WILEY & SONS, New York • London • Sydney • Toronto

Library of Congress Cataloging in Publication Data

Menzies, Robert James.
  Abyssal environment and ecology of the world oceans.

  "A Wiley-Interscience publication."
  Bibliography: p.
  1. Marine ecology. 2. Abyssal zone.
I. George, Robert Y., joint author. II. Rowe,
Gilbert Thomas, 1942–   joint author. III. Title.
QH541.5.S3M4   574.92   72-8780
ISBN 0-471-59440-7

Printed in the United States of America

10-9 8 7 6 5 4 3 2 1

Foreword

The enormous amount of progress in the field of abyssal oceanology during the last 10-year period required critical reconsideration of some of the old conceptions on the antiquity, the origin, and the peculiar biology of the abyssal fauna. There was also a need to synthesize clearly the continually increasing knowledge concerning this immense and important field of research.

These are the aims of the present volume, and we may anticipate that the authors have achieved both their tasks by presenting us with a clear, comprehensive, and convincing image of the various problems raised by the abyssal fauna, including paleogeography, ecology, adaption, migration, and division into zones.

Many papers concerning some or certain groups of abyssal animals have been published during the last 10 to 20 years (to mention here but Ivanov's mono-graph entitled "POGONOPHORA," 1960, in Russian; 1963 in English, or those on the "ISOPODA" by R. J. Menzies, R. Y. George, T. Wolff, and Ya. A. Birstein). Only a few papers treat the overall problem of abyssal fauna, namely: *Tiefseebuch* by Correns et al. (1934); Unesco's *On the Distribution and Origin of the Deepsea Bottom Fauna* (Naples, 1954); a chapter by A. Bruun in *Treatise on Marine Ecology and Palaeoecology* (1957), a popular volume telling of the voyage around the world of the research vessel *GALATHEA*. Finally, there is *The Bottom Fauna of the Greatest Depths of the Ocean* by G. M. Belyaev (1966), a first summation of the problem, based especially on the rich material collected by the Soviet expedition.

The present volume is written by three well-known experts in the abyssal fauna nearby the American coasts, and particularly by Prof. Robert Menzies, who led many expeditions in both the Northwest Atlantic and the Southeast Pacific oceans, as well as in the Antarctic Ocean, to explore the deep-sea abysses.

Prof. Menzies and his collaborators, Dr. R. Y. George and Dr. G. T. Rowe, are already well known as experts in abyssal Isopoda and epifauna.

From the very first chapter, the authors establish the concept of variable depths limiting the benthal deep-sea fauna, in contrast to any arbitrary and constant bathy-metric limit. The starting point of this deep-sea fauna cannot remain at the 2000-m isobath as old concepts established; rather, it oscillates according to the regions, depending on the water types and currents. The currents determine also the granulo-metry and the composition of sediments between 1000 and 3000 m. Neither the 10° C temperature (A. Bruun, 1957) nor the 4°C temperature (Medov) can char-acterize the beginning (or the upper limit) of the abyssal province.

The book aims also to establish an original bentho-abyssal scheme of the divi-sion into zones of the world ocean bentho-abyssal animals, correlated to many peculiar physical and biological phenomena (temperatures, currents, sedimentary requirements, biomasses, standing stock, etc.).

In the authors' original interpretations there are many varieties, such as emergence or polar submergence, the maximum density of the mesoabyssal species, and the relatively recent origin of the ultra-abyssal fauna derived from the archibenthal zone. These concepts are very different from the commonly accepted ones, and to demonstrate them the authors use a vast material of abyssal isopods. Most of their evidence comes from off the Carolinas, the Peru–Chile Trench, and the Antarctic regions; the material was often collected by themselves and compared with many samples from Arctic regions (collected by other explorers). Prof. Menzies and his colleagues have also used many submarine photographs that show the spreading and the density of the fauna *in situ*.

In the elaboration of the original points of view, which are distributed almost throughout the 14 chapters of the volume, the authors used both the elements of their numerous works already published and many unpublished data, particularly in connection with Isopoda.

The original observations Menzies made in 1965, in the Peru Trench, and more recently those of Menzies, George, and Rowe in North Carolina, serve as main bases of comparison and as sources for many of their conclusions.

The work begins with a very instructive and objective history of the evolution of abyssal-biology research, starting with the English (J. Ross, Forbes, Thomson, and Murray), the Norwegian (M. Sars and G. O. Sars), and the American pioneers (A. Agassiz), and ending with the present countries' endeavors in the research of this field, the United States and the Soviet Union being in first place.

This history is followed closely by the evolution of certain original concepts on the abyssal fauna, as well as other concepts presented in later chapters. The authors, for instance, not only share the conception of Murray and Agassiz of the migration of shallow-water fauna toward the depths but, by means of arguments and examples, they give the idea new dimensions and bases.

In the second chapter, the collecting methodology and the preparation of bentho-abyssal samples are critically analyzed. The important qualities of the Menzies trawl, which I have seen myself as a participant in the eleventh Cruise of the R/V *Anton Bruun,* must be also emphasized. The dredge is excellent for collecting of certain microfauna elements.

The opinions concerning the inefficiency of present types of quantitative sampling devices and of the washing methods are very objective indeed, and an ideal grab type is also proposed. As the only happy solution of such a grab, however, I would choose the adaptation of the Campbell or Okean grabs or the Băcescu-grab type of sampler.

Special attention was given to deep-bottom photography as a technique facilitating the research on the deep sea. The authors are practiced experimenters with this technique and have contributed to its improvement. A good photo of the bottom may indicate the macrofauna that several trawls are not able to collect.

We are given the first complete description of the deep-bottom trawling technique as well as a view of the disadvantages of the respective operations. The comparative data regarding the quality of various trawl types and of various sieve sizes are quite instructive.

Proper collecting of the abyssal microbenthos remains a problem for the future.

To define the abyssal division into provinces and zones, the authors adapted Ekman's point of view that the maximum change of fauna is much more concerned than the physicochemical conditions or the bathymetry. They elaborate a method to detect faunal change and also present Savage's difference coefficient. Based on extensive experience, the authors found the distribution of benthal isopods as ideal indicators for the bentho-abyssal zones.

Giving various examples, the authors convince us of the reality of emergence and submergence. For instance, 87% of the Peru-Trench isopod genera are present too, all along the continental limit of the Antarctic, and this is probably a question of Antarctic emergence.

The species living in the archibenthal or even in the abyssal zones in the tropics emerge toward the continental shelf in the poles. Similarly, Pogonophora appear to emerge in the northern climate and to submerge in lower latitudes.

In the last part of the book, in order to demonstrate the old character and the zoogeographic spreading of the abyssal fauna, the authors make a learned incursion in the climatological marine evolution, using the latest seismic data of sediment cover.

They establish as fundamental features of the changes that have occurred the extinction and the evolution of the various animal groups, further proving that the abyssal zone has changed its environment throughout the geological times and that the change of temperatures (+30° C in Carboniferous; +15° C in Tethys) and of sediments under the current actions or under the absence of oxygen led to the changing of communities.

Today the province of the continental Peruvian plateau is completely isolated from its abyssal province. The richer and more ancient Antarctic fauna differs from the poorer Arctic fauna; there are very few endemisms, and primary productivity and oligomixic biocenoses are reduced. The authors' explanation is that the Arctic is considered a recent environment, in contradistinction to the Antarctic, which is an ancient one.

Starting from Vinogradova's scheme, they propose a new scheme to delimit the abyssal provinces, and to define it they combine the temperature with the bottom topography, introducing new and judicious divisions into zones, such as the Peruvian zone and the Trench zone of North and Middle America.

Perhaps it would also be useful to consider as "C_2" one Southwest African province (i.e., the Benguela current zone), and the Afro-Malgache as another. Later on, the rich Arctic fauna of the Tethyan times disappeared and, unlike the Antarctic, endemic genera appeared no more. The Arctic fauna lacks the intertidal zone, and rich fauna extends like a belt around the continental margins of the Arctic. The Arctic fauna is a Cenozoic–Pleistocene fauna which migrated on a Paleozoic geosyncline from the Pacific Ocean and, more recently, mainly from the Atlantic Ocean.

Of special interest are the tables of the various animal groups found in the Arctic and Antarctic oceans compared with those of the neighboring oceans at the same depth, and reaching to the comparison of even the genera and the species of some animals.

The definition of the "Oceanic Trench" is completed by such ideas as seismic zones and negative gravity anomalies (due to Ewing and others).

The authors discuss also the criteria according to which the hadal fauna was separated—at 6000-m depth—from the abyssal fauna, considering as a reality only the absence of certain animal groups from the various trenches. The morphological differences between hadal and abyssal species, the eye-reduction, the size increase, and the oligomixic biocenoses are considered as relative and not clearly distinct. Otherwise, the deep-sea fauna should be a gradually differentiated derivative of the abyssal depths. Thus the authors avoid using terms like ultra-abyssal and hadal, placing fauna that might have been so described in the lower abyssal zone. I think it is not necessary to make this substitution because the authors are discussing Bruun's (1956) and Zenkevitch's (1968) peculiar hadal concept of the fauna, acknowledging the need of a proper term for the bottom fauna of deep-sea abysses.

The term "hadal" fauna is good indeed, even though in its delimitation the bathymetric criterion is an arbitrary one, as it was in the case of the start of the abyssal benthos (as compared with the start of the archibenthal).

This book puts special emphasis on the problem of variety (diversity) as an ecological entity; the variety of the dominance increases with the increasing depth of the abyssal, which is a measurable and important characteristic of the abysso-benthal in the establishing of communities and of their zoogeography.

In a very interesting section of Chapter 14—"Average Age of the Deep-Sea Fauna"—the authors prove that the abyssal fauna was derived from the Jurassic, warm fauna (of Tethys type) or even later—in any case, earlier than the great majority of the abyssal zone researchers wish it to be.

The earlier Paleozoic forms are decreasing with the depth in proportion to the increase of the Tertiary representatives. The authors' statements are based not only on the study of Isopoda, but also on the latest data of the studies concerning Echinodermata (Madsen) and Mollusca (Clarke). As such the ultra-abyssal zone, with only few exceptions (*Neopilina* species being one of them), was not a refuge of the Paleozoic fauna.

The archibenthal zone must be considered as a transition zone and a zone of origin for the abyssal fauna within its space. It contains the most ancient marine fauna known today. Consequently, as the authors have shown, the whole present marine fauna is much more closely related to the Tertiary than to the Cretaceous or even to earlier fauna.

The abyssal communities, the feeding, the breeding, and the sight of the animals composing these communities, were also analyzed by the authors, who bring again new and valuable viewpoints. Many species follow a seasonal, cyclical development, in spite of the uniformity of their abyssal life conditions, and the authors consider this as evidence of such species' recent origin from the littoral part of the abyssal fauna.

Several features ensure this volume a place among the best in the field of current knowledge about the bentho-abyssal fauna. These include the idea of employing, in the analysis of the various problems raised by the abyssal bottom fauna, a group of animals of such a large bathymetrical spreading as the Isopoda (from the intertidal to more than 10-km depth), as well as the richness of original examples and the use of the latest studies of other animal groups. The original, varied illustrations (more than 100 figures) and particularly the bottom photographs, which are almost all original, are also valuable.

The new ideas and the bold directions so prominent in the 14 chapters of this fluently and clearly written book will surely stimulate controversies among abyssal zone researchers and form a source of suggestions for future oceanographers, by showing the inadequacies of the present research techniques and the gaps in our knowledge of the latest field of oceanology—the fauna of the deep-sea abysses.

Mihai C. Băcescu, Director

Musée D'Histoire Naturelle, Bucuresti
Republica Socialista Romania,
1971

Preface

~~~~~~
~~~~~~

It is commonly recognized that water covers 70.8% of the earth's surface. But it is a little-known fact that this water has an average depth of 3 km and that 85% of the sea bed lies at the depth in excess of 2 km. The area covered by the abyssal oceans is twice that of exposed land on earth. Below 2000 m the water takes on special characteristics. There is no sunlight, and living green plants are absent. The water is cold, with an average temperature near 2° C, and the bottom temperature ranges from −1.8°C at the Poles to +13°C in the Mediterranean. The salt content, however, is nearly the same everywhere, at 35 parts salt per thousand parts water. These environmental conditions are encountered nowhere else on the Earth.

The abyss has often been referred to as a constant environment. It is quite clearly the largest uniform environment on earth. We are interested in defining the major environmental features that influence the distribution of animals along the abyssal margins of the ocean, as well as in its greatest depths. We have found that this little-studied abyss has not been constant throughout geologic time, either in its temperature or in its sediments. It is our notion that, similarly, the unusual fauna has not been constant in character. For example, it is not known today when the oceans achieved their present volume. The Atlantic is believed to have started its evolution in Jurassic times, and it is known that Paleozoic and Mesozoic seas once covered most of the earth's surface and later receded into the oceans. In our opinion, changes of such magnitude must have affected the fauna of the abyss.

The information assembled strongly suggests that there is a distinct deep-sea fauna, but its depth limits vary so much between regions that a constant upper depth limit becomes meaningless. Recognition of this fundamental feature in the distribution of life in the sea should do much to reduce the arguments about the point or depth of the start of the deep-sea fauna, and it should eliminate the obviously convenient but meaningless 2000-m depth limit that heretofore has been commonly employed as the start of the abyssal fauna.

As might be expected, we have found some exceptionally interesting and tantalizing relationships between the animal zonation we have described and other biological phenomena such as species diversity, stenobathy and eurybathy, polar emergence and submergence, biomass, temperature requirements, breeding cycles, eye development, and origin and antiquity.

This study, which is to a large extent based on original data for four regions of the world oceans (Figure 1) has necessitated inquiry into the evolution of marine climates and marine sediments as they affect the abyssal fauna and its zonation. We believe it has also provided insights into some fundamental features of zoo-geography and the origin and antiquity of the abyssal fauna. Unsurprisingly, more

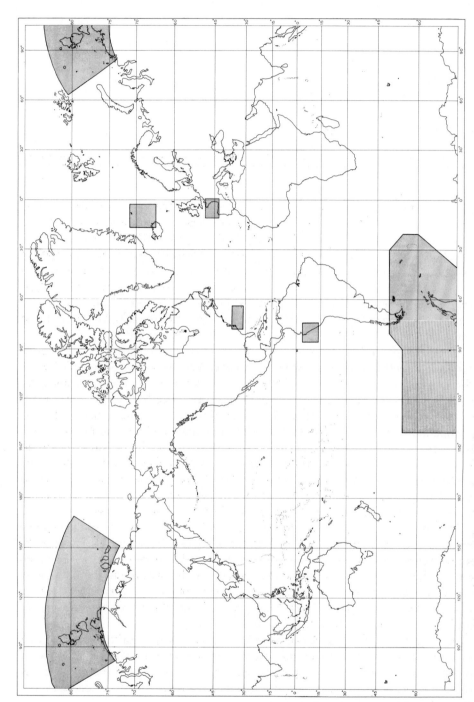

Figure 1. World map showing regions investigated in this study.

questions are raised than are answered, and quite clearly the data are not as comprehensive as such a massive task requires. But the significant trends are likely to remain regardless of the nomenclature used for them. At best we can only cite the quotation of Peterson (1914, p. 6) when he developed his thesis on level bottom communities:

> I venture to make this attempt well knowing how imperfect it must be, because it will have to be made sooner or later, and the sooner it is made, the sooner I trust, will it be repeated to better effect.

The original illustrations presented here of the large epifauna off North Carolina were prepared from dead animals captured in trawls and from underwater photographs showing the living animal *in situ*. In this presentation an enlargement of the photograph of the living animal is inserted onto the charcoal wash drawing, or vice versa. The drawings were made by a Duke University art student, Miss Marcia Johnson, under supervision of the writers. These *in situ* drawings are unique, and they constitute a far more exacting illustration of the deep-sea animals and their habitat than hitherto has been available in any text. Most earlier deep-sea habitat drawings are based on an artist's idea of the habitat and not on actual underwater photographs.

Underwater photographs were obtained from several sources. Among them were the United States Antarctic Research Program (USARP) and from Bullivant (1967) and Dearborn (1967). The photograph of Edward Forbes was furnished by the librarian of the Marine Biological Station, Port Erin, Isle of Man. Torben Wolff provided the photograph of Anton Bruun, and Lev A. Zenkevitch supplied information about the Soviet studies. Dr. Zina Filatova provided the photographs of the Soviet deep-sea scientists.

This book took its original sketchy form as a consequence of the several collaborative papers prepared in the 1960s by the senior author and Prof. Bruce C. Heezen at Columbia University. Unfortunately, continuing collaboration became impractical because of time and different interests, and the original plan was not realized.

We are especially indebted to Dr. Joel W. Hedgpeth of Oregon State University, who made a critical reading of the historical section and suggested several important corrections and additions. We are also indebted to Dr. John L. Mohr, of the University of Southern California, who appraised the chapters concerning the Arctic and the Antarctic. The critiques from our colleagues have done much to improve the text. Finally we are especially grateful to Dr. Mihai Băcescu and Mr. Gardner Soule, both of whom provided comments on the entire manuscript.

It is impractical to acknowledge everyone who contributed in some way to obtaining the data at sea; nevertheless, the senior author wishes to express his gratitude and esteem to Prof. Maurice Ewing, Director of the Lamont–Doherty Geological Observatory, Columbia University, for giving him the opportunity to learn firsthand about the abyss and its fauna. In many ways the stimulus provided by "Doc" Ewing kept alive the enthusiasm required for the generation of this text.

We realize that certain of our hypotheses are contrary to popularly held opinions about the abyss and its fauna. However, we hope that the data we present, if not totally convincing, will at least serve to stimulate the great amount of future research that is needed if man is to gain a cohesive body of knowledge of his greatest environment—the world's abyssal oceans.

ROBERT J. MENZIES
ROBERT Y. GEORGE
GILBERT T. ROWE

Tallahassee, Florida
Woods Hole, Massachusetts
June 1972

Contents

~~~~~~

CHAPTER 1

History of Deep-Sea Ecology (1815 to Present)

"What! 2,000 fathoms and no bottom!
  Ah, Doctor Carpenter, F.R.S!"
(Attributed to a pet parrot aboard the H.M.S. *Challenger*, according to Moseley.)

It is not known whether the parrot aboard the *Challenger* actually abbreviated "Fellow of the Royal Society" (F.R.S.). It is known that depths were determined by lead line which not infrequently, especially in deep water, failed to show bottom contact. This appears to have happened often enough to be translated into the vocabulary of the pet.

BEGINNINGS IN ENGLAND

The subject of deep-sea faunal zonation is of recent historical interest, being dependent entirely on man's ability to sample the sea floor. Our history is concerned with the men, their accomplishments, and the ideas they have developed. It has been most interesting for us to learn that each man who made a significant contribution was in direct contact with one of the others at some time of his life. The interconnections are shown in a diagrammatic way on Figure 1-1.

Edward Forbes

Edward Forbes (1815–1854; Figure 1-2), slightly more than a century and a quarter ago, initiated deep-sea studies without having taken a single sample from below 230 fathoms. At the age of 27 he virtually ceased his work on living marine animals and turned his interest to geologic problems, having contributed 65 published papers on the marine fauna (see Appendix). At the age of 39 he died of a fever contracted during his epochal study of the Aegean marine fauna and its zonation. His eight regions of depth in the Aegean Sea appear in Table 1-1.

In order to understand the important concepts elaborated by Forbes, it is essential to realize that organic evolution was not yet accepted. Darwin was aboard the H.M.S. *Beagle* between 1831 and 1835, and *The Origin of Species* was not published until after Forbes had died. Certain of Forbes's ideas were conflicting. He

1

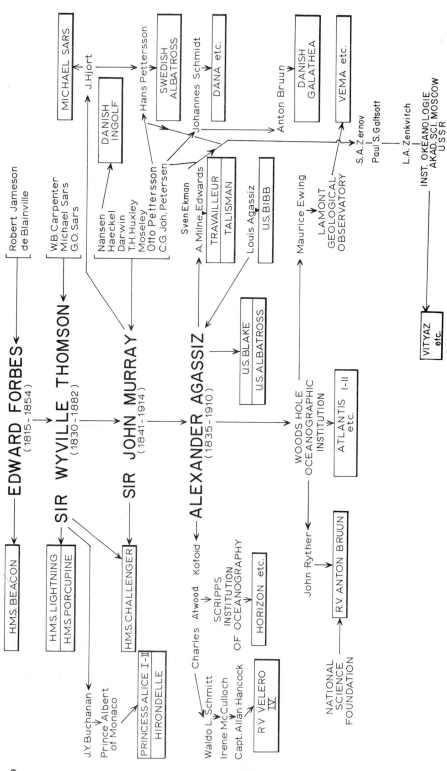

Figure 1-1. Diagrammatic history of deep-sea biological studies by men, ships, institutions, or expeditions between 1815 and the present. The ships are enclosed in blocks. The arrows indicate contacts between men.

2

Figure 1-2. Edward Forbes, from Wilson and Geike, 1961. Courtesy of the Librarian, Marine Biological Station, Port Erin, Isle of Man.

defined "representation" to mean that similar species living in different places were separately created. This is the antithesis of Darwinian organic evolution. However, Forbes' concept of "centers of origin" implies common descent of species. Because of the great interest in the azoic zone concept, certain other of Forbes' concepts have been more or less overlooked. Seven of them have significant bearing on the subject of animal zonation and the origin and distribution of marine life, and all are advocated by us with required contemporary changes. The concepts are cited here, often in the form presented by Forbes in his various writings (1846, 1854, 1859).

1. *Centers of origin:* All individuals composing a species descended from a single progenitor, or from two, according as the sexes might be united or distinct. Original pair created at a single spot, migrating to other localities. Discontinuous distribution due to formation of natural barriers between once common populations—or due to accidental transport.

2. *Representation:* In localities placed under similar circumstances, similar though specifically distinct forms were created.

3. *Maximum development of species:* "Every species has three maxima of development, in depth, in geographic space, in time. In depth, we find a species at first represented by few individuals, which become more and more numerous until

TABLE 1-1. DIAGRAM OF REGIONS OF DEPTH IN THE AEGEAN SEA*

Sea-Bottom = Deposits Forming	Region	Depth (fathoms)	Characteristic Animals and Plants
Extent—12 ft; ground various; usually rock or sandy (con-glomerates forming)	I	2	*Littorina coerulescens, Fasciolaria tarentina, Cardium edule* Plant: *Padina pavonia*
Extent—48 ft; muddy, sandy, rocky	II	10	*Cerithium vulgatum, Lucina lactea, Holothuriae* Plants: *Caulerpa* and *Zostera*
Extent—60 ft; ground mostly muddy or sandy; mud bluish	III	20	*Aplysiae, Cardium papillosum*
Extent—90 ft; ground mostly gravelly and weedy; muddy in estuaries	IV	35	*Ascidiae, Nucula emarginata, Cellaria ceramioides* Plants: *Dictyomenia volubilis, Codium bursa*
Extent—120 ft; ground nulli-porous and shelly	V	55	*Cardita aculeata, Nucula striata, Pecten opercularis, Myriapora truncata* Plant: *Rityphloea tinctoria*
Extent—144 ft; ground mostly nulliporous; rarely gravelly	VI	79	*Venus ovata, Turbo sanguineus, Pleurotoma maravignae, Cidaris histrix* Plant: *Nullipora*
Extent—156 ft; ground mostly nulliporous; rarely yellow mud	VII	105	*Brachiopoda, Rissoa reticulata, Pecten similis, Echinus monilis* Plant: *Nullipora*
Extent—750 ft; uniform bottom of yellow mud, abounding for the most part in remains of Pteropoda and Foraminifera	VIII	230	*Dentalium 5-angulare, Kellia abyssicola, Lingula profundissima, Pecten hoskynsi, Ophiura abyssicola, Idmonea Alecto*
Zero of animal life probably about 300 fathoms			Plants:
Mud without organic remains			

* From Forbes (1844), pp. 169–170.

they reach a certain point, after which they again diminish, and at length altogether disappear" (Forbes, 1854, p. 173).

4. *Zonation in the sea:* "Four marked zones of depth exist around all seaboards, each characterized by a distinct group of animals."

1. Littoral zone (space between the tide marks).
2. Laminarian zone (low water mark to about 15 fathoms—30 m).
3. Coralline zone (15–50 fathoms). Millipores, hydroids, and bryozoa.
4. Deep-sea corals: Forbes, 1859, Ed. R. G. Austen.

5. *Homoiozoic belts:* These are described as belts of animal life extending from one geographic region in shallow water to another region and hence connecting

the submerged fauna of one region with another. As described by Forbes, "A belt of nearly similar circumstances of climate extending through many degrees of longitude, but few of latitude."

6. "Extent of the range of a species in depth is correspondent with its geographical distribution" (Forbes, 1854, p. 171)—or, as stated today, eurybathyal species have the widest geographic distribution (Vinogradova, 1959).

7. "The alpine floras of Europe and Asia are fragments of a flora which was diffused from the North. The deep-sea fauna is in a like manner a fragment of the general glacial fauna" (after Forbes, 1846; see Herdman, 1923, p. 29). The marine "boreal outliers" of Forbes were assemblages of Northern species occupying the deeper areas of about 80 to 100 fathoms (Forbes, 1846).

THE AZOIC ZONE

The concept of an "azoic zone" was Forbes' most provocative contribution. Although it was incorrect, it has great historical interest because of its influence on marine explorations.

It is reasonable to believe that the azoic concept was a consequence of the inadequate collecting instrument available to Forbes and the impoverished nature of the marine fauna in the eastern Mediterranean. The survey ship H.M.S. *Beacon* was Forbes' research vessel for what was probably one of the world's first biological expedition to study the "deep-sea fauna."

Forbes used the naturalist's dredge first introduced by O. F. Müller in 1799 (Figure 2-1a). This little triangular, leather-mesh dredge was far from efficient, and its inadequacies as a collecting instrument must have contributed to Forbes's "discovery" of the "azoic zone" below 300 fathoms. Essentially, as he dredged deeper and deeper, he recovered fewer and fewer species. Extrapolating his findings, he

TABLE 1-2. LIST OF ANIMALS CAPTURED AT FORBES' DEEPEST AEGEAN STATION (AFTER FORBES, 1844)

Date: November 25, 1841
Locality: Southern extremity of Gulf of Macri, Asia Minor
Depth: 230 fathoms
Distance from shore: 1 mi (shore steep)
Ground: Fine yellowish mud
Region: VIII

Terebratula vitrea
Lingula profundissima
Arca imbricata
Dentalium quinquiangulare
*Hyalaea gibbosa**
*Cleodora pyramidata**
*Creseis spinifera**

"A grassy *Serpula* and the corneous tube of an *Annelide* adherred to the *Terebratula;* also a species of *Alecto* and a *Lobatula*."

---

* Now known to be planktonic, probably found as shells in sediments.

concluded:

> . . . within itself [8th region between 105 and 230 fathoms] the number of species of individuals diminishes as we descend, pointing to a zero in the distribution of animal life as yet unvisited. It [8th region; Table 1-2] can only be subdivided according to the disappearance of species which do not seem to be replaced by others (after Herdman, 1911).

It was, however, this singular statement that prompted many a successive deep-sea expedition up to and including the Danish *Galathea* (1950–1952). The Danish expedition settled the question with the recovery of animals from the greatest depth of the sea.

Forbes had used the dredge off the English coast earlier with such success that the British Association for the Advancement of Science formed a "Dredging Commission" to continue his work. The poem "Song of the Dredge" was one of Forbes' contributions to his Red Lion Clubbe (1839), a jovial congregation of colleagues.

SONG OF THE DREDGE

Hurrah for the dredge, with its iron edge, and its
    mystical triangle.
And its hided net with meshes set odd fishes to
    entangle!
The ship may move thro' the waves above, mid scenes
    exciting wonder,
But braver sights the dredge delights as it roves
    the waters under.

[First stanza, *Song of the Dredge*, E. Forbes,
Red Lion Clubbe, 1839 (After Herdman, 1923).]

Unknown to Forbes, Sir John Ross (1818) accidentally dredged up *Asterophyton linckii* Müller and Troschel (Figure 1-3) from a depth of 800 fathoms at 73°37′N lat. by 77°25′W long. while sounding the bottom. At this time (1818), Ross was engaged in his explorations for the Northwest Passage in the New World. In Norway, Michael Sars (1850) (Figure 1-4) published a list of 19 species living at depths greater than 300 fathoms. This listing was extended (1864) by his son G. O. Sars (Figure 1-5) to 92 species living between 200 and 300 fathoms. In 1860 the engineer Fleeming Jenkin retrieved a solitary coral *Caryophyllia borealis* Fleming (Figure 1-6); the coral was attached to a piece of submarine cable 40 mi from Sardinia in the Mediterranean at a depth of 1200 fathoms. The "azoic" zone had become "zoic"!

Sir Charles Wyville Thomson

The findings of Michael Sars (cited earlier) had a profound influence on the thoughts of Sir Charles Wyville Thomson (1830–1882; Figure 1-7). Thomson entered Edinburgh University at the age of 16. He knew Forbes and was familiar with the concept of an azoic zone. He visited Michael Sars at Christiania, Norway, and he saw firsthand the collections of marine animals from below 300 fathoms in the Lofoten fjords.

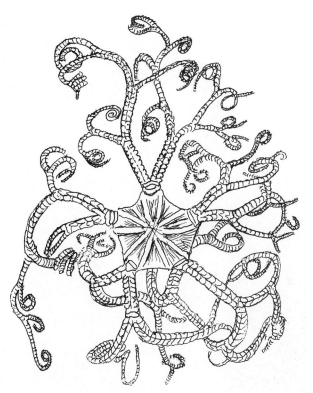

Figure 1-3.   *Asterophyton linckii* Müller and Troschel, the first deep-sea species captured by Sir John Ross (1819). Copied from Thomson (1874, Figure 1).

The crinoid *Rhizocrinus lofotensis* M. Sars (Figure 1-8) impressed him because it was a living fossil relict belonging to the Cretaceous family Apiocrinidae. This crinoid started Thomson on a lifetime search for living fossils in the sea. The lovely crinoid so stimulated Thomson that on his return to Britain he enlisted the aid of Dr. W. B. Carpenter, then Vice President of the Royal Society and his associate in the study of the embryology of the crinoid *Antedon,* to promote an expedition to explore deep water. Details of this event, resulting in the assignment of the H.M.S. *Lightning* to deep-sea exploration, were published by Thomson (1874).

LIGHTNING, PORCUPINE, AND CHALLENGER

Had the *Lightning* been awarded to men of lesser spirit and goals, it is doubtful that the subsequent events would have taken place, because as Thomson (*op cit.*) recorded, the cruise was no joyride. In his own words, the *Lightning* was a "cranky little vessel—perhaps the very oldest paddle steamer in Her Majesty's Navy." The boat leaked and the cruise was undertaken in foul weather. Nevertheless, the scientific crew consisting of Wyville Thomson, W. B. Carpenter, and his son Herbert Carpenter dredged to depths as great as 650 fathoms at station 15 southwest of the Faeroe Islands (Figure 1-9). Here the sea pen *Bathyptilum carpenteri* Kölliker was collected. Historically more important, negative bottom temperatures were recorded 10 to 30' north of 60°N, and positive temperatures were recorded south of 50°N. This was the first suggestion that a submarine ridge separates the North Atlantic from the

Figure 1-4. Professor Michael Sars. (After Murray and Hjort 1912.)

subzero Arctic bottom water. This finding regarding bottom temperature lead the Council of the Royal Society to apply for the use of another vessel to investigate water temperature and the gases of seawater. For this the H.M.S. *Porcupine* surveying vessel was assigned by the Admiralty, and three cruises were carried out in the summer of 1869 (Figure 1-9).

The first cruise of the *Porcupine* was under the scientific charge of Gwyn Jefferys, F.R.S., a malacologist and father-in-law of H. N. Moseley. On this cruise, between Scotland and Rockall, the depth of 1476 fathoms was successfully dredged at station 21 (Figure 1-9). The second cruise, led by Wyville Thomson, dredged to 2435 fathoms in the Bay of Biscay, station 37 (Figure 1-9). This sampling station was rich in animal life and was the deepest dredged at that time. The operation was repeated twice, with one failure. The collection at station 37 was described by Thomson (1874) as follows:

> On Thursday, July 22, the weather was still remarkably fine. The sea was moderate, with a slight swell from the northwest. We sounded in lat. 47°38'N, long. 12°08'W, in a depth of 2,435 fathoms (Station 37), when the average of the Miller-Casella thermometers gave a minimum temperature of 2.5°C.

Figure 1-5.   Professor G. O. Sars in his laboratory at Christiania University. Courtesy of Olaug M. Somme.

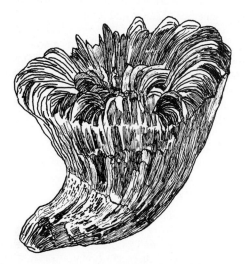

Figure 1-6.   Solitary coral, *Caryophyllia borealis* Fleming, obtained from a Mediterranean submarine cable at 1200 fathoms. Copied from Thomson (1874, Figure 4).

Figure 1-7. Sir Charles Wyville Thomson. From Hardman (1923).

As this was about the greatest depth which we had reason to expect in this neighbourhood, we prepared to take a cast of the dredge. This operation [was] rather a serious one in such deep water. . . . It was perfectly successful. The dredge-bag which was safely hauled on deck at 1 o'clock in the morning of the 23rd, after an absence of $7\frac{1}{4}$ hours and a journey of upwards of eight statute miles, contained $1\frac{1}{4}$ cwt. of very characteristic grey chalk-mud. The dredge appeared to have dipped rather deeply into the soft mud, as it contained amorphous paste with but a small proportion of fresh shells of *Globigerina* and *Orbulina*. There was an appreciable quantity of diffused amorphous organic matter, which we were inclined to regard as connected, whether as processes, or "mycelium", or germs, with the various shelled and shell-less Protozoa, mixed very likely with the apparently universally distributed moner of deep water, *Bathybius*.

On careful sifting, the ooze was found to contain fresh examples of each of the Invertebrate sub-kingdoms. When examined at daylight on the morning of the 23rd none of these were actually living, but their soft parts were perfectly fresh, and there was ample evidence of their having been living when they entered the dredge. The most remarkable species were:

MOLLUSCA—*Dentalium*, sp. n., of large size.
  Pecten fenestratus Forbes, a Mediterranean species.
  Dacrydium vitreum Torell; Arctic, Norwegian, and Mediterranean.
  Scrobicularia nitida Müller; Norwegian, British, and Mediterranean.
  Neaera obesa Lovén; Arctic and Norwegian.

Figure 1-8. The stalked crinoid *Rhizocrinus lofotensis* M. Sars. Copied from Thomson (1874, Figure 72).

CRUSTACEA — *Anonyx hölbollii Kroyer* (=*A. denticulatus* Bate), with the
    secondary appendage of the upper antennae longer and more slender
    than in shallow-water specimens.
    Ampelisca aequicornis Bruzelius.
    Munna, sp. n.

One or two annelides and gephyrea, which have not yet been determined.

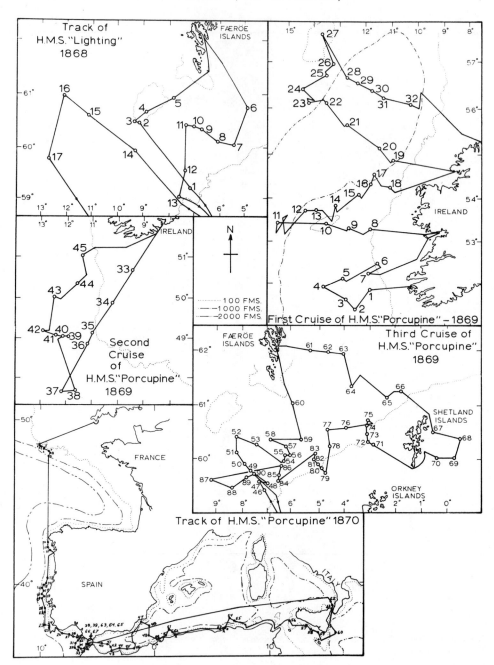

Figure 1-9. Cruise track and collecting stations of the H.M.S. *Lightning* and the H.M.S. *Porcupine*. (After Thomson 1874).

ECHINODERMATA—*Ophiocten sericeum* Forbes; several well-grown specimens.

Echinocucumis typica Sars. This seems to be a very widely distributed species; we got it in almost all our deep dredgings, both in the warm and in the cold areas.

A remarkable stalked crinoid allied to *Rhizocrinus,* but presenting some very marked differences.

POLYZOA—*Salicornaria,* sp. n.

COELENTERATA—Two fragments of a hydroid zoophyte.

PROTOZOA—Numerous foraminifera belonging to the groups already indicated as specially characteristic of these abyssal waters; together with a branching flexible rhizopod, having a chitinous cortex studded with globigerinae, which encloses a sarcodic medulla of olive-green hue. This singular organism, of which fragments had been detected in other dredgings, here presented itself in great abundance.

One or two small SPONGES, which seem to be referable to a new group.

On Friday, July 23, we tried another haul at the same depth; but when the dredge came up at 1:30 p.m. it was found that the rope had fouled and lapped right around the dredge-bag, and that there was nothing in the dredge. The dredge was sent down again at 3 p.m., and was brought up at 11 p.m., with upwards of 2 cwt. of ooze. We got from this haul a new species of *Pleurotoma* and one of *Dentalium; Scrobicularia nitida* Müller; *Dacrydium vitreum* Torell; *Ophiacantha spinulosa* M. and T.; and *Ophiocten kröyeri* Lutken; with a few Crustaceans and many foraminifera.

Note from Thomson's list of specimens that *Bathybius haeckeli* Huxley, which had been discovered in alcohol-preserved bottom muds collected by the H.M.S.

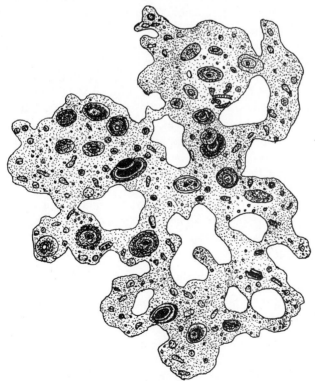

Figure 1-10. *Bathybius haeckeli* Huxley, the moner of the deep sea. Copied from Thomson (1874, Figure 63).

Cyclops in 1857, was considered to be an animal and was thought to permeate the ocean floors in ancestral dignity (Figure 1-10). The demise of *Bathybius* came during the *Challenger* expedition when the chemist J. Y. Buchanan found it to be a sulfate precipitate of alcohol and seawater.

Meanwhile, by the completion of the second cruise of the *Porcupine,* Thomson was satisfied that life would be found on the bottom even to the greatest depths. The third cruise, under Carpenter, investigated the fauna of the "warm" and "cold" areas. Besides confirming existence of life down to 2435 fathoms, the unusual temperature data of the *Porcupine* were confirmed, creating a perturbing puzzle regarding the cause of the negative temperatures adjacent to positive ones. To solve this problem the H.M.S. *Knight Errant,* with John Murray and Captain Tizard, conducted a topographic survey in the summer of 1880 to prove the existence of a submarine ridge rising to a depth of 300 fathoms and separating the Norway Basin of the Arctic Ocean from the North Atlantic.

In the summer of 1870 the *Porcupine* had cruised the Mediterranean (Figure 1-9), but the dredging results were so poor that Carpenter was led to believe, like Forbes, that the depths beyond a few hundred fathoms were nearly azoic. The benthic fauna, however, appeared to be more abundant near the African coast. These results encouraged the French expeditions of the *Travailleur* and the *Talisman* under the scientific guidance of Alphonse Milne-Edwards.

After the death of Thomson, Murray and Tizard, this time aboard the H.M.S. *Triton,* charted the submarine ridge that was named the Wyville Thomson Ridge (Faero–Iceland–Greenland Ridge). The discovery that it separates different communities of animals prompted several later expeditions, the most noteworthy being that of the Danish *Ingolf* (1895–1896). The contents of the *Ingolf* expedition reports are reproduced in the Appendix.

The correspondence of the Royal Society and the proposals for a circumnavigation of the world are recorded in great detail by Thomson (1874). The ship selected was the H.M.S. *Challenger* (Figure 1-11) and Mr. C. Wyville Thomson was appointed "chief scientist" of the cruise at an annual salary of £1000. Although he was quite obviously a guiding light in the various expeditions and a prolific writer, Wyville Thomson produced scarcely any lasting deep-sea ecologic concepts. His greatest scientific accomplishment, besides the conduct and organization of the *Challenger* expedition, lies in the discovery of the temperature differences across the ridge bearing his name.

The *Challenger* set sail from Sheerness on December 7, 1872, and returned to Sheerness on May 26, 1876 after $3\frac{1}{2}$ years circumnavigating the earth and sampling the deep sea. Her cruise track and stations are recorded in Figure 1-12. Wyville Thomson (in *Narrative of the Cruise,* 1885, p. 941) summarized the statistics:

> Between our departure from Sheerness on the 7th December 1872, and our arrival at Spithead on the 24th of May 1876 we traversed a distance of 68,890 nautical miles, and at intervals as nearly uniform as possible, we established 362 observing Stations.

There is no better means of conveying the scope and magnitude of the expedition in terms of its results than to cite the published Reports. The preparation of these was entrusted to Sir John Murray, who financed the publication of the Reports and directed the *Challenger* office after the death of Wyville Thomson. Work on the *Reports* commenced in 1876 and they were published three years later. The collec-

Figure 1-11. H.M.S. *Challenger*. (From Narrative, Vol. 1, Challenger Reports, Figure 1.)

tions are still being studied today. The citations to this great work, which has such obvious bearing on deep-sea animal zonation, are given in the Appendix.

Sir John Murray

Sir John Murray (1841–1914; Figure 1-13) entered the science of oceanography by "accident" when an appointed member of the *Challenger* expedition dropped out. Wyville Thomson had selected Murray as a naturalist, and he developed personal interests in plankton, deposits on the sea floor, and the origin and mode of formation of coral reefs. Murray, unlike Forbes, Thomson, and others, never became a professor in a university, but he made more scientific contributions to oceanography than did his advocate, Wyville Thomson, and like Sir Charles he was knighted in recognition of his works. Herdman (1923, p. 75) described him

> as the archmagician of the laboratory—a sort of modern scientific alchemist, bringing mysterious unknown things out of store-bottles, and then showing us how to demonstrate their true nature.

Murray knew personally Fridjof Nansen, Johan Hjort, Otto Pettersson, and C. G. Johannes Petersen, among the Scandinavians, and Alexander Agassiz of the United States; he was also known to such famous contributors to the *Challenger Reports* as Ernst Haeckel. His influence was felt by many men. With Hjort, Murray participated in the cruise of the *Michael Sars,* and he recorded the details of this voyage in his book written with Hjort, *The Depths of the Ocean* (1912).

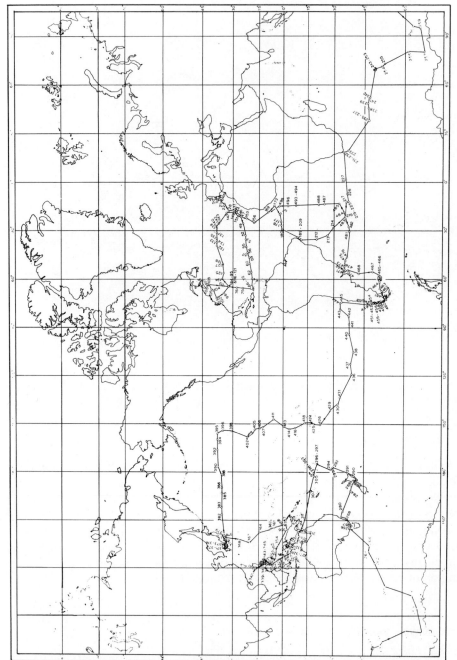

Figure 1-12. Cruise track and stations of the H.M.S. *Challenger*. Copied from Narrative, Vols 1 and 11.

Figure 1-13.   Sir John Murray. From Herdman (1923).

Murray's last expedition was on the *Michael Sars* with Hjort (Murray and Hjort, 1912). They were accompanied by Hans Pettersson, the son of Otto Pettersson, and this connection was culminated by the great Swedish deep-sea expedition of 1947–1948 (H. Pettersson, 1966), aboard the Swedish *Albatross* (Figure 1-14). Biological specimens collected by the *Michael Sars* were entrusted to several noted biologists. Among them was Prof. Carl Chun of Leipzig, who earlier led the German deep-sea expedition of 1898–1899 aboard the *Valdivia*.

Murray elaborated several interesting concepts (see Herdman, 1923):

1. The permanence of the continents.
2. Bipolar distribution of marine animals.
3. That the deep sea was not inhabited during the Cretaceous time when the surface seawater temperatures were 70 or 80°F.
4. That polar animals and deep-sea animals all have direct development. [Often attributed to Thorson.]
5. That the freshwater fauna is much more archaic than the deep-sea fauna.
6. That Darwin's subsidence theory of the formation of coral reefs was wrong.

EARLY AMERICAN BEGINNINGS: ALEXANDER AGASSIZ

Alexander Agassiz (1835–1910; Figure 1-15), the son of the renowned teacher and zoologist Louis Agassiz (1807–1873), was the American equivalent of Wyville Thomson. When he was young, his father introduced him to work at sea aboard the

Figure 1-14. The Swedish *Albatross*, 1947–1948. (After Pettersson.)

U.S. Survey Ship *Bibb*. Alexander Agassiz was well acquainted with the works of Forbes and collaborated on occasion with Alphonse Milne-Edwards. He met Murray at the *Challenger* office in London and was doubtless influenced by him as well as by Moseley. Both Agassiz and Murray were antagonists of Darwin's subsidence theory of coral reef formation, and both did a considerable amount of work on deep-sea deposits.

Agassiz had a great interest in deep-sea life and made many fundamental contributions regarding its origin, distribution, and nutrition. His persistent reference to "representation," a term constantly used by Forbes, implies a great familiarity with the works of Forbes. In close working arrangement with the U.S. Coast Survey, he managed one important deep sea investigation after another, from the *Blake* (Figure 1-16) to the U.S. *Albatross* (Figure 1-17). His book *Three Cruises of the Blake* (1888) is a standard reference for those interested in deep-sea exploration and biology, particularly off the United States. His contributions to deep-sea biological thought, which were based mainly on the collections of the *Blake* and the *Albatross*, cover the following general headings: food supply, pressure, bathymetric range and zonation, color, origin and antiquity, and distribution. Like his predecessor Wyville Thomson, who was directly responsible for the discovery of the Faero–Iceland–Greenland Ridge, Agassiz predicted the presence of the Mid-Atlantic Ridge.

It is unfortunate for deep-sea biology that both Agassiz and Murray ended their scientific careers with extensive investigations into the formation and distribution of shallow tropical coral reefs around the world. This proved to be a scientific blind

Figure 1-15. Alexander Agassiz. From Murray and Hjort (1912, p. 12).

Figure 1.16. The U.S. Coast Steamer *Blake*. (After Agassiz, 1888).

Figure 1.17.   The U.S. *Albatross* at anchor. (From C. H. Townsend, 1901, Pl. IV.)

alley, and the two men reinforced each other's opinions, failing to break new ground for scientific thought. The several concepts elaborated by Agassiz are given in his words.

>   *Bathymetrical range:* The bathymetrical range of these littoral species is small, and with few exceptions they do not live within the limits of either the continental or the abyssal areas, where flourish the fauna of the other two belts. The temperature of the continental belt is sufficiently low to crop out in high latitudes, and hence the large number of so-called arctic species which have been found to be quite common in moderate depths upon this so-called continental area; while the strictly deep-sea or abyssal forms rarely occur within the range of the continental, and still more rarely within that of the littoral belt [p. 302, Vol. I].

>   *Zones:* [Main outlines of bathymetrical faunal divisions.] They consist of a littoral fauna, all light, motion, and heat; a continental fauna, with a superabundance of food and an equable temperature; and a deep-sea fauna, having a cold, unvaried temperature, deriving its food largely from pelagic animals and plants. It is however impossible to determine zones of depth except in the most general way, because representatives of nearly all the principal groups characteristic of the deep-sea find their way up to higher levels, and vice versa [p. 162, Vol. I].

>   *Distribution:* There are many of the same types in the deep waters of the Atlantic, the Pacific, and the Southern Oceans—the survivors perhaps of an abyssal fauna, which thanks to the distributing agency of the equatorial currents, extended, in mesozoic times over the whole floor of the equatorial and part of the temperate zones [pp. 151–152, Vol. 1].

>   Under favorable conditions, animals having a wide geographic and geological distribution have been found in proximity to continental masses. The *Albatross* has dredged in deep water off Chesapeake Bay many deep-sea types first known from the European tertiaries, from the Arctic and Antarctic, from the shores of Europe and of South America, from the West Indies and even from India and the Pacific Ocean [p. 152, footnote, Vol. 1].

The fauna found at great depths in the ocean is peculiar, and appears to contain many species of extensive geographical range, and to be made up of a smaller number of representative species than is common in areas of lesser depth [p. 162, Vol. 1].

Origin: . . . the abyssal fauna has descended from the littoral and other shallow regions, to be acclimatized at great depths. The conditions of existence becoming more and more constant, or even in the deeper regions perfectly uniform, species of the most varied derivations, when they had once attained a certain zone, could spread everywhere. This explains at once how the deep-water fauna presents a very uniform composition in all regions of the globe, but at the same time includes various species the analogues of which lived in the sublittoral regions of both cold and hot climates, and may have sent an occasional wanderer into deeper waters [pp. 155–156, Vol. 1].

Antiquity: All the evidence thus far tends to show that the deep-sea fauna originated at the close of the paleozoic times [p. 151]. Since the temperature of the sea is nearly the same everywhere in deep water, we have a uniform cosmopolitan fauna, of considerable antiquity, at great depths, corresponding to that of high isolated peaks or mountain chains. They have preserved down to our time the remnants of a former fauna, which may once have been connected with a fauna at lower levels during an epoch of ice or of lower temperature [p. 164, Vol. 1].

We have (with color) a strong argument in favor of the gradual and comparatively recent migration of littoral forms into deep water in the fact that there are still so many vividly colored bathyssal animals belonging to all the classes of the animal kingdom, and possessing nearly all the hues found in types living in littoral waters [p. 311]. There is as great a diversity in color in the reds, oranges, greens, yellows, and scarlets of the deep-water starfishes and ophiurians as there is in those of our rocky or sandy shores. There is apparently in the abysses of the sea the same adaptation to the surroundings (color, commensalism) as upon the littoral zone. We meet with highly colored ophiurians within masses of sponges, themselves brilliantly colored, at a depth of more than 150 fathoms [p. 310].

Except for the book *Three Cruises of the Blake*, much of Agassiz' expedition material, especially that from the *Albatross*, is scattered widely in many scientific journals. Accordingly we have given a reasonably complete list of Agassiz' marine publications and have duplicated the list of publications resulting from the *Albatross* in the Appendix.

EARLY EUROPEAN CONTRIBUTORS

Certain of the early European works have already been cited with reference to the development of the *Challenger* expedition and the *Challenger Reports*.

French Naturalists

According to the cogent review of Le Danois (1948), Audouin and Milne-Edwards undertook the first oceanographic research when they used a dredge along the coast of France in 1830. After a lapse of half a century, work was continued on the *Travailleur* by a "Commission of Dredgers" composed of such notables as Perrier, Marion, Fischer, Vaillant, and of course Milne-Edwards. The *Travailleur* was replaced by the *Talisman* in 1883, and research was continued along the coasts of

Morocco, the Canaries, Spain, Portugal, and the Azores, and out into the Sargasso Sea.

Prince Albert I of Monaco

The career of Prince Albert I was dedicated to advancing the young science of ocean-ography. In several decades around the turn of the century he directed, and financed, numerous expeditions of the *Hirondelle* and her replacements, the *Princess-Alice I* and *II*. These ships concentrated on the eastern Atlantic, Mediterranean, and es-pecially in the Gulf of Gascogne. Most of the early notable work done by Albert I has been published in a series of monographs (*Res. Camp. Scient. Pr. Monaco*) similar to those resulting from the *Challenger*, the *Porcupine*, the *Blake*, the *Albatross*, and other vessels.

Listings of stations taken by the French and Monégasque explorers (as well as less well-known researchers from adjacent countries) have been published by Le Danois (1948). Unfortunately, copies of this thorough yet succinct publication are difficult to obtain. Had it not been printed in limited quantity and in paperback, it could perhaps rank with the works of Ekman, Murray and Hjort, and Agassiz.

The Netherlands' Siboga

The Dutch expedition of the *Siboga* (1899–1900) deserves mention because of the excellence and thoroughness of the monographs on the tropical Western Pacific fauna. Its reference to deep-sea species is only occasional (Appendix).

OUR IMMEDIATE PREDECESSORS

Johannes Schmidt (1877–1933), who is best known for his discoveries about the spawning grounds of the European eels in the Sargasso Sea, led many expeditions of the Danish Ships *Thor* and *Dana*. These were concerned mainly with pelagic work, and on the *Dana II* expedition (1928–1930) only two Petersen grab samples were taken at 25 and at 8m. The *Dana* materials, however, played an important role in the young life of Anton Bruun (1901–1960; Figure 1-18), who later became the scientific leader of the Danish *Galathea* expedition.

Hans Pettersson organized the Swedish *Albatross* expedition (1947–1948). The voyage had been the dream of his father Otto Pettersson, who had earlier promoted a round-the-world scientific expedition that had met with failure. His plan had been to use the *Hirondelle,* which was then for sale, for an international expedition (Pet-tersson, 1966, p.7). Regrettably, this did not materialize; however, it is a great trib-ute to his son that the Swedish *Albatross* expedition came about.

The voyage of the Swedish *Albatross* (Figure 1-14) marked the start of seismic explorations at sea under Wiebull, as well as the extensive use of the "vacuum corer" or piston corer invented by Kullenberg (Pettersson and Kullenberg, 1941). In addition, Kullenberg developed the technique of single-wire otter trawling, which was later practiced aboard the *Galathea* (Figure 1-19). The aim of the Swedish *Al-batross* was not primarily deep-sea biology; consequently, only a few deep-sea trawlings were taken. (The stations and cruise tracks of the Swedish *Albatross* and the

Figure 1-18. Anton Bruun, leader of the Danish *Galathea* expedition. Courtesy of Torben Wolff.

Danish *Galathea* are presented in Figure 1-20.) Nevertheless, the specimens were described in considerable detail in an outstanding series of papers by Swedish and other naturalists: Mortensen (Echinoidea), Silen (Bryozoa), Madsen (Asteroidea, Ophiuroidea, Holothuroidea), Kramp (Hydrozoa and Scyphozoa), Eliason (Poly-chaeta), Shellenberg (Amphipoda), Nordenstam (Isopoda), Millar (Ascidiacea), Gordon (Decapoda), Nybelin (bottom fishes), Broch (Pennatularians), and Odhner (Mollusca, Brachiopoda) (for titles see Appendix).

The Danish *Galathea* expedition made significant contributions to taxonomy of the deep benthos (Bruun, 1956b; Bruun et al., 1956; see Appendix). The concept of a hadal fauna peculiar to the deep trenches was first elaborated by Bruun (1956a) on the basis of *Galathea* results and by Zenkevitch (1958) on the basis of *Vityaz* results.

CONTEMPORARY EFFORTS

The men of our times who have contributed significantly to deep-sea studies are few in number. The list of those who have participated in deep-sea expeditions is smaller. Most investigators have made only one contribution based on a single expedition or a single series of samples, and most reside in the Soviet Union.

Figure 1-19. The Danish *Galathea*. From A. F. Bruun, 1957, Galathea Report Vol. 1, Figure I, p. 6.

Scandinavia

Among the Scandinavians the name of Sven Ekman (1953) stands out as the man with the most knowledgeable view of deep-sea ecology and the factors governing it. Anton Bruun (1901–1961), leader of the Danish *Galathea* expedition and ichthyologist, will always be remembered as the friend of all who tried to probe the abyssal fauna and for his term "hadal" fauna. Torben Wolff (1962) is a recognized worker on deep-sea asellote isopods. He has written several popular articles about the *Galathea* expedition and its general results, especially with reference to the hadal fauna. Lemche and Wingstrand's (1959) monograph on the anatomy of the *Neopilina* is outstanding.

United Kingdom

Most of the current benthic biological work in the United Kingdom is restricted to shelf depths or shallower with no identified, continuing deep-sea biological program. However, the contributions of Dr. A. D. McIntyre (quantification of grabs and trawls and underwater photography), Dr. and Mrs. Southward (Pogonophora), and

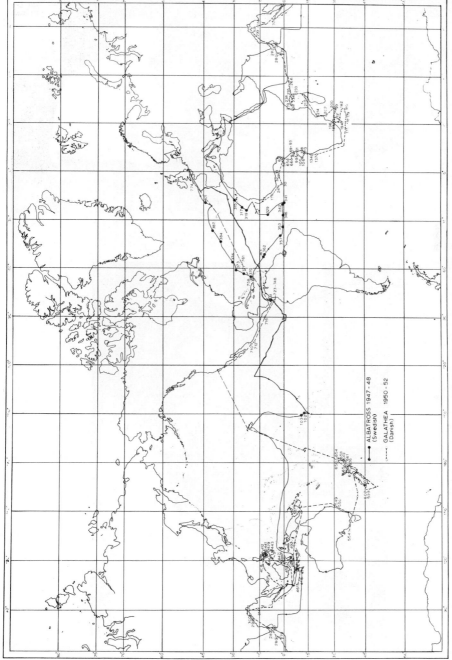

Figure 1-20. Stations of the Swedish *Albatross*, from Swedish Deep- Sea Expedition Report, Vol. 1, and benthic stations of the Danish *Galathea*, as well as cruise tracks of both.

Figure 1-21.   Deep-Sea biologists: (a) Dr. J. A. Allen, United Kingdom; (b) Dr. M. Horikoshi, Japan; (c) Dr. Mihai Băcescu, Rumania.

Dr. J. A. Allen (Figure 1-21a) (Molluscan soft parts, including deep-sea species) deserve special mention.

Germany

Contemporary studies on abyssal problems have been restricted in recent years almost entirely to studies on hydrostatic pressure by Schlieper (1968), his colleagues, and his students from the new German R/V *Meteor*.

Japan

Deep-sea studies commenced in Japan in the 1950s with the organization of the Japanese Expedition of Deep Seas (JEDS). The sampling gear and animals collected on JEDS-2 have been reported by Y. Suyehiro et al. 1960. The benthic biology has been studied by Masuoki Horikoshi (Figure 1-21b), whose main interest is in benthic community analyses using quantitative techniques. Many of the species collected from the Japan Trench remain to be identified and described.

France

French deep-sea biology, mainly dealing with zonation, is represented by the work of Pérès (1961). Additionally, the French pioneered the development of deep diving vehicles that have permitted first hand glimpses of the abyssal environment and biota by man, see also p. 70.

Rumania

Mihai Bǎcescu (1962), the leading expert on benthic ecology of the Black Sea (Figure 1-21c), has published definitive works on abyssal Cumacea and Mysidacea from the R/V *Vema* and the R/V *Anton Bruun*.

Soviet Union

As indicated earlier, the major effort in deep-sea biology has come from Russian efforts and especially through the late, world-renowned expert on the deep-sea Lev A. Zenkevitch, Director of the Institute of Oceanology in Moscow (Figure 1-22a). His book on the biology of the seas of the Soviet Union, which was published in English in 1963, is a classic and contains a wealth of information, especially about biomass distribution. His collaboration with the late Ya A. Birstein, Professor of Zoology at Moscow State University (Figure 1-22b), the Russian expert on deep-sea marine isopods (1963), resulted in the concept of abyssal giantism. The work of M. Vinogradov (Figure 1-22c, left) on vertical distribution of plankton and amphipods (1964, etc.) is paralleled by the definitive study on abyssal zoogeography by his wife Nina Vinogradova (Figure 1-22c, right) (1958). The monographic work on the Pogonophora, a group of shallow-to-deep tube-dwelling wormlike animals, by their discoverer A. V. Ivanov (Figure 1-22d) (1963), is a major contribution to the biology, systematics, and zoogeography of these animals. The studies on the communities and quantity of abyssal life in the North Pacific and added studies on deep-sea mollusc are basic works of Zina Filatova and N. G. Barsanova (1964). The work on

Figure 1-22a. Professor Lev Zenkevitch, Director of the Institute of Oceanology, Moscow, examining an abyssal sipunculid (deceased June 20, 1970).

Figure 1-22b. Professor Ya. A. Birstein, Moscow State University, at work on abyssal isopods (deceased July 1970).

Figure 1-22c.   Dr. M. Vinogradov (left) and his wife Nina Vinogradova rowing a boat. Vinogradov worked mainly with vertical distribution of zooplankton. Mme. Vinogradova became an expert on deep-sea zoogeography.

Figure 1-22d. Professor A. V. Ivanov, the world's foremost expert on *Pogonophora*.

Figure 1-22e. P. Ushakov, the Soviet expert on the Arctic and Antarctic deep-sea fauna and biomass.

Figure 1-22f. Georg Belyaev, the Soviet expert on the trench fauna.

Figure 1-22g. Mme. M. N. Sokolova, the Soviet expert on deep-sea trophodynamics.

Arctic and Antarctic quantitative distribution by the indefatigable P. Ushakov (Figure 1-22e) (1963) is classic. The contribution of Georg Belyaev (Figure 1-22f) (1966) to the trench fauna will be a lasting reference. The studies on trophic levels among deep-sea populations initiated by M. N. Sokolova (Figure 1-22g) (1959) represent an intriguing approach to levels of trophic distribution in the abyss. These are only a few of the outstanding Soviet deep-sea fisherman. Others should doubtless be included such as Kriss (1962), marine microbiology; Neiman (1966), benthic trophic distribution; and Pasternak (1958), Antipatharia.

SOVIET RESEARCH VESSELS

Soviet research vessels used in deep-sea biological studies include the *Vityaz* (Figure 1-23*a*) under the scientific leadership of Dr. Zenkevitch. It was from this ship that the mapping of the benthic fauna of the North Pacific was accomplished, as well as the detailed studies on the fauna of the Bering Sea and the Kurile–Kamchatka Trench. The biomass of the North Atlantic was investigated by Kusnetzov (1960) from the *M. Lomonosov* (Figure 1-23*b*). High Arctic studies have been made from the Soviet Ice Islands. Antarctic studies have been carried out mainly from the *Ob* (Figure 1-23*c*). Besides the foregoing, the *Ac. Kurtschatov* and the *D. Mendeleev* (Figures 1-23*d* and *e*, respectively) are now in operation.

United States

Contributors to deep-sea biology in the United States are fewer in number than those of the Soviet Union. This reflects most logically on the lack of emphasis accorded to the subject by the American government. Even taxonomists are scarce in this country.

J. L. Barnard (Figure 1-24*a*) has contributed significant works regarding the abyssal amphipods (1962); A. H. Clarke (1962*b*) (Figure 1-24*b*) has catalogued the abyssal molluscs. Olga Hartman has published several works concerned with abyssal polychaetes off California, the Antarctic, and Massachusetts, and on biomass

Figure 1-23. Soviet research Vessels: (a) *Vityaz*, (b) *M. Lomonosov*, (c) *Ob*, (d) *Ac. Kurtschatov*, (e) *D. Mendeleev*. Photographs courtesy of Dr. Zina Filatova, Institute of Oceanology, Moscow.

Figure 1-23. (Continued)

distribution off California with J. L. Barnard. Deep-sea isopods and trench phenomena have been a major interest of two of us (Menzies and George, 1967) (Figure 1-24c), and the distribution of abyssal epifauna in relation to currents, sediments, and water masses has been a major concern of Rowe (1968) (Figure 1-24d).

Studies on species diversity have occupied the attention of Hessler and Sanders (Figures 1-24e and f, respectively) and associates (1967) over a period of several years. The work of ZoBell (Figure 1-24g) on pressure effects (1953) is classic, and his students Oppenheimer and Morita have continued this work.

Beyond this brief listing there are naturally others who have studied one or more deep-sea animals and have contributed to the deep-sea effort. We beg forgiveness for our oversights. We exclude ourselves from among the major contributors but include

Figure 1-23. (Continued)

our photos for the curious. Unfortunately, we could not obtain photographs of all persons mentioned.

UNITED STATES RESEARCH VESSELS

Contemporary American research vessels which have engaged in deep-sea biological investigation include the *Velero IV* (Figure 1-25; Barnard, Hartman, and Menzies) working mainly off California and Mexico, the *Anton Bruun* (Figure 1-26; Menzies and Chin, 1966) in the eastern Pacific Ocean, and the *Horizon* (Figure 1-27; Parker, 1963). High Arctic studies have been conducted from the American Ice Islands and from Ice Stations Alpha, Charlie, Arlis I-II, and T-3 (Figure 1-28). Antarctic deep-sea studies have been conducted from the U.S.N.S. *Eltanin* (Figure 1-29).

Figure 1-24.   United States deep-sea biologists: (a) J. L. Barnard, Amphipods; (b) A. H. Clarke, Jr., Mollusca; (c) R. J. Menzies (left), isopods and zonation; R. Y. George (right), isopods and zonation; (d) G. T. Rowe, epifaunal zonation; (e) R. R. Hessler, isopods and community diversity; (f) H. L. Sanders, community diversity; (g) C. E. ZoBell, Microbiology.

Figure 1-24. (*Continued*)

Figure 1-25. Research vessel *Velero IV* with the late Captain G. Allan Hancock on the flying bridge. Courtesy of the Allan Hancock Foundation, University of Southern California.

Figure 1-26.   Research vessel *Anton Bruun,* belonging to the National Science Foundation. Photo courtesy of Captain Rothrock, *Anton Bruun* Cruise 11, South Eastern Pacific Expedition.

Figure 1-27.   Research vessel *Horizon,* belonging to the Scripps Institution of Oceanography.

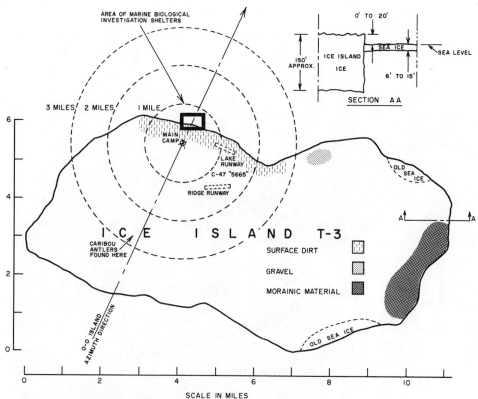

Figure 1-28. U.S. Drifting *Ice Island T-3*. Courtesy of John L. Mohr.

The R/V *Vema* (Figure 1-30) has taken extensive samples from the Atlantic, the Pacific, the Indian, and the Antarctic oceans, as well as from the Caribbean and the Mediterranean seas. The R/V *Eastward* (Figure 1-31) has sampled mostly in the vicinity of the Carolinas. The vessels of the Woods Hole Oceanographic Institution, including the *Atlantis II,* the *Chain* and the *Gosnold,* have worked the North Atlantic, the Caribbean, and the Mediterranean (Sanders and his associates).

Commonly no single United States vessel has devoted its efforts entirely to deep-sea biological studies. This opportunity has been rare in the history of deep-sea biology, regardless of the sponsoring country. Usually deep-sea biology has been conducted on an adventitious basis whenever time allowed. There have been few exceptions to this practice, and the net result has been a curtailment of the work that might have been accomplished. The *Galathea* expedition and the *Anton Bruun* South Eastern Pacific expedition are notable exceptions. Few scientists can afford the luxury of riding a research vessel for weeks and months waiting for the chance to do some deep-sea fishing.

ERAS IN THE DEVELOPMENT OF
DEEP–SEA BIOLOGY

The majority of the deep-sea expeditions between 1873 and 1960 were not concerned with deep-sea biology and took deep-sea samples only occasionally. Wüst (1964) divided the history of deep-sea physical explorations into three eras. The first

Figure 1-29.   R/V U.S.N.S. *Eltanin* during shakedown cruise.

was an era of exploration between 1873 and 1914, the second was characterized by national systematic and dynamic ocean surveys (1925–1940), and the third is a time of international research cooperation (since 1957).

As far as accomplishments are concerned, biological studies are still in the exploratory stage, and history may not offer again the opportunity for ships that will allow biologists to attain the technological level of their colleagues in the physical sciences. It is possible that the Swedish *Albatross*, the Danish *Galathea*, and the Russian *Vityaz* indicate entry of deep-sea biological studies into the second era of Wüst (*op. cit.*).

It should be obvious that a great distance has to be traveled historically in the accumulation of basic biological data before biologists can profit from an international era. Biologists have not yet asked the important questions that can be answered by international teamwork, and they are stuck, like it or not, with a dearth of knowledge of the fauna.

The techniques for data gathering are archaic and so time consuming that the task sometimes seems to exceed the lifespan of the investigator himself. Such a dismal picture is not the kind to attract new students; it relates to the great difficulty of collecting raw data, to the diversity of the data themselves, and to the difficulty of reducing them into meaningful terms.

The physical scientist, for example, can ask meaningful questions regarding a series of water temperatures. His data reduction requires only that the temperatures

Figure 1-30.   R/V *Vema* entering New York harbor, 1953. Courtesy of Lamont Geological Observatory.

Figure 1-31.   R/V *Eastward,* owned by Duke University. Photograph courtesy of Sturgeon Bay Ship-
building and Drydock Company.

be corrected or converted into some other form. It takes less time to do this than it does to collect the temperatures, and the investigator's data, expressed as a few digits on a piece of paper, can be meaningful to many persons shortly after collection.

"Not so with the biologist." An animal or plant, often represented by one specimen which may not be reduced to its scientific name for 10 to 20 years after collection, becomes available long after other sciences have answered their questions and have moved on to new and more fruitful fields or are able to enter new eras of investigation. A promising facet of biological data collecting lies in quantitative samples or photography, which can be reduced into terms of standing stock or crop and utilized widely. There remains then the problem of using uniform techniques of sampling and reporting.

Biologists interested in the deep sea should take heart in the undeniably high probability that the future problems of mankind relating to the seas will have strong biological orientation. Our limited knowledge about the marine biota today prevents an accurate ecosystem model except in broadest terms. An inventory of the biologic resources must be completed before even the simplest ecosystem model can be tested and evaluated. The physical and geological inventories of the seabed have largely been completed. Future emphasis on ecologic understanding requires the acquisition of biological information.

In historical perspective, it must be evident that biological investigations of the seas have not had and will not have the same history as the physical explorations. Studies on how, where, and when the fauna of the deep sea originated are fascinating, but they contribute very little directly to the economy of any nation and hence have no claims to immediate "national" priority. However, they are required for a basic knowledge of the biotic potential of the seabed for use by mankind, to assay the effects of ocean waste disposal, and to determine the interdependence among the biota of the various components of the hydrosphere.

CHAPTER 2

Deep-Sea Sampling

> Deep-sea expeditions could certainly have arrived at corresponding results if their methods of dealing with bottom materials had been more satisfactory; it may be considered quite certain that hundreds of species of small crustacea, etc., lived in the bottom materials hauled up by the *Challenger* and the later great European and North American expeditions and were flushed into the sea again. (H. J. Hansen, 1913, p. 3, with reference to *Ingolf* collections.)

Hansen's comments emphasize the importance of one major feature of deep-sea sampling, namely, mesh size. Other features may be equally important.

Two types of samples, quantitative and qualitative, are generally accepted as necessary when approaching problems involving the benthos, whether at shallow or deep levels. To gain precise knowledge of the abundance of animals on and in the bottom requires a quantitative sampler that can procure a known area or volume of sediment. In doing this, an investigator must be satisfied with a relatively small volume or area simply because it is so difficult to raise the mud from the sea floor. Unless numerous samples of this nature are taken, only the most common species can be captured. To overcome this obstacle, large qualitative samples are usually taken in order to collect the less common forms. In the deep sea as well as in the shallow water, these two general categories of sampling must be employed, but complications involved in sampler design and the resulting use at great depths are manifold.

Some of the early trawls and dredges used in deep-sea sampling aboard the *Porcupine* and the *Challenger*, including O. F. Müller's dredge, the naturalist's dredge, and the double-beam trawl (Sigsbee or Blake trawl) are shown in Figure 2-1.

The major activity of many scientists during a deep trawling station consists of wire watching (Figure 2-2), while the sampler makes one or more miles of excursion from the surface to the seabed and back.

QUALITATIVE SAMPLING DEVICES

Qualitative sampling devices were generally designed to capture as much of the fauna from as large an undefined area as possible. To meet this requirement, naturalists have often used modifications of the gear of commercial fishermen. Otter trawls are the largest, and are characterized by a set of heavy boards that act to spread the

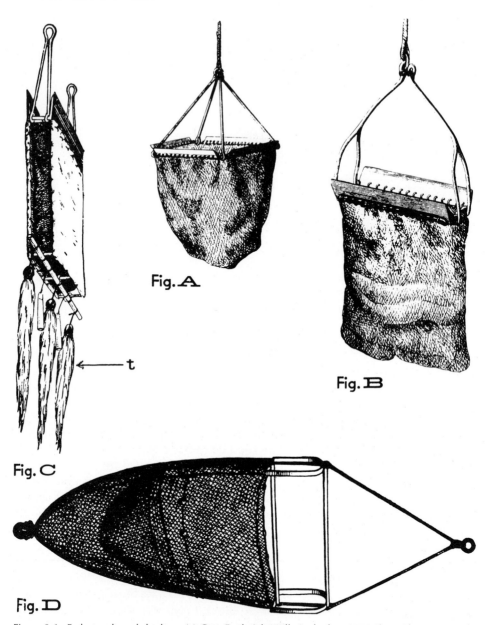

Fig. A

Fig. B

Fig. C

← t

Fig. D

Figure 2-1. Early trawls and dredges: (a) Otto Frederick Müller's dredge, 1779 (from Thomson, 1874, Figure 44, p. 239); (b) Ball's dredge or the naturalist's dredge (from Thomson, 1874, Figure 45, p. 240); (c) Blake dredge with Calver's swabs or tangles (t) (from Agassiz, 1888, Figure 23, p. 24); (d) Blake trawl–double-beam trawl, Sigsbee trawl (from Agassiz 1888, Figure 26, p. 26).

net over the bottom. Shallow-water bottom fishermen use similar nets up to a mouth diameter of 40 to 60 feet (20 m), and these have occasionally been used in the deep sea with limited success. The Herring Otter Trawl (HOT) used aboard the *Galathea* measured 100 ft (30 m) from board to board. On this trawl the float line that served on smaller nets to hold the upper half of the net above the bottom was replaced by a third board. Although difficult to use, the trawl occasionally takes tremendous

Figure 2-2.   Men "wire-watching" and not wire watching aboard the U.S.N.S. *Eltanin*.

samples in areas of high biomass. The *Galathea* expedition samples taken off Costa Rica (Wolff, 1961) form one such example.

The other configuration can be called a double-beam trawl or Sigsbee (Figure 2-3, a Russian model), and it has gained wider use in the deep sea, mainly because it is smaller and much easier to handle. It consists of a metal frame that supports the orifice of the net. To prevent the net from tangling in the wire, the frame has often been continued around the rear of the net as in Menzies' Small Biology Trawl (SBT) or "Menzies trawl" (Figure 2-4), or the larger "Blake" trawl used by Rowe at the Woods Hole Oceanographic Institution (WHOI) and by Rowe and Pilkey on the *Eastward*. A much larger 10-ft model was used by Menzies off Peru from the *Eltanin* (Figure 2-5).

Another qualitative sampler finding wide use is the modification of the Riedel or WHOI epibenthic sled (Figure 2-6) described and used by Sanders et al. (1965) at WHOI. Designed to skim the surface rather than to bite deeply, it has a blade that can be adjusted to skim up the top few centimeters of mud. Although its percentage of success is unknown to us, the catch by successful trawls is remarkable, and totals of 2000 to 3000 individuals have been taken (Sanders et al. 1965). The Russians have added a meter wheel to a double beam trawl, called a trawlograph, to aid in quantitative assessment of trawls (Figure 2-7).

Mesh Size

Regardless of configuration of frame or method of use, the characteristic of a trawl that principally determines its function on the sea floor is the size of the mesh of the net. The function of the net itself is to allow the easy filtration of water and sediment

Figure 2-3. "Caged" Sigsbee-type trawl utilized aboard *Vityaz,* where the spring accumulator is built into the "A" frame on deck. (From Filatova, 1960, Figure 4.)

while retaining the animals. The Herring Otter Trawl has large meshes several centimeters across, decreasing in size to less than 1 cm toward the cod (rear) end, whereas the SBT generally is lined with 550 μ mesh. Menzies (1964) compared numbers and kinds of animals captured by these two devices from off Costa Rica and illustrated that a diverse assemblage of small benthos is taken with the small mesh, whereas numerous large organisms unrelated in taxa and habit are taken with the HOT's large mesh (Table 2-1). The SBT, rather than filtering as hoped, probably fills quickly with sediment when a moderate fraction of foraminiferan tests clog the mesh (Menzies and Rowe, 1968). The larger mesh of the epibenthic sled (1.5 mm) probably improves filtering, but it is doubtful that this is done with any degree of efficiency. When large "bodies" or several gallons of "foraminiferal sand" are retained in the net, it can be assumed that filtering was less than optimal. The Blake trawl has been used extensively by Rowe at all depths with 0.5 in. (1.25 cm) stretch mesh,

TABLE 2-1. COMPARISON OF CATCHING CAPACITY OF THE *GALATHEA* HOT AND THE *VEMA* SBT USED AT SAME STATION (FROM WOLFF, 1961)

Animal Groups	Galathea Station 716 (3570 m) HOT — Number of Species	Number of Specimens	Relative Order of Abundance (1–10)	Vema Stations 128–129, 131–139 (3290–3700 m) SBT — Number of Specimens	Relative Order of Abundance (1–10)
Porifera	3	ca. 41		21	
Coelenterata	16	ca. 137		141	
Hydroidea	3	4		5	
Scyphozoa (Stephanoscyphus)	1	13		101	6
Alcyonaria	1	8		15	
Pennatularia	3	17		0	
Coral	0	0		3	
Actiniaria	8	95	6	17	
"Vermes"	ca. 23	ca. 317		1142	
Aschelminthes	0	0		52	10
Polychaeta	ca. 20	ca. 310	3	1090	1
Hirudinea	1	1		0	
"Gephyrea"	2	6		0	
Crustacea	ca. 32	ca. 208		742	
Copepoda	1	3		6	
Ostracoda	0	0		15	
Cirripedia	2	32		1	
Nebaliacea	0	0		8	
Cumacea	1	1		41	
Tanaidacea	7	60	8	221	4
Isopoda	7	50	10	198	5
Amphipoda	5	12		250	3
Decapoda	ca. 9	ca. 50	10	2	
Pycnogonida	3	18		1	
Mollusca	14	132		539	
Monoplacophora (Neopilina)	1	10		1	
Solenogastres	1	15		80	8
Scaphopoda	1	55	9	33	
Gastropoda	6	18		51	
Bivalvia	5	34		374	2
Brachiopoda	0	0		1	
Bryozoa	0	0		89	7
Echinoderma	30	1171		104	
Holothuroidea	14	512	1	25	
Crinoidea	1	3		2	
Asteroidea	7	210	4	7	
Ophiuroidea	5	315	2	67	9
Echinoidea	3	131	5	3	
Pogonophora	0	0		3	
Ascidiacea	2	10		16	
Pisces	9	66	7	0	
Total	ca. 132	ca. 2100		2799	
Average per trawling				256	

Figure 2-4.   Small biology trawl (SBT auct. Menzies' trawl), 1 m wide at mouth, showing weight and corer attached, on deck of USNS *Eltanin*.

which seems to be an appropriate size for filtering sediment. Used in conjunction with camera surveys, it offers optimum efficiency in taking those dominant large species that are visible in photographs.

QUANTITATIVE SAMPLING DEVICES

Quantitative samplers are generally clamshell-type grabs which bite out a specific area of the bottom down to a given depth in the sediment, depending on their size and design. Of those invented, only the Petersen, the Okean, the Campbell, the Van Veen, and the Smith-McIntyre grabs (Figures 2-8 through 2-11) have gained wide use. The combined Ekman grab and box-corer called "Sonda" (Figure 2-12), developed by Băcescu (1957), has a 100-cm² sampling area.

At best, however, grabs are horribly inefficient collecting devices and sample far too small an area. The largest is the Campbell grab, which covers 0.6 m² and works about 50% of the time. The large undisturbed bottom sampler, or LUBS (Menzies and Rowe, 1968; Figure 2-13) holds much promise as a useful quantitative sampler of the infauna as well as the epifauna, but it has not had enough use to win wide acceptance. Nevertheless, it does appear to meet most of the following criteria suggested by Holme (1964) and Hopkins (1964).

Figure 2-5. Caged 10-ft double-beam trawl after successful trawling of Milne-Edwards trench.

1. *Penetration:* To at least 30 cm.
2. *Surface sediments:* Not disturbed.
3. *No sample loss:* Retrieval should entail no sample loss, retaining entire sample with overlying water.
4. *Sample removal:* Must be relatively easy to remove.
5. *Area:* As wide as possible, preferably 1 m².
6. *Sturdy construction:* Enough to endure repeated handling on deck and impacts on bottom.
7. *Minimum of working parts:* Delicate working parts are unsuitable, especially if exposed to sea water.
8. *Corrosion resistant:* Disparate metals should not be used together.
9. *Orientation:* Should orient correctly before contacting bottom.
10. *Handling:* Easily handled on deck and over the side of the research vessel; should not be overly dangerous.
11. *Bottom contact:* Must be recognizable even in deepest water.
12. *Frontal water shock waves:* Water shock waves at the mouth of the sampler must be avoided. This is often accomplished in part by cutting holes in the top of the open-grab sampler. Otherwise, much fauna is lost through this imperfection of design.

All the points cited are important to success. However, we would like to emphasize one of them — namely, that the sampler should be designed in such a way

Figure 2-6. WHOI epibenthic sled. Courtesy of Howard Sanders, Woods Hole Oceanographic Institution.

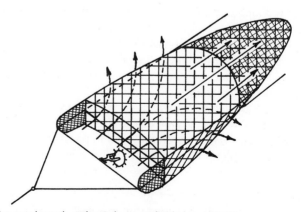

Figure 2-7. Russian trawlograph. (After Belyaev and Sokolova, 1960.)

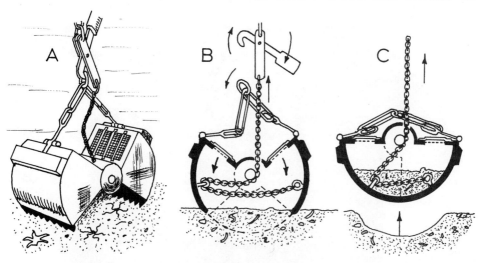

Figure 2-8. Idealized sketch of a Petersen grab taking a sample on the seabed. (After Holme 1964; redrawn from Hardy, 1959.)

that impact is not preceded by a "shock wave." In other words, the sampler should have a minimal frontal (impact) area, thus reducing the tendency to push a wall of water in front of it. This problem can be eliminated by allowing water to pass through the sampler as it descends. If this design feature is not satisfied, then criterion 2 cannot be met because the wave of water will wash the more vagrant epifauna from the sampling area. This failure appears to be a common source of error in quantitative sampling in deep water.

An anchor dredge (Figure 2-14) has been used to take what are presumed to be quantitative samples in several recent investigations (Sanders et al., 1965; Griggs et al., 1969). Useful only in mud and along smooth bottom, it bites to a specific (10 cm) depth, ideally fills to capacity, and then rejects additional material. The behavior of the dredge on the bottom, when observed in shallow water by George

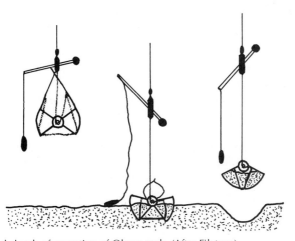

Figure 2-9. Idealized sketch of operation of Okean grab. (After Filatova).

Figure 2-10. Open surface of Campbell grab with bottom camera in place; insert shows grab closed after sampling Peru–Chile Trench on *Anton Bruun* Cruise II. (After Menzies, Smith, and Emery, 1963.)

Hampson using scuba equipment, appeared to follow the above mentioned sequence properly (Sanders et al., 1965). Unfortunately, the density of what can be classified as "mud" on the basis of particle size, can vary considerably. On relatively hard "mud" bottoms, such as in shallow water just off Gay Head, Mass. or along the walls of the Hudson Canyon or the Milne Edwards Deep, the dredge has a tendency to "skim" the surface, collecting only from the top several centimeters. If not recognized, this behavior could lead to vast overestimation of population densities. Off Peru (Rowe, personal observation), for example, the dredge brought up several hard blocks of dense sample. Moreover, there is little likelihood that it can ever effectively sample much of the epifauna when the substrate is soft, although some epifauna is usually taken.

The anchor dredge, however, has found wide usage, as indicated by the reports

Figure 2-11.   Van Veen grab rigged for lowering (on land).

just cited. There are several good reasons for this. First, it can take a large sample. A dredge 50 cm wide and 10 cm deep, can sample, depending on the bag volume, more than one square meter, and this is important when dealing with a sparse fauna. Perhaps more important is its reliability. Although the efficiency of the anchor dredge may be questionable as indicated, its simple design (there are no moving parts) gives it a great advantage in use at sea, especially in foul weather. For example, a Campbell grab succeeding in 50% of its efforts over the side would gather considerably less data than an anchor dredge with 90% success. When working time at sea is at a premium despite the weather, the anchor dredge is put to work.

Internal agreement between samples from the same areas (Griggs et al., 1969; Sanders et al., 1965) suggests that although anchor dredging may not be as precise as a clamshell grab or the LUBS, the sampling efficiency is adequate to uncover many ecological variations manifested in infaunal biomass. Comparisons have also been made of the other various sampling devices in shallow water (Lie and Pamatmat, 1965; Wigley and McIntyre, 1964), but few of them are germane here because deepsea sampling presents so many additional difficulties.

PHOTOGRAPHY

The development of deep-sea photographic techniques has progressed at a rapid rate in the last two and a half decades. The competence and reliability of the systems in use are no longer questioned, and numerous applications have added considerably

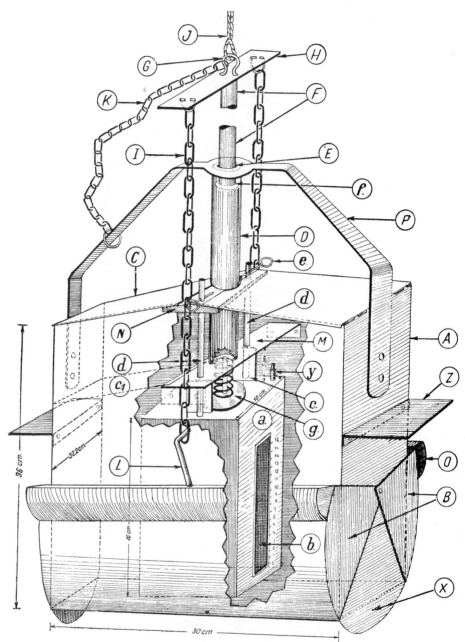

Figure 2-12.　"Sonda" grab as developed by Băcescu (1968).

to knowledge of the deep sea. The major contributions, however, have been in geology (Hersey, 1968). There is, however, some question regarding the efficacy of photographs in the biological investigations. Whereas Owen et al. (1967) doubted the usefulness of camera techniques in estimating population densities, Fell (1967) discussed and demonstrated their application. In agreement with earlier suggestions (McIntyre, 1956; Czihak and Zei, 1960), photography appears to give

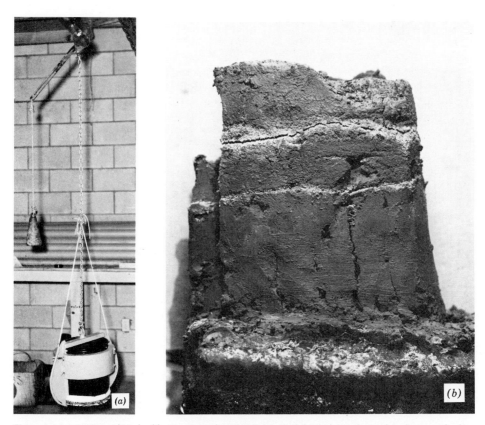

Figure 2-13. Large undisturbed bottom sampler (LUBS) (a) minilubs with 5-gal sampler photographed in closed position; (b) sample from minilubs showing stratified sediments.

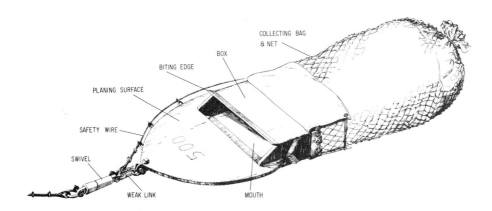

Figure 2-14. Anchor dredge. Courtesy of Howard Sanders, Woods Hole Oceanographic Institution.

data on the abundance of large epibenthonts that is more reliable than that accrued from conventional samples. This is mainly because a much greater area of the bottom can be recorded on film.

Deep-sea photography has probably not been used to its full capacity in the study of deep-sea animals because of the limited information available on the identity of the animals seen in the photographs. Menzies et al. (1963) developed the grab-camera (Figure 2-10) in order to solve this identification problem, and Emery et al. (1965) continued this technique in shallow water. Rowe (1968) utilized camera lowerings with good results by following each with a trawl.

We have generally used (except as cited otherwise) a multishot (unperforated Plus X, 35-mm film), Ewing-Thorndike shutterless camera (Figure 2-15). A compass attached to the camera frame and in the focal plane allows the determination of animal orientation with respect to magnetic north; likewise, the compass is useful in describing the direction of bottom currents (Rowe and Menzies, 1968). Between 20 and 25 exposures are taken at each station, although some stations require a greater number of exposures to yield reliable information on epifaunal abundance in relation to space. The setting of the focal distance determines the area photographed, and although there is some distortion at the margin of each frame, animals can be counted in each exposure. In time, as the epifauna becomes better known, underwater photography should allow mapping of relatively wide areas of seabed, such as that done by Rowe (1968). As previously stated, one good photograph can be equivalent to several trawl attempts, because a 12-ft focal distance usually allows the coverage of an area of 6.5 m² with each photograph.

The development of wide-angle lenses and their use in cameras towed from underwater sleds, in conjunction with transponders and side-scanning sonar, has permitted precise positioning and the development of photographic mosaics of large areas of the deep-sea floor, as well as excellent indications of biological activity (Brundage et al. 1967).

SAMPLING WORK AT SEA

Bottom sampling of any kind, be it trawling, grabbing, or photographing, is much more difficult in deep water than in shallow water. Menzies (1964) has determined the success and failure of bottom sampling of various deep-sea expeditions. Failure to hit bottom was probably the major reason for not getting a sample. Looping and tangling the wire is the second reason. The first mishap results from not allowing enough wire to be paid out, while the vessel drifts or steams too rapidly; the latter is caused by paying wire out too fast or paying out too much of it when the ship is virtually hove-to, relative to the bottom. To overcome these two difficulties, we have concluded that a most important factor is knowing exactly when a piece of gear hits the bottom. Several techniques have been developed as aids in recognition of bottom contact.

The most promising modern technique (Rowe and Menzies, 1967) involves the use of an acoustic pinger of the same frequency as the depth recorder to visually position the dredge or trawl (Figure 2-16). Lacking such sophisticated instrumentation, it is possible (Menzies 1964) to locate bottom during trawling by watching the spring accumulator and recording the wire tension (Figure 2-17). In this case, a sud-

Figure 2-15. Multishot underwater camera being lowered over the side from the *Anton Bruun,* cruise 11, Dr. Menzies and technician Mr. Yano stabilize camera.

den drop in tension indicates bottom contact (*H*-1 to *H*-7 in Figure 2-17). This occurrence is more simply determined by transduction of the spring tension signal to a strip chart recorder (Figure 2-18), as described by Rowe and Menzies (1967) and Wall and Ewing (1967). The resulting record of wire tension during an operation provides invaluable information about the activity on the bottom and about actual contact. It is a permanent record of each station which can be related to the success of each attempt.

Little more than imagination has been involved in past trawling evaluations; common trawl failures have been attributed to mysterious anomalies of the sea floor, now proved to be nonexistent, or to animal attacks. In fact, the most probable cause of deep-sea trawling failure is not some untoward feature of the sea floor but poor winch operation, such as overspeeding the wire payout (Figure 2-19).

Trawling at sea depends on many variables, with the seabed character being (in deep water, at least) the least significant of these. Winch operation, weather, surface currents, ship speed, ship maneuvering during trawling, and bottom-contact recogni-

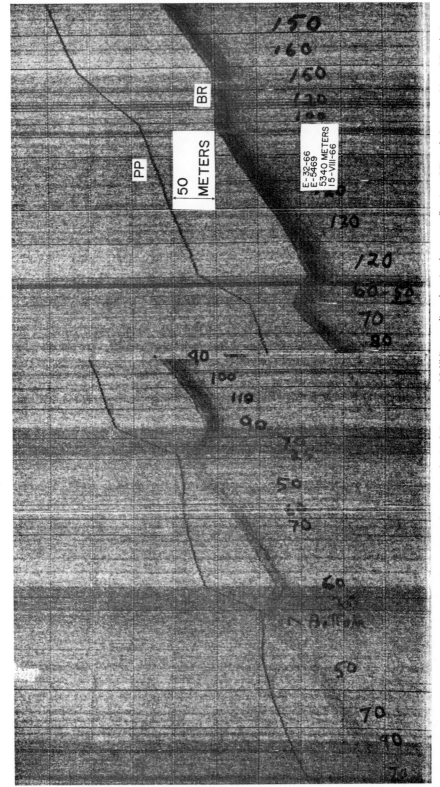

Figure 2-16.   Precision echo-sounding recorder (PESR) (Alpine Geophysical Company, 12 KH) recording showing bottom reflection (*BR*) and pinger position (*PP*) relative to bottom, scale in meters. (From Rowe and Menzies 1967, Figure 2.)

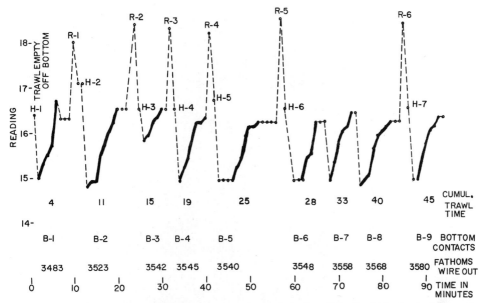

Figure 2-17. Deck (winch) spring accumulator readings taken aboard the U.S.N.S. *Eltanin* Station 35 (USC). Heavy line indicates on-bottom trawling, *H-1*; holding indication from empty trawl at depth above bottom; *R-1* to *R-6* are raisings of trawl from bottom to obtain an off-bottom accumulator reading; *H-1* to *H-7* are readings of the accumulator with winch stopped and trawl off bottom. Bottom contacts are indicated by accumulator readings of 16 units or less.

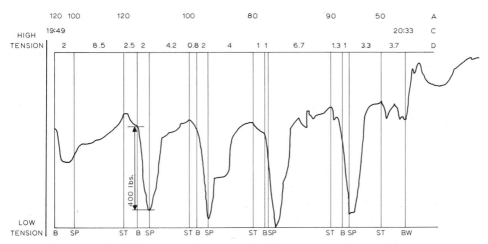

Figure 2-18. Strip-chart graph of tension record (corrected for ship's roll): A = pinger distance from bottom (m), B = begin payout, BW = begin wind in, C = time, D = elapsed time (min), SP = stop payout, ST = stop trawling. The 400-lb decrease in tension represents the weight hitting the bottom. (From Rowe and Menzies, 1967, p. 272.)

Figure 2-19. A "moderate" kink in 0.5-in. steel cable after it was allowed to lie slack on the bottom because of failure to observe bottom contact during R/V *Eastward* cruise.

tion are much more important. A change in any of these can make or break a station. The weather is one variable that cannot be controlled, but adjustment to its lack of cooperation is possible.

Winch speed should be only fast enough to allow the trawl to sink more rapidly than the trawl wire. Wire payout should be at a constant rate, especially when bottom contact is near. Surface currents, subject to change in velocity and direction during a long station, must always be considered. Ship maneuvering is a crucial matter, and sudden changes in surface wire angle through ship maneuvering can destroy all attempts to recognize bottom contact.

A great deal of attention has been given to wire payout and wire angle during deep trawling. Kullenberg (1956) has elaborated formulas and nomograms for wire payout during trawlings of the massive HOT utilized from the *Galathea*. It is quite practical to use much less wire in deep trawling than the commonly accepted ratio of 3 lengths of wire out to water depth. This was demonstrated (Figure 2-20) by Rowe and Menzies (1967), using the pinger technique for bottom positioning. Wall and Ewing (1967) employed a similar technique for determining piston-corer bottom contact.

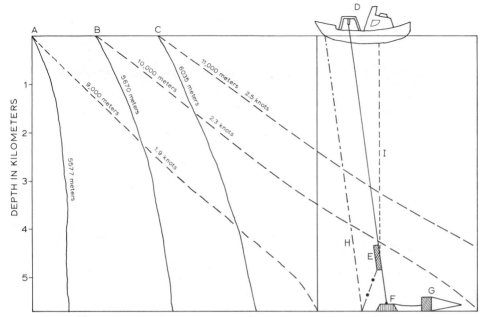

Figure 2-20. Comparison of wire length required for trawling as used aboard the R/V *Eastward* with wire angle: (a) 40° surface wire angle, speed 3 to 5 knots; (b) 45° surface wire angle, speed 3 to 4 knots: (c) 48° wire angle, speed 3 to 4 knots—line is wire amount calculated by Kullenberg at various speed and comparable surface wire angles in (a), (b), and (c). Also included is a diagrammatic sketch of the described technique, where D is the ship and E is the sonic pinger 100 m above F. This location is 100 m from G, the trawl. I is the transmitted sound from the pinger, and H is the sound reflected off the bottom. (After Rowe and Menzies, 1967.)

Failure to ascertain contact of the trawl with the bottom has accounted for about 50% of the deep-sea trawl failures (Menzies, 1964), and often the joy of work at sea is dissipated in removing kinks from wire that has been paid out on the bottom (Figure 2-19). Failure has also resulted from the entrapment of the net in the wire (Menzies, 1964), thus tearing it off the frame.

TRAWLING TECHNIQUES

The general methods used by the men of the *Porcupine* for getting a successful trawl sample are about as good as any devised to date (Figure 2-21). First the depth was measured, and 500 fathoms of line and a heavy weight (W) were attached to the dredge (D). Then line equal to the water depth was paid out, plus an additional 500 fathoms. Then the ship streamed back on its course (A–C) and collected a good sample more often than not.

On the *Challenger* (Figure 2-22), the single modification was to first pay out the line and then to slide a weight (G) down the rope to a stop located 500 fathoms from the dredge (B–F). The added weight caused the line to sink to the bottom and some time was saved. Both the *Challenger* and the *Porcupine*, it should be noted, used hemp rather than wire rope, and on the *Porcupine* coiling was done by hand!

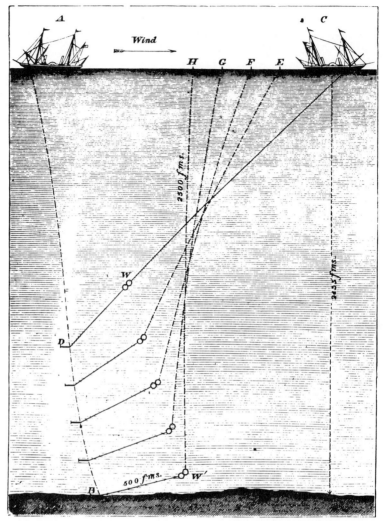

Figure 2-21. Diagram of relative positions of the H.M.S. *Porcupine:* (A–C) = weight, W—W', dredge = D = D' while dredging in deep water. (From Thomson, 1874, Figure 50, p. 253.)

Deep-Sea Trawl Behavior

Heretofore there has been no visual record of how a trawl was behaving on the bottom in the deep sea (depths greater than 2000 meters), although divers have recorded the action of trawls in shallow water (Holme, 1964). On the R/V *Eastward,* a 5-ft Blake trawl was outfitted with a time-activated Edgerton, Germeshausen, and Grier underwater camera that was placed inside the trawl looking directly out the mouth (Figure 2-23). It took pictures at 30 to 50 sec intervals, and the photographs tell approximately what the trawl was doing on the bottom. The Loran "A" fixes suggest that the trawl was towed between lat. 32.31°N by lat. 32.34°N (or .03 degrees to the north), and between long. 67.41°W by long. 67.45°W (or 0.04 degrees to the west), trawling a rough distance of .30 nautical mile, over the bottom during the total

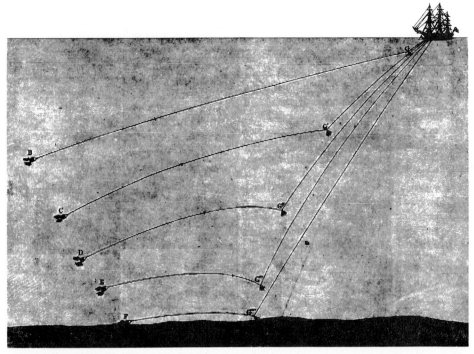

Figure 2-22. Deep-sea dredging from the H.M.S. *Challenger:* A = ship, B–F = dredge, G = weight being slid down rope. (From Tizard et al., 1885, Figure 22, p. 77.)

of 1 to 1.5 hr on bottom. Bottom contact was observed using a pinger and a strip-chart tension recorder.

The camera took pictures first at 30-sec intervals but by the end of the run, 1 hr, 7 min later, it had slowed to 50 sec per exposure. In all, 192 frames were taken during the trawling. In order to understand the information presented in the photographs, it is essential to know the rigging of the trawl and the placement of the camera and strobe light (Figure 2-23).

Photographs of the descent showed a clean trawl pipe with the weight in front of the apex of the trawl bridle. Obviously, the trawl wire was moving toward the bottom (Figures 2-24a and b). A cloud of mud (Figures 2-24b and 2-23f) is visible in front of the trawl. Figure 2-24c reveals muddy water but no mud clouds, suggesting that the camera face was covered with a thin layer of mud. Figures 2-24d and 2-23d show the bridle and weight being lifted from the bottom with a cloud of mud coming off the junction of the bridle and the weight. Photos of mud clouds and no bridle within the photo frame suggest trawling (Figures 2-24b and 2-23f). These figures, not arranged in time sequence, furnish information characteristic of deep trawling that is now available through photographs, and also information that formerly was based on imagination.

Of what was supposed to have been a 1.5-hr work period, the trawl spent a cumulative period of 252 sec (roughly 4 min) doing what it should have been doing. The bottom trawling efficiency (not its catching efficiency) amounts to less than 10%. It would appear from the photographic evidence that most of the time was

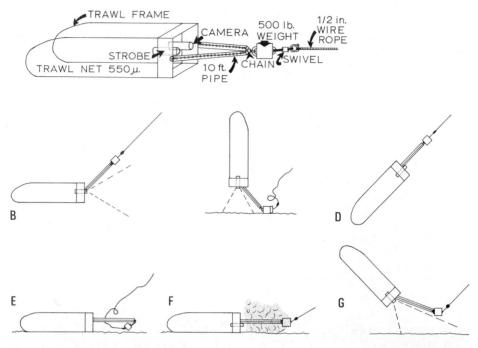

Figure 2-23.   Deep-sea trawl behavior; (a) Diagram of 5-ft beam trawl showing position of camera and strobe; (b) trawl being lowered, photos taken show clean water and not bridle in pictures; (c) trawl upset, camera faces bottom, takes photo perpendicular to bottom; (d) trawl on way up, bridle and clean water or mud falling off weight and bridle are visible; (e) trawl on bottom and weight drops toward trawl mouth; (f) trawling, mud clouds in photographs, no bridle; (g) trawl oblique, camera takes oblique view of bottom.

spent twisting, flipping and flopping over the sea bed, spilling its contents back onto the sea floor, or being buried mouth first in the mud. This trawl captured 0.33 gal of the manganese-encrusted pebbles, debris, and animals as indicated below:

RV *Eastward* Station 7799

Porifera	2	Brachiopoda	1
Nematoda	0	Mollusca	8
Alcyonaria	1	Polychaeta	3
Actiniaria	8	Cumacea	0
Tanaidacea	1	Isopoda	7
Bryozoa	3	Amphipoda	20
Echinoidea	0	(some pelagic)	

Quantifying Trawl Samples

Certain approaches can be used to estimate the area that trawls cover while on bottom and also on how much they effectively sample. Walkers attached to the frames which meter the distance traveled by the trawl have been used and have met with some success (Figure 2-7). Another method of determining the distance traveled is to take accurate navigational fixes at the beginning and the end of a drag. As we have seen from the statements regarding the behavior of one deep-sea trawl (p. 61), the distance covered may bear little relation to the area that was effectively fished. In addition, a fine-mesh trawl similar to the Menzies SBT (0.55-mm mesh) usually clogs and therefore samples a much smaller area than would have been indicated from time on the bottom and by any "before-and-after" navigational fixes.

To prove this point, we cite the findings of Menzies and Rowe (1968), who took a LUBS and a SBT trawl at the same position and depth. The indications were that the

Figure 2-24.   Underwater trawl camera photographs showing trawl behavior on bottom at 5050 m (R/V *Eastward* Station 7799): (a) Trawl descending (depth 900 m); (b) mud cloud, trawl probably working (depth 5000 m); (c) camera face covered with mud, no photo of bottom; (d) trawl leaves bottom, note mud falling from bridle (depth 5000+ m).

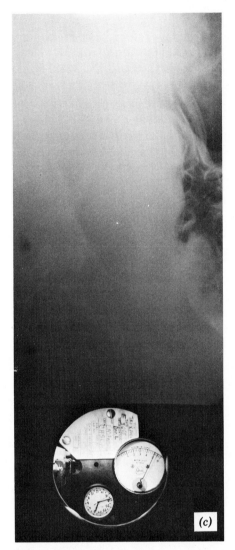

Figure 2-24. (*Continued*)

SBT captured 180 animals and the LUBS, although covering an area only 0.25 m², captured 69 animals (Table 2-2). This means that the SBT, providing it was not selectively sampling only certain members of the fauna, effectively sampled only a 0.6 m² area, even though it was on bottom for 2.17 hr. This figure must be considered to be a rough estimate at best, since it was assumed that the animals were evenly distributed numerically, but it does give an idea of how little a trawl can do. Although use of larger meshes overcomes the clogging problem, the smaller and more numerous animals are lost. If only large epifaunal animals are considered, the simultaneous use of camera and trawl can allow accurate quantification of trawl catches.

Even though methods of arriving at quantification of trawl catches as described previously seem to be of dubious value, they do provide relative indications of abundance. When several trawls, photographic stations, and quantitative grabs are taken in the same area, statistical tests can be applied assuring reliability of abundance.

TABLE 2-2. CATCHING CAPACITY OF LUBS AND SBT IN DEEP WATER* (FROM MENZIES AND ROWE, 1968)

Animal Group	Mega-LUBS 7813 (3065 m)	SBT 7814 (3155 m)	$f=4$ N/m²	Mini-LUBS 7609 (775 m)	SBT 6241 (975 m)	$f=16$ N/m²
Isopoda	3	57	12	1	26	16
Amphipoda	6	18	24	0	11	0
Tanaidacea	1	11	4	0	0	0
Echinoidea	2	2	8	0	0	0
Pogonophora	3	1	12	0	12	0
Sponges	3	8	12	0	5	0
Gastropods	25	1	100	12	0	192
Bivalves	12	54	48	14	12	224
Stephanoscyphus	8	5	32	4	0	64
Polychaeta	0	6	0	12	42	192
Ophiuroids	0	4	0	1	0	16
Alcyonaria	1	1	4	0	0	0
Solenogasters	2	3	8	0	10	0
Decapoda	0	2	0	0	1	0
Actiniaria	0	3	0	0	0	0
Bryozoa	0	1	0	0	0	0
Nematoda	3	0	12	3	45	48
Cumacea	0	3	0	0	8	0
Corals	0	0	0	0	0	0
Groups	12	17	—	7	10	—
Specimens	69	180	276	47	172	752
Depth (m)	3065	3155	3065	775	975	775

* Diameter mega-LUBS = 56.2 cm, penetration 12.5 cm, volume 3735 cm³ sample.
Area = 2490 cm² = ca. 0.25 m²; free-fall 10 ft.
Diameter mini-LUBS = 28 cm.
Area 617 cm² = ca. 0.062 m².
f = Conversion factor to square meters.

An alternate and probably valid procedure would be to compare the quantity of particles above the mesh diameter of a trawl with similar size particles captured in a corer or grab. Particles such as pelagic foraminiferal tests would provide a useful comparison because these, unlike the benthic Forminifera, are more likely to be evenly distributed over and in the sediments, unless winnowing by bottom currents has caused them to accumulate selectively. Winnowing occurs in some areas of the seabed, but it is unusual.

If trawls are poor collecting devices and grabs take too small an area of bottom for the work required, then what are the alternatives? Obviously a program of selective sampling using a deep-submergence research (rescue) vessel (DSRV) offers the best hope for truly quantitative sampling in the future. Until better times, there is the frightful thought that trawls and grab samples with or without underwater photographs will constitute the prime source of information about deep-sea bottom life for many years to come.

Treatment of the Sample

Once on board, a successful sample faces a critical period. The indiscriminate investigator, by allowing the loss of water from the mouth of an anchor dredge or from out of the flaps of a LUBS or grab, can lose a significant fraction of his animals.

For example the quotation at the beginning of this chapter reveals how Hansen (1913) was appalled at the treatment of mud samples aboard most ships, and on the Danish *Ingolf* expedition he used a fine-mesh sieve in order to recover small animals.

The sieving of fresh samples has been done in various ways by different workers, and Reish (1959) has shown (Figure 2-25) the importance the method bears on the kind and amount of specimens recovered. In practice, the largest sieve size used should be equivalent to that of the net used in the trawl and no larger. With grab samples the sieve size should be small enough to capture the smallest animal desired by the investigator. The writers have used 550-μ mesh so that SBT and LUBS catches could be compared (Menzies and Rowe, 1968).

Sanders and his associates at Woods Hole use 0.42-mm mesh for sieving anchor-dredge samples at sea, and this size has been adopted by others (Griggs et al. 1960 at Oregon State University, Hessler at Scripps Institution of Oceanography, and Southward at the Plymouth Laboratory in England). When dealing with biomass it is acceptable to use a sieve of much larger mesh, at least in shallow water, since most biomass is retained on a much larger screen (Figure 2-25). This is probably not true in deep water due to the paucity of the fauna and the small size of the individual animals. A compromise size on the order of 1.0 mm rather than 0.42 mm may not affect biomass measures appreciably, but this hypothesis has not been tested. If effective, the proposed change would affect numbers of animals recovered appreciably (Figure 2-25).

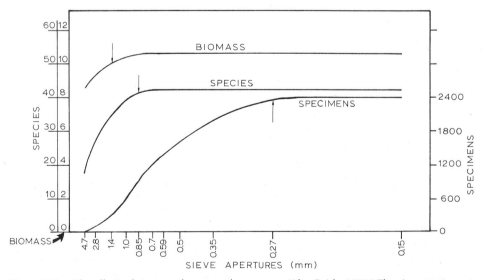

Figure 2-25.   The effects of sieve mesh on sample recovery. (After Reish, 1959.) The sieve aperture at which more than 90% of the total was retained is indicated in each instance by an arrow. (Redrawn from Holme, 1964, Figure 36, p. 236.)

Water-shaker tables with several levels of screens have been used in shallow water, but their value is limited when dealing with the tenacious clays from deep water. Care must always be taken that the small, delicate animals are not crushed when water flow is too forceful. Patience is essential, and several hours may be required to sieve a full anchor dredge, Campbell grab, or any other large sample.

In most cases (e.g., with grabs and the anchor dredge), it is necessary to measure the volume of the mud taken in order to estimate area covered. With LUBS the depth of penetration is an accurate measure of volume taken. Other devices require a dumping of the sample into some other container before volume estimates can be made.

Fixing the Sample

A recommended practice after determining the sample volume is to freeze an aliquot for sedimentary analyses and meiofauna investigations. Shortly thereafter, the sieved samples should be fixed in 10% neutral formalin. The formaldehyde (40%) may be neutralized using the organic buffer hexamethylinamine in a saturated solution, roughly 1 lb per quart of formaldehyde, or a saturated solution of borax (sodium borate).

If the samples are to be used by systematists, it is necessary to follow their advice in fixation closely. Calcareous animals should be left in formalin for as short a period as possible to prevent the dissolution of the calcium carbonate. Generally, these specimens should be changed to ethyl alcohol (ca. 70%) after several days; in certain cases, such as samples including some of the echinoderms, specimens can be dried without destroying taxonomic characteristics.

Unfortunately, both formalin and alcohol make the specimens unsuitable for various chemical analyses. In fixation, formalin coagulates the protein, as well as tending to dissolve carbonates through its slight acidity. Alcohol extracts various organic constituents, and this is manifested in a visible loss of color from the animals, which in turn leads to a discoloration of the preservative. Freezing, a customary alternative to the routine prescribed, frequently has the unfortunate effect of destroying the morphological qualities most required by systematists. It is often fruitful to induce specimens to relax by narcotizing them with barium sulfate prior to fixation. Similarly, many specimens, such as anemones, need to be injected with fixative.

SAMPLE SORTING

Separating the individual animals from the remaining sediment and detrital particles is another critical step; indeed, if done improperly it can negate a sample's value. The most efficient method is to suspend aliquots in a large container and then decant off the small species, which usually sink more slowly than the heavier sediment particles. The supernatant removed then must be "picked" using a binocular dissecting microscope at low power, in order to extract the individuals. The suspension technique is continued until no more animals appear in the supernatant. To extract the large heavy animals such as bivalves, thin layers of the sediment itself must be scanned under the microscope. A dilute solution of rose bengal stain can be used to

TABLE 2-3. BENTHIC BIOMASS ANALYSIS SHEET

Research Vessel _____ Cruise No. _____ Station No. _____

Date _____ Time: _____
Gear _____
Depth _____

Taxon	Number of Specimens	Displacement Volume or Weight	Taxon	Number of Specimens	Displacement Volume or Weight
Foraminifera			Cephalopoda		
Porifera			Scaphopoda		
Hydrozoa			Polychaeta		
Scyphozoa			Pycnogonida		
Stephanoscyphus			Copepoda		
Actiniaria			Ostracoda		
Gorgonaria			Mysidacea		
Alcyonaria			Cumacea		
Turbellaria			Decapoda		
Nematoda			Isopoda		
Nemertina			Amphipoda		
Ectoprocta			Tanaidacea		
Brachiopoda			Cirripedia		
Pogonophora			Ascidiacea		
Kinorhyncha			Enteropneusta		
Sipunculoidea			Cephalocordata		
Phoronida			Cyclostomata		
Priapuloidea			Chondrichthes		
Echiuroidea			Osteichthyes		
Asteroidea			Squid beaks		
Ophiuroidea			Shark teeth		
Echinoidea					
Crinoidea					
Holothuroidea					
Aplacophora					
Monoplacophora					
Polyplacophora					
Gastropoda					
Bivalvia					

Volume _____
O_2 (ml/liter) _____
Temperature (°C) _____
Sediments
<500 μ
<250 μ
>250 μ
Salinity (‰) _____

aid in separation. This substance, which stains only "living" tissue, can be extremely useful with animals such as benthic Foraminifera. During this step, separation can be made to phylum, class, family, and so on, depending on the knowledge of the investigator. Table 2-3 represents the data form we used at this step.

Deep-sea biology, in its era of exploration, must contend with the fact that the fauna is relatively unknown. In most deep-sea samples a large fraction of the small individuals have not been described, and the large species, if described, are still known only by a few well-trained specialists. Unfortunately, the dearth of competent systematists creates a bottleneck in the advancement of understanding ecological problems in the sea. The immediate future does not promise relief because little support is available for training students who are oriented to taxonomy. Hence benthic biologists, regardless of prime interest, should learn at least one group of animals in order to fill the void.

There is no doubt that the state-of-the-art of benthic sampling in the deep sea is archaic. Except for wire rope, photography, and the pingers and echo-sounders used to position gear on the bottom, there have been no major advances since the *Challenger,* a century ago. Quantitative grabs seldom work well in deep water and almost never function after inadvertent slamming against the ship. In rough weather they become lethal if not used with caution.

Figure 2-26.   Research submersible vessel *Trieste I* now used by the U.S. Naval Electronics Laboratory, San Diego. Official photograph, U.S. Navy.

At first glance the answer would seem to be simply to hire engineers to design and construct an ideal sampler that fits a specific list of criteria. In our experience, however, this suggestion, has seldom produced satisfactory results because it leads to "overengineering"—the constructed gear is too complex, too well-machined, and has too many moving parts. After several successes it becomes clogged with abyssal ooze or is dropped on the deck too hard and broken. At this point, if it is too "well-engineered" to be fixed at sea by the biologist, it is lost for the rest of the expedition.

Except for a few interesting exceptions (Barham et al., 1967) and several French accounts (Laborel et al., 1961; Vaissiere and Carpine, 1964; Reyss, 1964a; Reyss, 1964b; Guille, 1965; Reyss and Soyer, 1965), deep submergence research (or rescue) vessels (DSRV) have not been used by biologists. As the problems of managing our ocean ecosystem become more pressing, we no doubt will find them necessary.

The new avenues available to the study of the science of the benthos will be worthwhile, especially since *in situ* biology will allow many experimental approaches that have heretofore been impossible. It could be argued that these innovations essentially make the subjects of this chapter obsolescent, but the writers se-

Figure 2-27. U.S. Navy photograph of parts of the submarine *Thresher* on the bottom of the Atlantic Ocean, 220 mi east of Cape Cod, Mass. Courtesy of the U.S. Naval Photographic Center. Photograph taken aboard the *Trieste:* note the extreme clarity.

Figure 2-28. Shrimp on bottom off Florida coast, photo taken by Dr. Harvey Bullis, Base Director, U.S. Department of the Interior, Fish and Wildlife Service, from a view port on the submersible *Aluminaut*. Note the great detail.

riously doubt it. Rather, we envision DSRV's used as complements to conventional research. They should allow us to place limits on the actual capabilities of grabs, photography, and other devices, while allowing visual confirmation of zonation, clumping behavior, and other interesting phenomena. Photos taken by submersibles such as the *Trieste* (Figure 2-26) are outstanding in quality compared with the results of conventional tethered underwater cameras (Figures 2-27 and 2-28).

CHAPTER 3

Defining Faunal Zones

The boundaries of the abyss have been variously designated:

> 3000 to 6500 to 7000, according to Pérès (1957, p. 131).
> 600 to 6000, according to Zenkevitch and Birstein (1956, p. 55).
> 2000 to 7000, according to Bruun (1957) and Wolff (1960).
> "An upper limit needs to be agreed upon," according to Menzies (1965, p. 196).

Knowledge of the bathymetric distribution of the benthic marine fauna has been little improved at depths greater than the intertidal and shelf since the works of Agassiz (1888) and Murray and Hjort (1912), the synthesis of Le Danois (1948), and the investigations of Ekman (1935, revised 1953) and Vinogradova (1962). It is difficult to select one scheme of zonal nomenclature over another mainly because they are based not on faunal change, but instead on isobaths or isotherms. Hedgpeth placed the problem in perspective:

> There has been less argument—and less confusion—about high seas and the deeper parts of the ocean, perhaps because our information to date is too scanty to tempt the coiners of terms. Archibenthal (or archibenthic), continental slope, bathyal, and archibathyal have frequently been applied to the environment of the continental slope down to about 1000–2000 meters, but their usage has not gone beyond the diagrams in textbooks. (Hedgpeth, 1957, p. 21.)

The most definitive reference to the general problem of zonation in the sea beyond the intertidal is that of Ekman (1953). He stated

> It is impossible to fix at a certain depth the generally valid boundary between the deep-sea fauna and that of the shelf. In nature, this boundary is quite indistinct and consists of a mixed or transitional region. Its position must be determined by reference to the *distribution of the animal species* and not primarily with regard to physical conditions such as the temperature or illumination. It must not be placed at the lower limit for the distribution of certain shelf animals or the upper limit for certain deep-sea animals if it does not also coincide with the level for the greatest faunal change [p. 266]. . . . Investigations into the position of the greatest faunal change seem not to have been undertaken and will have to wait until our knowledge of the deep-sea fauna is more complete [p. 267] (emphasis supplied).

There have been some attempts to give a stabilized nomenclature for the biotic zones of the sea, even though the "zones" were of dubious reliability as Ekman noted (*op. cit.*).

National committees as well as international commissions [Committee on the Treatise of Marine Ecology and Paleoecology of the U.S. National Research Council (Ladd et al., 1949, 1951) and the Commission Internationale pour l'Exploration de la Mer Mediterranée (Pérès, 1957)] met with little success, largely because they too lacked data on the fauna. Several schemes (Table 3-1) are recognized today; it is

TABLE 3-1. COMPARISON OF FAUNAL ZONE NOMENCLATURE AND DEPTH (VARIOUS AUTHORS).

Hedgpeth (1957)	Pérès (1957)	Ekman (1953)	Vinogradova (1962)	Zenkevitch (1959)
Supralittoral (above high water)	Supralittoral (upper level) Springtide to intertidal (0–5 m) [upper level mean tide]			Surface zone littoral (0) Surface zone sublittoral (0–200 m)
Littoral (intertidal) (between high and low water)	Mediolittoral (upper mean to lower mean tide) Infralittoral (lower mean tide to lower spring tide)			
Sublittoral– inner (low water: 75 m) Sublittoral– Outer (75– 250 m)	Circalittoral shelf (50– 200 m)			
Bathyal (250– 4000 m?) Abyssal (4000 [?]– 6500)	Bathyal (200– 300 m) a) Epibathyal b) Mesobathyal c) Infrabathyal Abyssal (3000– 6500 m)	Archibenthal Abyssal	Abyssal (3000– 6000 m) Upper abyssal subzone (3000– 4500 m) Lower abyssal subzone (4500– 6000 m)	Transition zone (200–600 m) Abyssal (600– 10,000 +m) Upper abyssal (600– 2000 m) Lower abyssal (2000– 6000 m)
		Abyssal lower zone		
Hadal (6500– 10,000 +m)	Hadal (6500– 11,000)		Ultra-abyssal zone (6000– 10,000 +m)	"Super-Ozean-ische Tiefe" (6000– 10,862 m)

obvious that the zones are not all of equivalent rank, nor are they always at the same depth.

To approach the problem of defining deep-sea zones, we follow the advice of Ekman (*op. cit.*). That is, we determine those regions on the sea floor where the fauna changes at a maximum rate. Two sets of information are prerequisite to the recognition of a rate of faunal change. First, the species comprising the fauna must be known to the investigator. They are the only basis for the differentiation and recognition of zones, and therefore errors in identification could lead to erroneous results as readily as errors in sampling. Second, and of equal importance, the collections must have been made with sufficient intensity, as well as sufficient variation in depth, to allow the reliable determination of a rate of faunal change. Preferably the collections should have been taken from the same ship, by the same investigator, and with the same instruments and techniques.

However, we do acknowledge that, from a philosophical point of view, any data that truly represent the fauna would do, no matter how they were acquired. Of course, to avoid the mistake of indiscriminate pooling of data, as has been done by previous workers, the collections used in any calculation of faunal change should cover a single geographic region alone and should not be syntheses of many areas. We believe that our regional treatments satisfy the foregoing criteria as well as can be expected today.

In order to aid in selection of terms and the placement of faunal boundaries, we have developed criteria for the recognition and the classification of provinces and zones. The zones selected are based mainly on isopod Crustacea, but data from large epifaunal animals suggest that the scheme has general applicability.

DETERMINATION OF FAUNAL CHANGE

The method of determining the homogeneity or distinctiveness between sampling points and depth intervals is exceptionally simple and straightforward. The total taxa in common (T_c) between any two points was subtracted from the total taxa (T) at the two points, divided by the total (T) and multiplied by 100 to gain the percentage of distinctiveness (D). With this formula, the more taxa in common between two points then the lower the percentage distinctiveness, and vice versa. Here is a hypothetical case:

Depth Point or Interval

Taxa Genus	A	B	C	D	E	F
1	×					
2	×					
3	×					
4	×					
5	×					
6	×	×				
7		×	×	×	×	×
8		×	×	×	×	×

9   ×   ×   ×   ×   ×

10   ×   ×   ×   ×   ×

$$\frac{T - T_c}{T} \, 100 = 0\%$$

Distinctiveness D

Between A and B	90
Between B and C	16
Between C and D	0
Between D and E	0
Between E and F	0

Conclusion: The maximum rate of faunal change is between points A and B. Points B to F inclusive represent a single faunal unit. It would be our practice then to combine points B to F into a single faunal unit or depth range. In this case B to F inclusive represents sampling stations, collectively called depth interval 1, 2, 3, etc.

Using this formula, we can arrive at the values of percentage distinctiveness along a bathymetric scale. Abrupt changes in the slope of the line connecting these values indicate differences in faunal composition between points, whereas little slope in this line means there is little change in faunal composition. Depths at which the rate of change is great in comparison with points in between are presumed to be the borders between faunal units. The larger the number of taxa and depth intervals involved the greater is the reliability of the measurement.

This approach is similar to the "coefficient of difference" developed by Savage (1960). The authors have also used his technique and arrived at comparable results. His formula for province recognition, $CD = 1 - C/N_2 \times 100$,* has the advantage of reducing the error induced by small samples. However, it does not take into account the distinctiveness of zones as indicated by taxa starting or stopping at one depth, nor does it consider the endemics between the two depth regions that are compared. Once faunal units are identified both in depth range and composition, we express the data in graphic form. The proportional distribution of taxa from one province or zone is plotted with reference to the fauna on the entire bathymetric scale. The plot is a linear expression of the trellis diagram utilized by Sanders (1960). Once a zone or province is identified and named, its characteristics are described in quality and quantity, including endemism (bathymetric), number of contained taxa, species diversity/genus (not total fauna), morphological characteristics of the animals, proportional group distribution, and large animal composition.

Selection of Depth Intervals

Depth intervals of sampling were ascertained to a great extent by convenience and chance collection, rather than by predetermined or arbitrary depth units. Those stations in close proximity and those having an identical fauna were combined even

---

* CD = Coefficient of difference for faunal provinces.
  C = Number of species common between two points.
  N_2 = Number of species at point having largest number of species.

though there may have been considerable difference in depth between sampling points. Quite often many stations occurred within a limited depth range (see Figure 4-1 and Tables 4-1 to 4-3) but from different points horizontally on a chart and hence several kilometers apart, within the general region investigated.

Few of the studies that have been made have been characterized by adequate sampling intensity in any restricted geographic location. The usual approach has been to integrate data from many different geographic regions with the hope of obtaining meaningful results. Our comparison of faunal zones, based on data collected from small regions of several major parts of the oceans, is a significant departure from the usual method of pooling data on species depth distribution from all oceans. Although the pooling technique is available to those with sufficient library resources as well as a comprehensive knowledge of the taxonomy of each group of animals, there is a mixing of the depth ranges of species from various places.

The result is loss of zonal character because species are distributed at different depths, depending on the special hydrological and ecological features of each geographic region. Vinogradova (1958) applied the pooling technique with most interesting results. Alternately, studies have been based on samples taken along a narrow line (Sanders et al., 1965; Sanders and Hessler, 1969). Both methods yield valuable information, but neither yields data that are entirely suitable for comparing the rate of faunal changes within a single geographic region and hydrologic regime.

Isopods as Valid Zone Indicators

Benthic isopods are ideal indicators of zonation for several reasons. First, the animals develop directly from the parental brood-pouch and thus have no planktonic or migratory larval stage; second, most species are obligatory benthonts without any means of swimming and are generally mud dwellers in and on the surface sediment (Hult, 1941); third, they are among the most abundant of the deep-sea invertebrate Crustacea; and fourth, their remains are not subject to transport from shallow water to deep water by turbidity currents or slumps, as is the case with foraminiferan and mollusc shells. The animals sampled therefore are living and provide a real picture of the fauna at a given locality and time that is not confounded by secondary biotic or abiotic mixing processes.

Other Animals Characteristic of Zones

Throughout this work certain large invertebrates and vertebrates have been found in the various zones that were determined first on the basis of isopod data. Off North Carolina and off Peru, where sampling intensity has been comparatively great, it has been possible for us to recognize the rate of change of the fauna of large animals and to associate this with isopod zonation. The coincidence between these two different kinds of data prompted us to conclude that the zones determined from Isopoda would in all probability apply equally well to many other members of the marine fauna. Drawings based on underwater photographs, as well as the preserved specimens and bottom photographs, have been used to demonstrate the fauna of various zones. A classification has been arrived at that is more objective than that of convenience or that based on subjective intuition, such as has been used in the past.

CLASSIFICATION SYSTEM FOR ZONATION

The major environmental subdivisions of the sea are the benthic and the pelagic. The pelagic has been divided into photic and aphotic realms based on the presence or absence of sufficient sunlight for photosynthesis (Sverdrup et al., 1942, p. 275, and others). An application of the same reasoning to the benthic division results in photic and aphotic benthic realms which are equivalent to Ekman's (1953) shallow-water and deep-sea systems and to the phytal and aphytal realms of Pérès (1957). Thus the hierarchy of faunal classification proposed starts with the benthic division composed of the photic (phytal), and Aphotic (aphytal) realms. These realms are divided further into provinces and these, in turn, are split into zones (Figure 3-1).

Faunal Realm (Distinguished at Kingdom Level)

As suggested previously, two inclusive faunal realms constitute the benthic division. These correspond with the presence or absence of sunlight and the presence or absence of living sedentary members of the plant kingdom along a vertical profile of depth. Zoogeographers have tended to minimize the importance of these categories in the past (Hedgpeth, 1957) because of the limited correlation between temperature and the lower boundary of the photic zone, and probably also because benthic

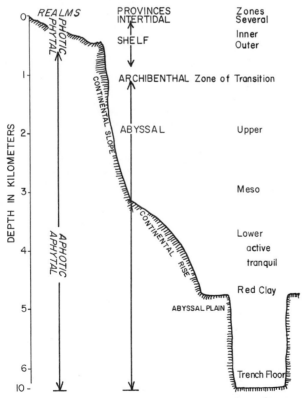

Figure 3-1. A semidiagrammatic view of the components of the Benthic division of the oceans from realms to provinces and zones, arranged according to depth on an idealized profile of a continental margin.

animals do not follow the vertical distribution of living plants closely. Obviously, the only animals with exact coincidence in distribution with living plants would be the plant parasites or the obligatory herbivores. Few (by number of species) benthic animals are parasites or herbivores and thus limited precisely to the distribution of living plant matter. Deep-sea animals are not directly dependent on living plants, although some live in or near the photic zone in polar regions. This relative independence, however, does not invalidate the broad concept of photic and aphotic realms in the sea, because they are the two most salient characteristics associated with the distribution of life in the sea in the pelagic as well as in the benthic division.

Faunal Province

A faunal province is a major division of a realm. It may be distinguished from other provinces by the rate of change of genera along a bathymetric scale within a realm. The assemblage of genera within a province constitutes its fauna (Figure 4-12 to 4-13, 4-16, 4-29).

Faunal Zone

A faunal zone is a subdivision of a province and is distinguished by the rate of change of species along a bathymetric scale within a faunal province. The species contained within a faunal zone constitute the members of the zone (Figures 4-11, 4-18).

Utilizing the foregoing definitions and the method already described to determine the depth points of greatest faunal change (for realms, provinces, and zones), we have been able to distinguish the upper and lower limits of a variety of faunal units and to recognize and compare their special characteristics in several regions of the world oceans. Little attention has been paid here to the details of the well-documented zones of the intertidal or the shelf in the photic realm. Major effort has been expended on zones within the aphotic realm. That the shelf province and the intertidal province can be subdivided into one or more zones, as indeed others have done, is abundantly clear. Our shallow-water data are too scanty to recognize the subzones except those off North Carolina and off Peru, where distinctive subzones of the shelf province are evident.

Start of the Abyssal Faunal Province
(Deep-Sea Fauna)

Our data on zonation reveal one obvious feature—namely, that the abyssal fauna is located at different depths in different parts of the world, as suggested by Menzies (1965) and as implied much earlier by many others. That this zonation appears to be directly related to the temperature structure of the oceans has often been suggested before, but most recently by Anton Bruun (1956, 1957), who located the abyss at the 10°C isotherm. From our zonation picture it is evident that the 10°C isotherm is not the start of the abyssal fauna in any part of the world that has been investigated. The next several chapters discuss the results of applying the approaches presented here to data from the various regions of the world ocean.

CHAPTER 4

North Atlantic

> The fauna of the European Atlantic . . . is in some respects the best known in the world. But still how recent is our knowledge! When the 18-year-old Georges Cuvier travelled in 1788 to the coast of Normandy to study its fauna, it was the more than 2000-year-old writings of Aristotle which he used as a reference book. (Ekman, 1953).

We realize immediately that Ekman's comment on the "recentness" of the acquisition of knowledge is even more applicable to the American Atlantic shores.

THE NORTHWESTERN ATLANTIC: CAROLINAS TO GEORGIA

General Features

The region of the North Central Western Atlantic (Figure 4-1a) is characterized at the surface by the nearshore shelf water, the offshore Gulf Stream (Florida current), and beyond that, the Sargasso Sea. The intertidal and shelf nearshore fauna is seasonally transitional between the warm-temperate fauna of the Virginias and the subtropical fauna down to Florida.

In the discussion of each province and zone, various correlated physical factors are mentioned. It is useful here to review them briefly. Figure 4-1a demonstrates where the provinces and their zones are located. It is evident that there are multiple correlations in all instances. Thus the zones of the abyssal faunal province seem to be most powerfully associated with current velocity and current direction, and, because sediments are also influenced by these factors, the correlations between sediments and provinces are good also. Conventional isobaths and conventional isotherms do not define provinces or zones. The intertidal province, the shelf province, and the archibenthal zone of transition are all under the influence of seasonal changes in currents and temperature. The abyssal faunal province starts where fine muds with less than 15% sand commence and where seasonal temperature variation is no longer apparent. The 4°C isotherm defines the upper boundary in this region.

Figure 4-2 is a topographic planar reconstruction of the seabed off the Carolinas. The illustration also shows the position (a–h) of the actual depth record profiles of Figure 4-3 and bottom photographs Figure 4-4. Sediments were obtained along this area and have been plotted and contoured (Figure 4-5). Photomicrographs of smears

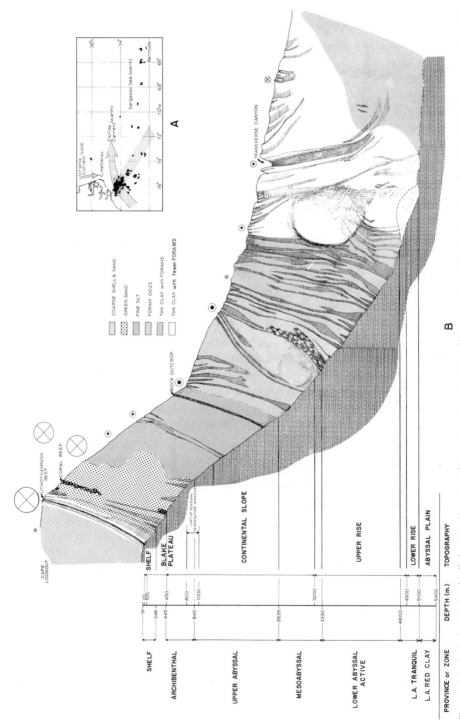

Figure 4-1. Area investigated off the Carolinas, North Western Atlantic: (a) Collecting stations (having isopods) plotted; (b) planar reconstruction of vertical profile of area showing zones, topography, sediments, current direction (as indicated by symbols ○ south and ⊕ north, and relative current velocity by the diameter of circles). Figure 4-5 provides greater detail regarding sediments.

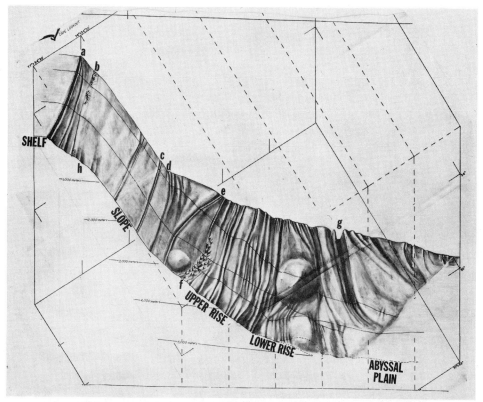

Figure 4-2. Topographic reconstruction of sea floor off Cape Lookout, North Carolina: coordinates in latitude and longitude, vertical exaggeration 1 : 39; scale 1 : 111,000. Letters refer to PESR traces and submarine photographs in Figures 4-3 and 4-4, respectively.

of sediments corresponding to the sediment types that are contoured appear in Figure 4-6. In addition to these presentations, it is useful to have an idea of the currents along the profile (Figures 4-1 and 4-7c), where the arrow size indicates relative strength of the currents. The reader may gain a visual impression of the seabed and its dominant features by studying Figure 4-4, which presents bottom photographs of the features (Figure 4-6, A–H inclusive). These are referred to in the discussion of correlative features and faunal zonation off the Carolinas. We also present (Figure 4-7a) the distribution of isotherms and the distribution of dissolved oxygen (Figure 4-7b). The information on topography, faunal provinces, and currents is summarized in Figure 4-1a.

The intertidal province coincides with the tidal range and associated factors such as desiccation and sand or rock, as it does elsewhere in the world (except at the poles).

The shelf faunal province, with its inner and outer zones, does not coincide with the shelf break and instead extends from below the tides to a depth of 246 m. The shelf faunal province shows neither sediment nor depth correlations even though the faunal zones within the shelf province are sediment affiliated (e.g., the submerged tropical zone over the rocky *Lithothamnion* bank and the inner shelf zone affiliated with coarse sands and rubble or mud embayments. The lower boundary of this prov-

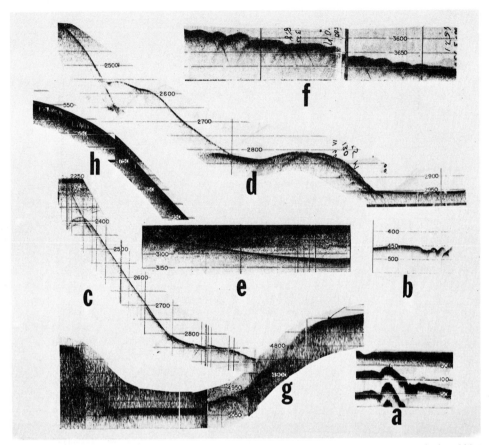

Figure 4-3. PESR traces of lettered features in Figure 4-2; depth in meters: (a) *Lithothamnion* algal reef-like structure, scale 0 to 750 m (60 m); (b) *Lophohelia* coral bank, scale 0 to 750 m (458 m); (c) outcrop, scale 2250 to 3000 m (2400 to 2800 m); (d) contact, continental slope—upper continental rise, scale 2250 to 3000 m (2950 m); (e) outcrop upper continental rise, scale 3000 to 3750 m (3120 m); (f) hyperbolic hillocks, continental rise, 33°22.9′N × 75°08′W, scale 3000 to 3750 m (3700 m); (g) Hatteras submarine canyon, E-20-66, scale 4500 to 5250 m (5250 m); (h) outer edge of Blake Plateau, scale 0 to 750 m (540 m).

ince does correlate most evidently with the start of the well-sorted, foraminiferan-rich glauconitic sands that commence near this depth.

The archibenthal zone of transition (445–940 m) corresponds to glauconite-rich foraminiferal and pteropod oozes (marked *G* and *S* in Figure 4-5) under the influence of the Florida current. It embraces the rapid change in isotherms with depth and an intermediate zone of no motion just above the western boundary undercurrent (WBUC) between the Florida current and the WBUC. Its temperature ranges from 10 to 5°C. The water is subject to seasonal changes in temperature in excess of 4°C.

The abyssal faunal province (1000–5315 m) commences with the start of the true abyssal sediments (*F, RC, RG*) and with the temperature isotherm of 4°C it embraces all the bottom water including that of the WBUC and the Antarctic bottom water. Topographically it includes most of the continental slope, all the continental rise, and the abyssal plain.

Figure 4-4. Underwater photographs at locations (a) to (h), inclusive, of Figures 4-2 and 4-3.

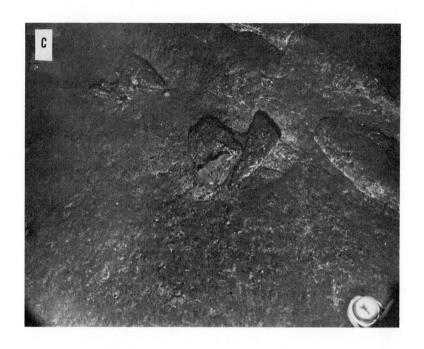

Figure 4-4. (Continued)

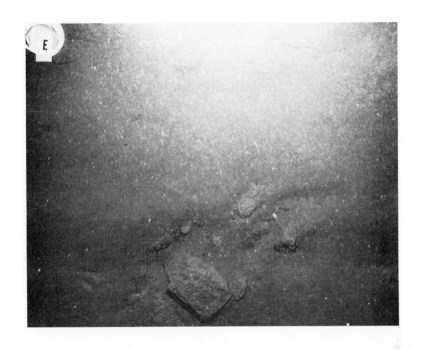

Figure 4-4. (*Continued*)

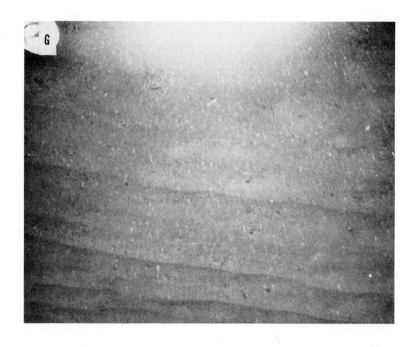

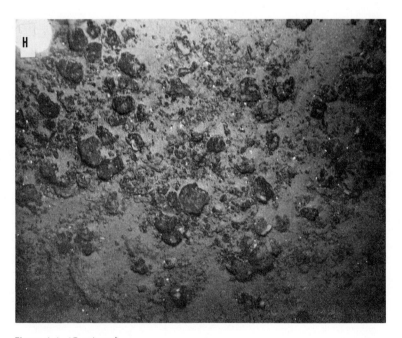

Figure 4-4. (*Continued*)

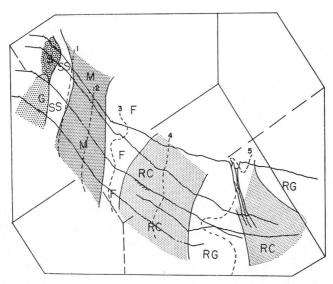

Figure 4-5.   Distribution of sediment types off North Carolina (data from Menzies and Rowe, in SS): G represents well-sorted glauconite–foraminiferal sand, S is poorly sorted glauconite–foraminiferal sand, SS is sandy silt, M is fine silt and clay, F is foraminiferal ooze, RC is clay and silt with a small fraction of sand composed of foraminiferan tests, and RG is lutite with almost no (less than 1.0%) sand fraction. (After Rowe and Menzies, 1968, Figure 9, p. 717.)

The upper abyssal zone (940–2635 m) starts with the abyssal muds and embraces almost exactly the width of the band of M-category sediments. These are mainly silt and clay, with much less sand than the neritic sediments of the archibenthal. The grain size is much smaller (compare Figures 4-6a, b, and c with Figures 4-6d, e, f, and g) in the M category than in the G, S, or SS category of the archibenthal and shelf. The currents in this region are rapid and to the south; the percent of sand is generally less than 10% and does not exceed 15%.

The mesoabyssal zone (2640–3300 m) lies at the juncture of the continental slope and continental rise. It commences at the lower limit of the M sediments and the upper limit of the F sediment embracing the F sediments almost entirely. The temperature range is less than 0.1°C. The WBUC has a relatively low velocity in the mesoabyssal zone, and particulate organic matter in the form of Thalassia fragments reach their maximum abundance here.

The lower abyssal (active zone) (3340–4100 m) is found within the RC sediments but is not confined to them. It is, however, located in a region of high current activity.

The lower abyssal tranquil zone (4800–5080 m) is located between the lower edge of the WBUC and the start of the Antarctic bottom water flowing north. There is scarcely any current in this "no-motion" zone.

The lower abyssal red clay zone (5000–5300 m) covers much of the Hatteras abyssal plain under the influence of northward-flowing Antarctic bottom water. The sediments are red clay with little organic matter, a small fraction of sand, and manganese nodules of pebble size.

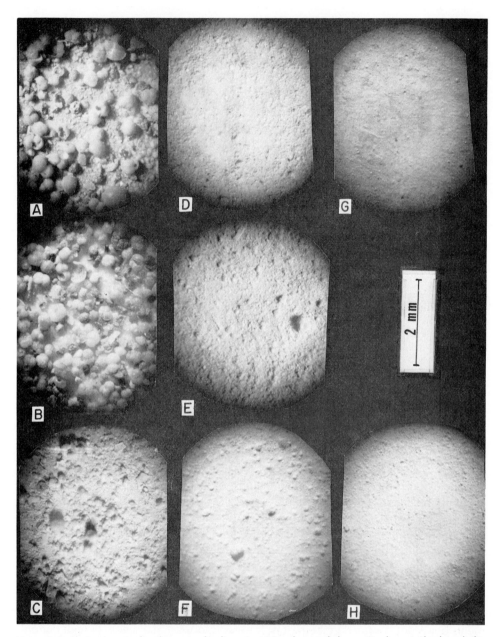

Figure 4-6. Photomicrographs of smears of sediments: (a) Poorly sorted glauconite–foraminiferal sand, (b) well-sorted glauconite–foraminiferal sand, (c) sandy silt, (d) fine silt and clay, (e) foraminiferal ooze, (f) clay and silt with a small fraction of sand composed of foraminiferan tests, (g) lutite with less than 1.0% sand fraction, (h) lutite. Dark particles of A and B are green glauconitic particles; others are pelagic foraminiferan skeletons.

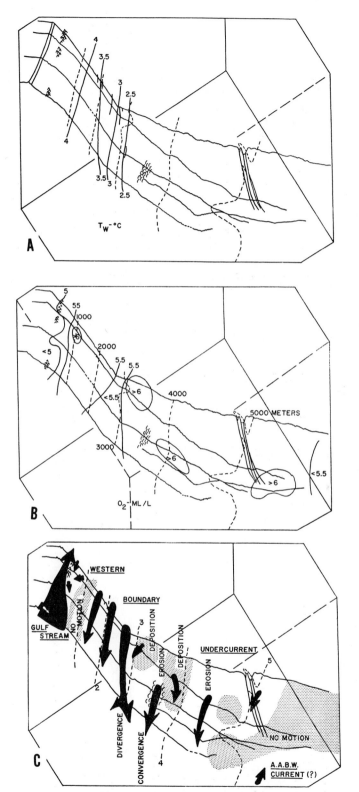

Figure 4-7. (a) Distribution of bottom-water temperatures; (b) distribution of dissolved oxygen in the bottom water; (c) summary of bottom currents. Relative current velocities indicated by arrow size. Dashed lines represent the 1 to 5 km isobaths. Stippled areas are apparent zones of no motion. (From Rowe and Menzies, 1968.)

89

TABLE 4-1. DATA ON DISTINCTIVENESS, COEFFICIENT OF DIFFERENCE, AND ENDEMISM OF ISOPOD GENERA AND SPECIES FROM CAROLINAS AND GEORGIA (NORTH ATLANTIC)

	Interval Depth (m)	Endemics		Starts		Stops		Total		Endemism, (%)		Coefficient of Difference		Distinctive- ness (%)		Faunal Zones
		G.	sp.	G.	sp.	G.	sp.	G.	sp.	G.	sp.	G.	sp.	G.	sp.	
1.	0–3	7	10	7	5	0	—	15	15	47	66	50*	71*	65*	82*	Intertidal
2.	5–20	2	6	6	6	2	2	16	17	13	35	25	47	29	53	Shelf (innershelf)
3.	21–70	0	0	1	2	2	1	13	11	0	0	16	10	15	17	Shelf
4.	80–90	0	1	0	0	0	0	11	11	0	9	8.3	8.3	8.3	17	Shelf
5.	91–150	0	1	1	1	3	7	12	12	0	8	25	67	31	73	Shelf
6.	200–250	1	3	0	0	5	3	10	7	10	43	60*	86*	67*	92*	Archibenthal
7.	450–550	0	4	2	2	1	0	6	7	0	57	17	57	17	62	Archibenthal
8.	600–650	0	1	0	0	0	2	5	4	0	25	74*	95*	74*	95*	Upper-abyssal
9.	900–1500	5	13	9	7	1	1	19	21	26	62	32	66	38	72	Upper-abyssal
10.	2000–2400	1	3	1	1	0	3	15	11	7	33	7	55	7	64	Upper-abyssal
11.	2400–2635	0	2	0	1	0	0	14	8	0	25	22	82*	22	83*	Meso-abyssal
12.	2640–3100	3	25	1	2	4	2	18	33	17	71	39	82*	42	83*	Lower-abyssal (active)
13.	3300–3600	1	2	0	0	0	1	12	8	8	25	8.3	64	15	70	Lower-abyssal (active)
14.	3800–4100	1	7	0	2	4	3	12	14	8	50	42	72*	50	83*	Lower-abyssal (tranquil)
15.	4800–5100	1	10	1	0	1	4	9	14	11	71	22	82*	30	83*	Lower-abyssal (tranquil)
16.	5200–5400	1	6	0	0	7	0	8	8	13	100	22	100*	30	100*	Lower-abyssal (red clay)

*Significant points of distinctiveness or coefficient of difference indicating faunal break.

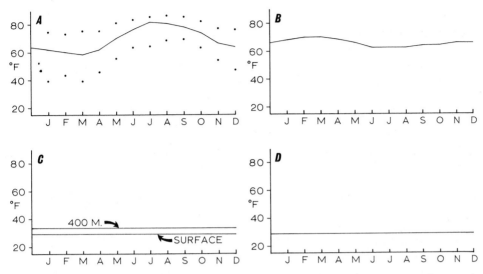

Figure 4-8. Seasonal surface seawater (or near surface) temperature curves for: (a) Beaufort, North Carolina; (b) Callao, Peru; (c) Arctic; (d) Antarctic.

Distinctiveness of Provinces and Zones

The data for the determination of faunal provinces, zones, percentage distinctiveness, and coefficient of difference are given in Table 4-1 and are discussed with reference to each zone. The proportional relationships of faunal provinces and zones appear in Figures 4-10 and 4-18.

VERTICAL DISTRIBUTION AND ZONATION

Intertidal Faunal Province (0–3 m)

The tidal range from the Carolinas to Georgia differs from north to south, being around 1 m in North Carolina and approximately 3 m in Georgia. The shallow-water environment consists of estuaries, bays, sand beaches, and offshore sandbars or banks. There is no natural rocky shore. A firm substrate is only found at artificial jetties and pilings. Because of this, the intertidal fauna and flora is considerably reduced in comparison with other regions such as California, Maine, or southern Florida, where a hard intertidal substrate predominates. The monotony of the sandy seashore, seasonal nearshore dilution from runoff, and the wide range of the annual surface sea temperature (Figure 4-8) from near 0°C in the winter to 30°C in the summer in the Carolinas, accounts for a comparatively impoverished seashore fauna as far as the diversity of taxa is concerned. Wave action varies from moderate to severe during tropical storms or "northeasters." The water bathing the seashore comes periodically from Virginia to the north with the cold Virginia current and from the south and east from the warm Gulf Stream (Figure 4-1a). Northern species such as *Mytilus edulis* (Wells and Gray, 1960) become established in North Carolina when the Virginia current enters the area. There are few, if any, species of invertebrates restricted in horizontal distribution to the intertidal of North Carolina, and among the isopods there are no endemic species in the shallow waters.

The genera and species considered to be intertidal are those found along the shore between the tide marks. This province has at least 15 genera of isopods, with 47% endemic to this province, and it is distinct from the upper part of the shelf province in 65% of the genera and 82% of the species. Apparently most intertidal and shelf genera do not mix, and less than 12% of the species are in common. Generic endemism in the intertidal compared with greater depths is 47% and species endemism is 65% (Table 4-1). A characteristic assemblage of intertidal animals in the Carolinas is presented in Figure 4-9.

At the generic level, 53% of the intertidal fauna enter the shelf province between 3 and 20 m, 40% between 21 and 70 m, 27% by 85 m, 27% by 120 m, 20% by 230 m, and zero by 445 m (Figure 4-10). Thus the intertidal generic representation decreases rapidly with increasing depth, and no contribution whatever is made to the generic fauna of the archibenthal zone.

Seven major groups of isopods (Figure 4-11) constitute the intertidal; the Sphaeromidae are dominant at 40%. The intertidal endemic group, the Oniscoidea, constitutes 13% of the species. The endemic isopod genera are *Ligia, Arma-*

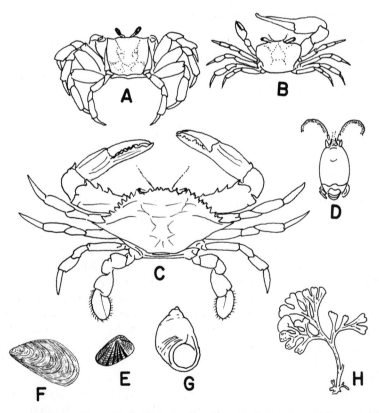

Figure 4-9. Assemblage of typical intertidal organisms from North Carolina: (a) Ocypode, (b) Uca, (c) Callinectes, (d) Emerita, (e) Donax, (f) Mytilus, (g) Littorina, (h) Fucus. (Figures e–g from Abbott, 1954; 1963.) Scales various.

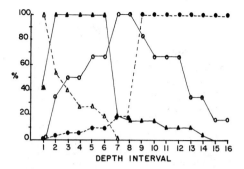

Figure 4-10. Proportional distribution of isopod genera within the four identified faunal provinces off the Carolinas: open triangle, intertidal; solid triangle, shelf; open circle, archibenthal; solid circle abyssal.

dilloniscus, Limnoria, Cassidinidea, Dynamenella, Sphaeroma, and *Munna* (Figure 4-12). All these have eyes, most are heavily pigmented, and most belong to the Sphaeromidae. There is good reason to believe that intertidal genera belonging to other invertebrate groups would show a similar distributional pattern with depth. Among them the barnacles *Chthalamus* and *Balanus,* the gastropod *Littorina,* the clams *Donax* and *Mytilus,* the crabs *Emerita* and *Ocypode,* and the plant *Fucus* are typical examples of large organisms which are generally endemic to the intertidal (Figure 4-9). Of course the plants *Spartina, Ulva,* and *Zostera* also typify the intertidal.

The intertidal province is divisible into several zones, but no attempt has been

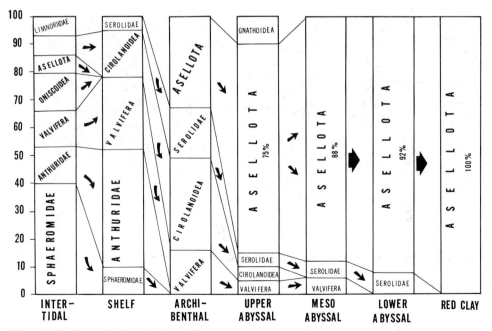

Figure 4-11. Proportional major isopod group representation at various provinces and zones in the Carolinas.

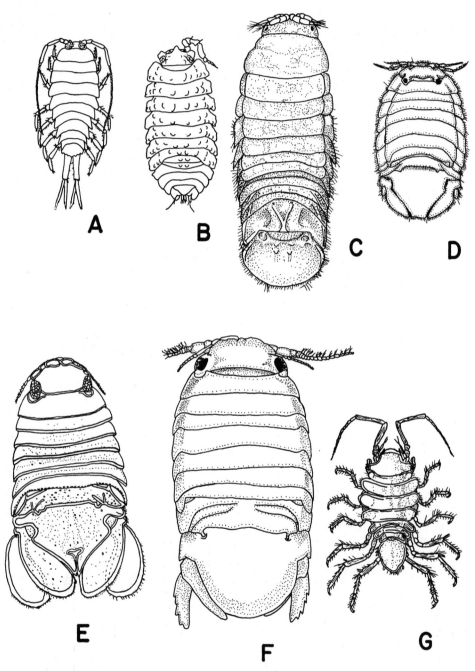

Figure 4-12. Assemblages of intertidal isopod genera characteristic of Carolinas, Intertidal: (a) *Ligia*, (b) *Armadilloniscus*, (c) *Limnoria*, (d) *Cassidinidea*, (e) *Dynamenella*, (f) *Sphaeroma*, (g) *Munna*. Scales and sources various.

made here to illustrate them. Reviews of intertidal zonation are available in Hedg-peth (1957).

Shelf Faunal Province (5–246 m)

The continental shelf off the Carolinas slopes gently for 40 nautical miles (nmi) from the offshore banks to a depth of 80 to 100 m at the shelf edge. The sediments consist of relict coarse sand and shell hash, and the bottom is generally smooth. The water on the shelf is variable seasonally, being subject to winter cooling and summer warming, dilution and mixing from the rivers, and mixing with the Virginia and Florida currents.

The water over most of the shelf out to the edge of the Florida current is turbid and rich in phytoplankton, and the visibility at the bottom is poor, due to high amounts of suspended matter. This is evident from the observation by the senior author that clear underwater photographs (focal distance 12 ft) could not be taken from any part of the shelf until the "reef" at the edge of the Florida current was reached at about 80 to 100 m depth, at least during winter months.

The surface temperature in April can change from 11 to 19°C in a period as short as 10 days (Menzies et al., 1966). This sudden change is caused by replacement of the cold winter water with warm Florida current water. Subtidal embayments often have a mud floor, whereas the reef (Figure 4-4a) at the edge of the shelf break consists of Lithothamnion rocks with an age of 20,000 years (B.P.) (Menzies et al., 1966). These two features of mud and rocks, which are quite distinct from the coarse sand bottom of the shelf proper, provide logical sites for zones and subzones within the province. Beyond the Lithothamnion reef from 100 to 246 m the slope of the seabed increases, crossing isotherms under the Florida current. The action of this current is apparent also from ripple marks in the bottom sediments parallel to the isobaths.

Thus throughout the topographic shelf and inclusive of the shelf faunal province, there is considerable variation in underwater climate and sediments. It is to the seasonal variation in marine climate that the faunal province appears to have responded with regard to distribution and diversity.

The shelf faunal province contains 19 genera of isopods, and nine of these are bathymetrically endemic. Two such endemic genera are located in the depth range of 5 to 20 m (subtidal) and one is limited to greater depths beyond the Lithothamnion reef (200–250 m). Over the 40 nmi expanse of the uniform topographic shelf below 20 m, there is not one endemic genus of isopods, either horizontally or vertically. The shelf province is 65% distinct in genera from the intertidal province and is 67% distinct in genera from the next deeper province, the archibenthal, which starts somewhere between 246 and 445 m. The genera across the shelf to a depth of 246 m are reasonably homogeneous in their distribution, whereas the species are not. This discrepancy suggests that there is added zonation of the kind repeatedly observed by those concerned with animal distribution on the shelf in many parts of the world [Cerame-Vivas and Gray (1966), North Carolina; Gilat (1964), Israel; Parker (1963), Gulf of California, etc.].

An inner shelf zone (auct. subtidal) is obvious at 3 to 20 m, and an outer shelf zone is represented between 200 and 246 m (Table 4-1). It is also apparent that the Lithothamnion reef between 80 and 90 m is a distinct ribbonlike zone which we call a submerged tropical shelf faunal zone within the shelf province (Menzies et al.,

1966). Cerame-Vivas and Gray (1966) found only 16% of the species of large animals in common between the inner shelf and the *Lithothamnion* reef, but they did not sample greater depths and were concerned only with large epifaunal organisms.

The shelf faunal province contributes (Figure 4-10) 42 per cent of its isopod genera to the intertidal and 21% to the archibenthal, but at a species level the shelf contributes 20% of its species to the intertidal and less than 5% to the archibenthal. It contributes 4% of its genera but not one species to the abyssal faunal province.

The single genus that is represented both on the shelf and in the abyssal is *Serolis*. By our method of defining faunal provinces, this genus was assumed to be found between its shallowest (5 m) and its greatest depth (3840 m). This does not appear to be the case with the two species of *Serolis* here, since the two, *S. mgrayi* Menzies and Frankenberg, and *Serolis* n. sp., are not closely related morphologically, and neither has been collected between 300 and 3500 m.

Isopod species characteristic of the shelf faunal province belong to the Valvifera (26%), the Anthuroidea (42%), and the Cirolanoidea (16%) (Figure 4-11). The Oniscoidea, the Limnoriidae, and the Asellota are absent, but because the Seroloidea is present, the shelf has only two major groups fewer than the intertidal.

The eight isopod genera characteristic of the shelf are *Cleantis* and *Edotea*, (Valvifera), *Horolanthura*, *Panathura*, *Accalathura*, *Apanthura*, *Calathura*, and *Xenanthura* (Anthuroidea) (Figure 4-13 and Table 4-2). These genera are pigmented and eye-bearing as a rule.

Large animals characteristic of the shelf are the crabs *Acanthocarpus*, *Cancer* (Figure 4-14), and *Ranilia*, the starfish Astrocyclus and *Astropectin* (Figure 4-15), the echinoids *Encope* and *Mellita*, and a host of mollusc genera such as *Aequepectin*, *Spisula*, *Echinochama*, *Mercenaria*, *Plicatula*, and *Chama*. Representative animals appear in Figure 4-16. One of the most characteristic animals of the shelf is the commercial shrimp *Penaeus* (Figure 4-16a). Pigmentation is the rule here, even among the echinoderms. The invertebrates that normally bear eyes, such as the Crustacea, have well-pigmented eyes. The shells of the molluscs are often thick and heavy, and the echinoderms and decapods are generally stout and heavily calcified in comparison to their deep-sea counterparts.

Archibenthal Zone of Transition (445–940 m)

Topographically, the archibenthal zone of transition (AZT) starts just a few kilometers beyond the *Lithothamnion* reef and extends to a depth of about 1000 m (Figures 4-2a, 4-3a, 4-4a). Toward the north near Cape Hatteras it is only about 5 km wide, but to the south it crosses the northern extension of the Blake plateau and is closer to 20 km across. It extends to the uppermost limit of the cold and uniform temperature water of the WBUC. The water on the continental margin in the depth range of the AZT consists of the bottom of the Florida current (at the southern border) and a zone of no motion between 800 and 1000 m just above the upper limit of the WBUC.

Underwater photographs (Figure 4-17) show ripples and dune marks caused by the Florida current, especially along the southernmost border. These ripples and dune marks disappear in the northern border of the transect, indicating that the Florida current has left the bottom as it bends seaward north of Cape Hatteras (Figure 4-1).

The water in the AZT is subject to seasonal changes in temperature. The 5°C

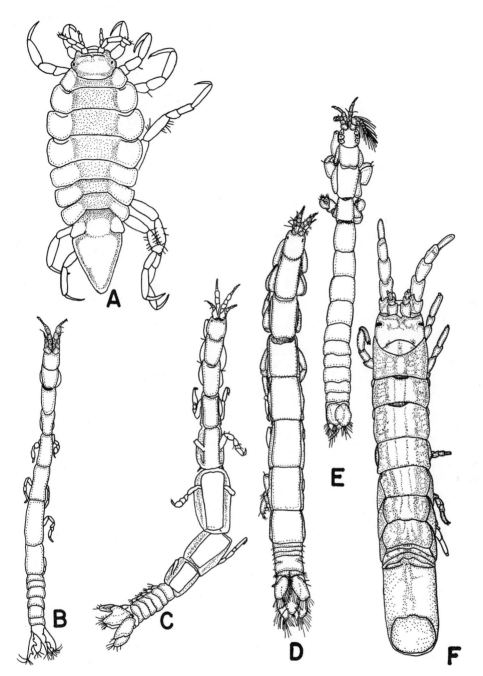

Figure 4-13. Assemblage of isopod genera characteristic of the North Carolina shelf: (a) *Edotea*, (b) *Horoloanthura*, (c) *Panathura*, (d) *Apanthura*, (e) *Xenanthura*, (f) *Cleantis*. (After Menzies and Frankenberg, 1966.) Scales various.

TABLE 4-2. BATHYMETRIC DISTRIBUTION OF ISOPOD GENERA OFF CAROLINAS IN THE NORTHWESTERN ATLANTIC; DEPTH IN METERS

Sampling Intervals

Isopod genera	1	2	3	4	5	6	7	8	9	10	11	12	13	14	15	16
	0–3	5–20	21–70	80–90	91–150	200–250	450–550	600–650	900–1500	2000–2400	2400–2635	2640–3100	3300–3600	3800–4100	4800–5100	5200–5400
Dynamenella	×															
Cassidinidea	×															
Sphaeroma	×															
Limnoria	×															
Ligia	×															
Armadilloniscus	×															
Munna	×															
Ancinus	×	×														
Edotea	×	×	×													
Erichsonella	×	×														
Chiridotea	×	×	×													
Paracerceis	×	×	×	—	×											
Cyathura	×	×	×	×	×	×										
Ptilanthura	×	×	×	×	×	×										
Eurydice	×	×	×	×	×	×										
Cleantis		×														
Horoloanthura		×	—	×	×											
Panathura		×														
Accalathura		×	—	—	—	×										
Apanthura		×	×	—	×	×										
Serolis		×	×	—	×	—	—	—	—	—	—	—	—	×		
Xenanthura		×	×	—	×											
Cirolana		×	—	×	×	×										
Astacilla			×	—	×	—	—	—	×	—	—	×				
Conilera				×		—	×	×	×							
Calathura						×										
Munnopsis							×	×	—	—	—	×				
Eugerda							×	×	—	—	—	—	—	—	—	×
Gnathia									×							
Dendromunna									×							
Desmosoma									×	×	—	×	×	×	×	×
Ilyarachna									×	×	—	×	×	×		
Macrostylis									×	—	×	×	×	×	×	×
Eurycope									×	—	—	×	×	×		
Notoxenoides									×							
Dendrotion									×	—	—	×				
Acanthocope									×	—	—	×				
Haploniscus									×	—	×	×	×	×	×	×
Ianirella									×	—	—	×	×	×	×	
Paramunna									×							
Asellota n. gen.									×							
Heteromesus									×	×	—	×	—	×	×	×
Haplomesus										×	—	×	—	—	×	×
Bathygnathia									×							
Rhacrura												×				
Antennuloniscus												×	—	×		
Hydroniscus												×				
Syneurycope												×				
Nannoniscus													×			
Echinothambema														×		
Storthyngura															×	
Ischnomesus															×	×
Abyssianira																×
Identified Faunal Province	Intertidal		Shelf			Archibenthal				Abyssal						

isotherm was located at about 450 m in November, and the temperature decreased to only 4°C by 1000 m. But during March, the bottom temperature at 500 m had increased to 8°C. During June, the temperature at 500 m had risen to 10°C. The seasonal changes in ripple mark orientation and the distribution of the density gradients in the water suggest that water cascades downslope during March. Dissolved oxygen concentrations vary from more than 5.0 ml/liter during March to less than 3.0 ml/liter during June.

Figure 4-14. Shelf crab *Cancer borealis* Stimpson: (a) Underwater photograph, 420 m; (b) habitat sketch, 320 m. Diameter of specimen, 10 cm.

The combination of characteristics of the water led Rowe and Menzies (1968) to conclude that the Gulf Stream intermittently impinged on the bottom during the time of their measurements, being on bottom during June but remaining somewhere offshore during March and November. In any case, it can be concluded the environmental temperature and the oxygen content are quite variable and somewhat unpredictable in the AZT.

The mean particle size of sediments and degree of sorting (composed of foraminiferan test sand) decreases seaward and north of the region where the Florida current impinges on the bottom. At the outer margin of the AZT at a depth of approximately 1000 m the sediment is a silty clay. Very little change in sediment particle size occurs at greater depths (Figure 4-6h). The water and the sediments indicate that this area is transitional in marine climate and influenced by seasonal changes. Similarly, the fauna appears to be of a transitional nature.

The archibenthal generic distinctiveness of marine isopods between 246 and 445 m amounts to 67%, and between 620 and 900 m it is 74% (Table 4-1). The genera of the AZT show a high affinity to the shelf but none to the intertidal, and a higher affinity to the abyssal down to the greatest depth (Figure 4-10). This pronounced generic representation in both shallow and deep water is also indicative of the transitional nature of the archibenthal fauna. At a species level (Figure 4-18) there is only a 12% representation in the lower part of the shelf (up to 120 m) and the same low percentage in the upper part of the abyssal (1270 m), whereas beyond these

Figure 4-15. Shelf sea-star: (a) drawing of *Astropecten americanus* diameter 11 cm (b) photo taken at 310 meters.

depths at both higher and lower elevation the relationship at a species level is zero. Because the genera show a higher percentage affinity to the abyssal province than to the shelf province, it may safely be concluded that the archibenthal fauna, although a mixture of shallow as well as deeper components, has its greatest affinities with the abyssal, as Ekman (1953) had earlier suggested for the archibenthal of world oceans.

The AZT contains only 4 genera of isopods, one of the lowest generic representations along the entire depth profile. These are *Eugerda, Munnopsis, Conilera,* and *Cirolana* (Figure 4-19), and not one of these is endemic to the depths of the AZT. *Eugerda* and *Munnopsis* enter the abyss, and *Conilera* and *Cirolana* occur on the shelf.

The AZT has only eight species of isopods, and within this zone their endemism ranges between 25 and 57%. The five species endemic are *Munnopsis typica, Cirolana gracilis, C. impressa, Eugerda* sp. A., and *Eugerda* sp. B (Table 4-3).

The archibenthal zone has only four major groups of isopods; the shallow-water group, Cirolanoidea (33%) and the deep-water group Asellota (33%) dominate (Figure 4-11). The Sphaeromidae and the Oniscoidea are lacking, and the Serolidae and the Valvifera are only assumed to be present. They were not collected from this depth range.

Large animals occurring in the AZT are the quill worm *Hyalinoecia artifex* Verrill (Figure 4-20) (non. *tubicola*); the burrowing anemone *Ceriantheomorpha braziliensis;* the hermit crabs *Parapagurus pilosomanus* (Figure 4-21) and *Catapagurus sharreri* with their commensal anemones; *Epizoanthus paguriphilus;* the bat fish *Dibranchus atlanticus* (Figure 4-22); the decapod crustaceans *Munida valida* (Fig-

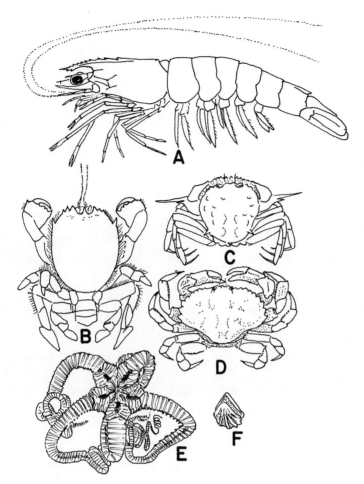

Figure 4-16. Assemblage of typical large shelf organisms from North Carolina: (a) *Penaeus duorarum*, (b) *Ranilia muricata*, (c) *Acanthocarpus alexanderi*, (d) *Cancer irroratus*, (e) *Astrocyclus*, (f) *Plicatula gibbosa*. Scales and sources various.

ure 4-23), *Rochinia crassa* (Figure 4-24), *Bathyplax typhla*, and *Geryon quinquedens;* the anemones *Actinauge rugosa (longicornis)* (Figure 4-25), *Bolocera tuediae*, and *Actinoscyphia saginata;* the solitary coral *Flabellum goodei* (Figure 4-26); and the fish *Nematonurus* sp. (Figure 4-27).

The large stalked and pigmented eyes of the decapods indicate the transitional nature of the archibenthal zone. Certain genera have most of their species in shallow water (shallow origin), whereas others have most of their species in deep water. In addition, some species show wide range in depth either onto the shelf (north of Cape Hatteras), like *Cancer borealis* (Figure 4-14) and *Ceriantheomorpha braziliensis*, or into the abyssal province, like *Parapagurus pilosomanus* and *Hyalinoecia artifex;* but many of the species are confined nevertheless to the AZT.

Abyssal Faunal Province (1000–5315 m)

The abyssal faunal province (AFP) commences 64 km from the shore, starting at 940 m depth and extending 200 km seaward. The top of the AFP appears to be

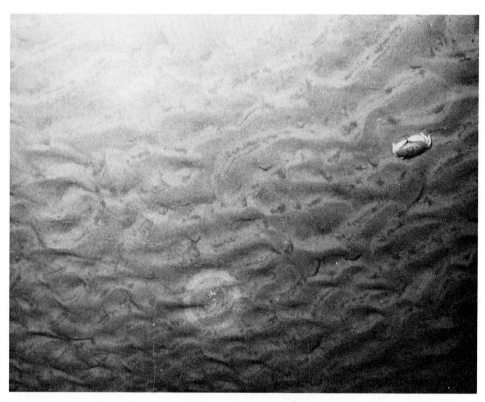

Figure 4-17. Cross-bedded ripple marks in the archibenthal zone off the Carolinas caused by the action of the Florida current at 502 m. Note also *Cancer borealis.*

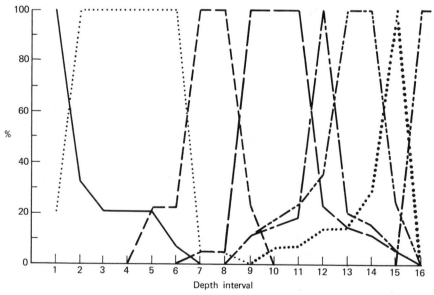

Figure 4-18. Proportional distribution curves of isopod species within the identified faunal zones off the Carolinas:——, intertidal; ....., shelf; ----, AZT; — —, upper abyssal zone; — - —, mesoabyssal zone; ------, lower abyssal active zone; •••••, lower abyssal tranquil zone; ---, lower abyssal red clay zone.

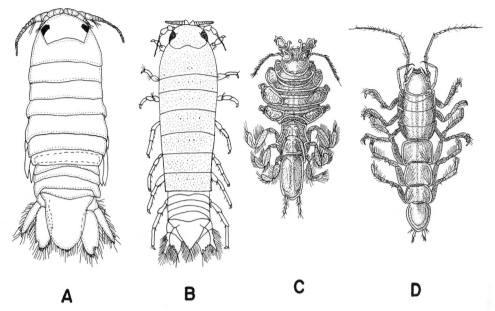

A　　　　　B　　　　　C　　　　　D

Figure 4-19. Assemblage of characteristic archibenthal genera of isopods from the Carolinas and Georgia: (a) *Cirolana*, (b) *Conilera*, (c) *Munnopsis*, (d) *Eugerda*. [(a) from Menzies and Frankenberg, 1966; (b) after Richardson, 1905; (c) and (d) G. O. Sars.] Scales various.

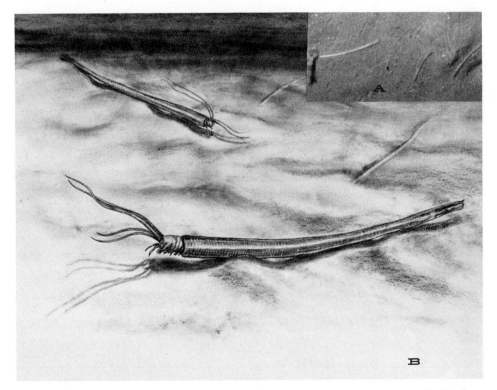

Figure 4-20. *Hyalinoecia artifex* Verrill, length of animal 6 to 7 cm: (a) Underwater photograph, (b) habitat sketch at 475 m.

TABLE 4-3. BATHYMETRIC DISTRIBUTION OF ISOPOD SPECIES OFF CAROLINAS (NORTH ATLANTIC): DEPTH IN METERS

Sampling Intervals

Genus	Species	1 0–3	2 5–20	3 21–70	4 80–90	5 91–150	6 200–250	7 445–550	8 600–650	9 900–1500	10 2000–2400	11 2400–2635	12 2640–3100	13 3300–3600	14 3800–4100	15 4800–5100	16 5200–5400
Dynamenella	dianae	×															
Ancinus	depressus	×	×														
Cassidinidea	lunifrons	×															
Chiridotea	caeca	×															
Chiridotea	sp. A	×															
Cyathura	carinata	×															
Limnoria	tripunctata	×															
Sphaeroma	quadridentatum	×															
Munna	reynoldsi	×															
Ligia	exotica	×															
Armadilloniscus	sp.	×															
Erichsonella	f. filiformis	×	×														
Cleantis	planacauda		×														
Accalathura	crenulata		×														
Chiridotea	nigrescens		×														
Cirolana	parva		×														
Edotea	montosa		×														
Panathura	sp. A		×														
Xenanthura	brevitelson	×	×	—		×											
Serolis	mgrayi	×	×	—		×											
Horoloanthura	irpex	×	—		×	×											
Eurydice	littoralis	×	×	×	×	×											
Chiridotea	stenops	×	×														
Paracerceis	caudata	×	×	×	—	×											
Ptilanthura	tricarina	×	×	×	×	×	×										
Apanthura	magnifica	×	×	—		×	×										
Cyathura	burbancki	×	—	—		×	×										
Eurydice	piperata				×	—	×										
Panathura	formosa					×											
Astacilla	laufi				×	—	×										
Cirolana	polita					×											
Calathura	sp. A						×										
Conilera	sp. A						×	—	×	×							
Cyathura	sp. A						×										
Accalathura	carinata						×										
Cirolana	borealis							×									
Cirolana	impressa							×									
Cirolana	gracilis							×									
Conilera	sp. B							×	—	×							
Eugerda	sp. B							×									
Munnopsis	typica							×	×								
Eugerda	sp. A								×								
Dendromunna	sp. A									×							
Desmosoma	sp. A									×	—	—	—	×	×		
Eurycope	sp. I									×							
Gnathia	sp. A									×							
Macrostylis	sp. J									×							
Acanthocope	sp. A									×	—	—	×				
Astacilla	caeca									×	—	—	×				
Dendrotion	sp. B									×	×						
Dendrotion	sp. C									×	×						
Haploniscus	sp. B									×							
Haploniscus	sp. C									×							
Heteromesus	sp. E									×							
Ianirella	sp. A									×							
Ilyarachna	sp. B									×	×						
Macrostylis	sp. D												×				
Macrostylis	n. sp.									×							
Macrostylis	n. sp.									×							
Macrostylis	sp. F									×	—	—	×	—	×		
Notoxenoides	sp. A									×							
Paramunna	sp. A									×							

TABLE 4-3. CONTINUED

Sampling Intervals

Isopoda (Genus)	Species	1	2	3	4	5	6	7	8	9	10	11	12	13	14	15	16
		0–3	5–20	21–70	80–90	91–150	200–250	445–550	600–650	900–1500	2000–2400	2400–2635	2640–3100	3300–3600	3800–4100	4800–5100	5200–5400
Asellota	n. gen., n. sp.									×							
Bathygnathia	sp. A									×							
Desmosoma	sp. C									×	—	—	×		—	×	
Haplomesus	sp. B									×							
Heteromesus	sp. D									×							
Haploniscus	n. sp.										×						
Haploniscus	spatulifrons										×						
Macrostylis	sp. E										×	×	×				
Antennuloniscus	sp. A											×					
Dendrotion	sp. A										×						
Desmosoma	sp. D										×						
Desmosoma	sp. E										×						
Eugerda	sp. D										×						
Eurycope	sp. A										×						
Eurycope	sp. C										×						
Eurycope	sp. D										×						
Eurycope	sp. E										×						
Eurycope	sp. F										×						
Eurycope	sp. G										×						
Eurycope	complanata										×			×	×		
Haploniscus	sp. D										×						
Haploniscus	n. sp.										×						
Heteromesus	sp. A										×						
Hydroniscus	sp. A										×						
Ianirella	lobata										×						
Ianirella	sp. A										×	—	—		×		
Macrostylis	sp. G															×	
Macrostylis	sp. I											×					
Macrostylis	sp. M											×					
Macrostylis	sp. N											×					
Macrostylis	sp. O											×					
Munnopsis	sp. A											×					
Rhacrura	pulchra											×					
Syneurycope	parallela											×					
Eurycope	sp. J												×				
Nannoniscus	sp. B											×					
Nannoniscus	sp. A													×			
Antennuloniscus	sp. B														×		
Echinothambema	sp. A														×		
Haploniscus	n. sp.														×	×	
Heteromesus	sp. B																×
Ilyarachna	sp. A														×		
Macrostylis	sp. C														×	×	
Macrostylis	sp. K														×		
Macrostylis	sp. L														×		
Serolis	n. sp.														×		
Haplomesus	sp. A															×	
Haplomesus	sp. C															×	
Haplomesus	sp. E															×	
Haploniscus	n. sp.															×	
Heteromesus	sp. C													×			
Ischnomesus	sp. A															×	
Ischnomesus	sp. C																×
Macrostylis	sp. A															×	
Macrostylis	sp. B															×	
Macrostylis	sp. H															×	
Storthyngura	sp. A															×	
Abyssianira	dentifrons																×
Eugerda	sp. C																×
Haploniscus	n. sp.																×
Desmosoma	sp. B																×

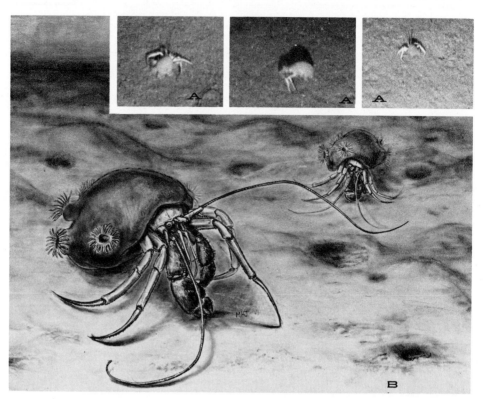

Figure 4-21. *Parapagurus pilosomanus* Smith, diameter of specimen 10 cm with commensal *Epizoanthus paguriphilus* Verrill: (a) Underwater photograph at 792 m, (b) habitat sketch at 645 m.

Figure 4-22. Archibenthal bat-fish *Dibranchus atlanticus* Peters: (a) Underwater photograph at 601 m, (b) habitual sketch at 620 m; length of animal 5.7 cm.

106

Figure 4-23. Archibenthal "shrimp" *Munida valida* Smith: (a) Underwater photographs at 420 m, (b) habitat sketch at 500 m; length of body, 5.7 cm.

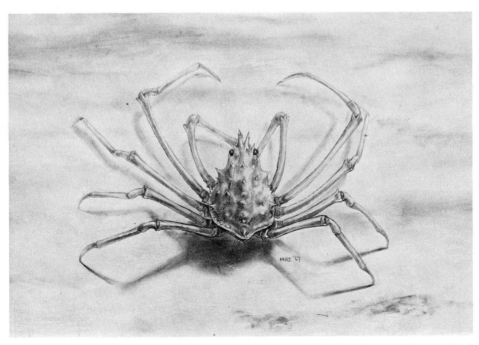

Figure 4-24. Archibenthal spider crab *Rochinia crassa* Milne-Edwards, habitat sketch at 475 m; width of carapace, 3 cm.

Figure 4-25. Archibenthal sea anemone *Actinauge rugosa* (= *longicornis*): (a) Underwater photograph at 420 m, (b) habitat sketch at 620 m; length of specimen, 6.5 cm.

Figure 4-26. Archibenthal–abyssal solitary coral *Flabellum goodei*; (a) Underwater photographs at 1045 m, (b) habitat sketch at 500 m; height of specimen, 4 cm.

Figure 4-27. Archibenthal–abyssal–rat-tailed fish *Nematonurus* sp.: (a) Underwater photograph at 1075 m, (b) habitat sketch at 500 m; length of specimen, 28 cm.

located at the top of the deep WBUC moving south, and the temperature of 4°C is not subject to seasonal variation. The lower margin of the province is located in the Antarctic bottom water (Figure 4-28) moving north over the red clay of the Hatteras abyssal plain. Between these extremes in depth and geography there is considerable variation in water current velocity and topography as well as in sediment type, from *Globigerina* ooze to red clay. The temperature in this depth range is 4.0 to 3.5°C on the upper border, dropping to 2.1°C at the base. It then rises to about 2.4°C on the Hatteras abyssal plain. It is unlikely that the restricted animal distributions are a result of these small differences.

The AFP contains 30 genera of isopods, and 84% of these are endemic (Table 4-2). It is distinguished from the archibenthal zone in 74% of the genera and 95% of the species at 940 m depth (Table 4-1). From this level of depth onward, the generic distinctiveness varies between 7 and 50% at sampling intervals within the province, whereas the species distinctiveness varies between 40 and 100%, suggesting strongly that the abyssal province can be divided into zones. The 25 isopod genera endemic to the AFP are presented in Table 4-2, and in Figure 4-29.

The isopod genera in the abyssal faunal province are limited mainly to the AFP and do not partake strongly of any lesser depths. Thus only 16% of the abyssal genera enter the archibenthal and less than 10% enter the shelf. Not one enters the intertidal (Fig. 4-10). The limited generic representation in the shelf is caused by the discontinuous distribution of *Serolis* and *Astacilla*, which was discussed earlier. Without these two unusual cases, the generic representation of AFP in the shelf would have been zero.

The abyss consists of 75 to 100% Asellota (Figure 4-11), and no other major

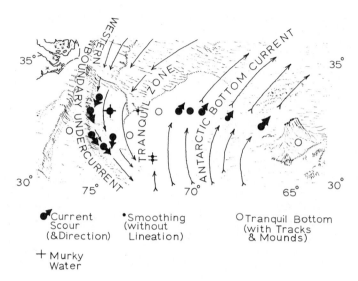

Figure 4.28. Western North Atlantic bottom currents. (Modified from Heezen et al., 1966, with required additions.)

isopod group occupies more than 10% of the total. Thus the abyss may be characterized as the province of the Asellota and also as a province with limited diversity at the family level. Its diversity at taxonomic levels below the family is great.

Large animals characteristic of the AFP belong mainly to the Echinodermata, the Coelenterata, and the Porifera; ophiuroids and holothurians dominate (Rowe, 1968; Rowe and Menzies, 1969). The number of genera of isopods increases in the AFP. Large animals show a progressive increase in abundance from 1000 m to a maximum at 2000 m. Below 2000 m the large epifaunal animals then exhibit a progressive decrease in density and in standing stock. Large animals as well as the isopods in the AFP are characterized by a lack of functional eyes and by a considerable reduction in pigmentation—most are pale in color. There is also an observable reduction in the degree of calcification in comparison with shallow-water species, especially among the molluscs.

The size of abyssal animals measured in body length is quite variable, and some are larger than their shallow-water counterparts. For example, the pycnogonid *Colossendeis*, the holothurians *Benthodytes* and *Psychropotes*, and the echinoid *Phormosoma* have species in the AFP which are among the largest known for these genera and even for their families.

Zonation Within the Abyssal Faunal Province

UPPER ABYSSAL ZONE (940–2635 M)

The upper abyssal zone starts the abyssal faunal province at 940 m. It extends 11 km to sea and down to 2635 m; thus it is located almost entirely on the continental slope within the limits of the WBUC (Figure 4-28). The sediments consist of a uniform *Globigerina* ooze. It is in this region that WBUC reaches its greatest velocity (Rowe and Menzies, 1968; Barrett, 1965), and hence the upper abyssal zone correlates well with a high-velocity tongue of the WBUC between 1000 and 3000 m.

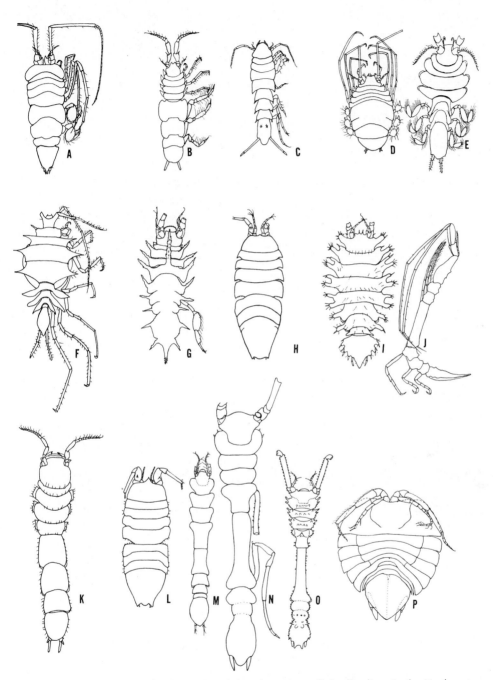

Figure 4-29. Isopod genera from the abyssal faunal province off the Carolinas in the Northwestern Atlantic Ocean: (a) *Ilyarachna*, (b) *Desmosoma*, (c) *Macrostylis*, (d) *Eurycope*, (e) *Munnopsis*, (f) *Dendrotion*, (g) *Acanthocope*, (h) *Haploniscus*, (i) *Ianirella*, (j) *Astacilla*, (k) *Eugerda*, (l) *Antennuloniscus*, (m) *Ischnomesus*, (n) *Heteromesus*, (o) *Haplomesus*, (p) *Serolis*.

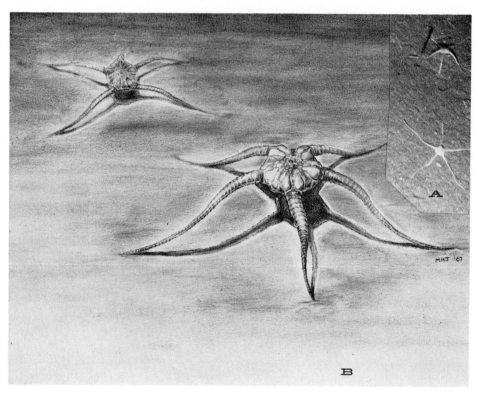

Figure 4-30. Upper abyssal brittle starfish *Ophiomusium lymani:* (a) Underwater photograph at 2000 m, (b) habitat sketch at 2000 m, diameter of specimen, 3 cm.

Figure 4-31. Upper abyssal brittle starfish *Ophiacantha simulans:* (a) Underwater photograph at 2000 m, (b) habitat sketch at 2400 m; diameter of specimen, 1 cm.

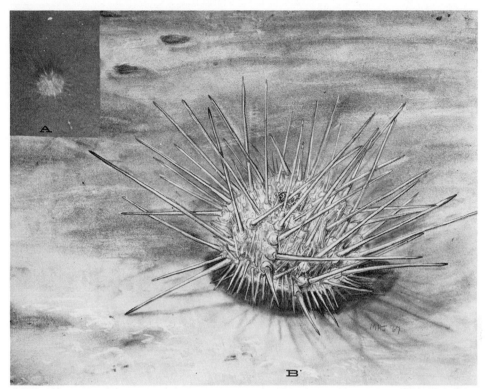

Figure 4-32. Upper abyssal sea urchin *Echinus affinus:* (a) Underwater photograph at 2400 m, (b) habitat sketch at 2300 m; diameter of specimen, 3.5 cm.

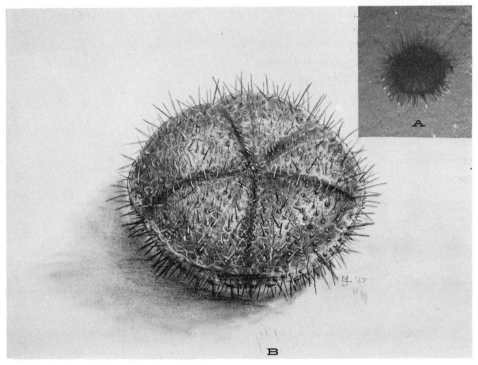

Figure 4-33. Upper abyssal sea urchin *Phormosoma placenta:* (a) Underwater photograph at 2440 m, (b) habitat sketch at 2460 m; diameter of specimen, 12 cm.

113

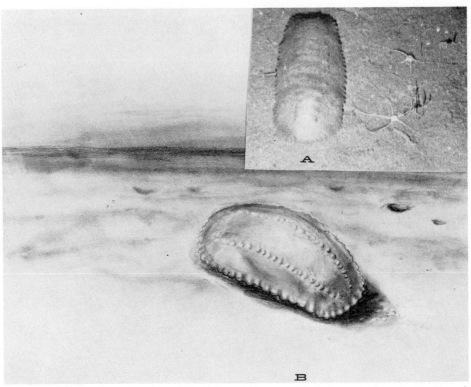

Figure 4-34. Upper abyssal holothurian *Benthodytes gigantea:* (a) Underwater photograph at 2000 m, (b) habitat sketch at 2500 m; length of specimen, 20 cm.

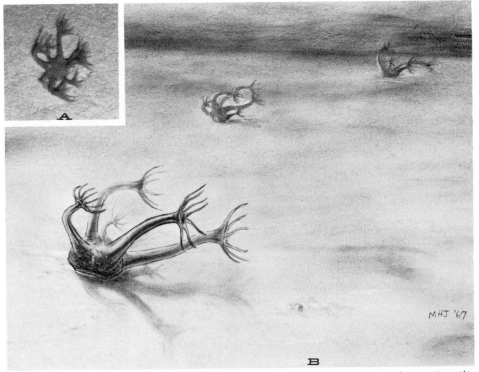

Figure 4-35. Upper abyssal sea pen *Anthomastus grandiflorus:* (a) Underwater photograph at 2440 m, (b) habitat sketch at 2600 m; diameter of crown, 2 cm.

Figure 4-36. Upper abyssal sponge *Cladorhiza sp.*: (a) Underwater photograph at 3062 m, (b) habitat sketch at 1940 m; height of specimen, 20 cm.

Of the 21 genera of isopods (Table 4-2) contained in the upper abyssal zone, 33% are endemic. They are *Notoxenoides, Paramunna, Bathygnathia, Gnathia, Dendromunna,* and a new genus of Asellota (Table 4-3), and each is represented here by a single species. The upper abyssal zone is distinct from the AZT in 95% of the species and from the next lower zone, the mesoabyssal, in 82% of its species. Between 940 and 2635 m two sampling intervals showed species distinctiveness ranging between 64 and 72%, suggesting that there may be added zonation within the upper abyssal.

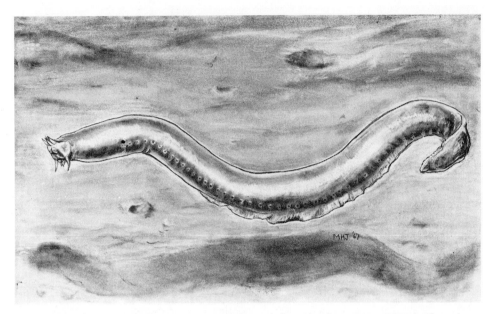

Figure 4-37. Upper abyssal hag fish *Myxine glutinosa:* habitat sketch at 1500 m; length of specimen, 34 cm.

Figure 4-38. Upper abyssal–mesoabyssal sea pen *Pennatula aculeata:* (a) Underwater photograph at 2440 m; (b) habitat sketch 2400 m, length of specimen 3.0 cm.

Figure 4-39. Mesoabyssal glass sponge *Euplectella suberea*: (a) Underwater photograph at 3062 m, (b) habitat sketch at 3100 m; length of specimen, 6.0 cm.

Only 3.5% of the species of the upper abyssal zone enter the AZT to 498 m, and none extends upward beyond 498 m. This range indicates that the relation of the upper abyssal to the AZT is at an exceptionally low species level. In contrast, the species enter the mesoabyssal and the lower abyssal at the values of 21 and 14%, respectively, and the greatest depth is less than 5%, showing how strongly the upper abyssal is allied and linked to the other zone of the AFP (Figure 4-18).

The upper abyssal zone has only five major groups of isopods, and the Asellota constitutes 75% of the fauna. The Gnathiodea, the Cirolanoidea, the Serolidae, and the Valvifera each contributes 5%, with one species each (Figure 4-11). The Gnathioidea made their first appearance in the upper abyssal zone, whereas the Sphaeromidae and the Oniscoidea are lacking. Blindness and reduced pigmentation are the main features (Figure 4-29).

Large animals within the upper abyssal zone are dominated by the ophiuroid *Ophiomusium lymani* (Figure 4-30), which is distributed from approximately 1400 to 3300 m. Maximum animal densities and a maximum number of species of large animals within the AFP are near middepth of the upper abyssal zone. Other animals here are the ophiuroids *Ophiomusium spinigerum*, *Ophiacantha simulans* (Figure 4-31) *and Bathypectinura heros*; the urchins *Echinus affinus* (Figure 4-32), *Phormosoma placenta* (Figure 4-33), *Hygrosoma petersi*, and *Pleisiodiadema antillarum*; the holothurians *Pelopatides gigantea-Benthodytes gigantea* (Figure 4-34), and *Molpadia musculus*, as well as the pennatulid *Anthomastus grandiflorus* (Figure 4-35), the sponge *Cladorhiza* (Figure 4-36), and the slime eel, *Myxine glutinosa* (Figure 4-37). As pointed out earlier (Rowe and Menzies, 1969), the large animals form a concise group within the swift tongue of the WBUC and this total assemblage appears to correlate well with the upper abyssal zone in isopods.

There are, however, several species which have invaded from above or have penetrated to greater depths. A burrowing anemone reaches from 400 m obliquely across the area to 3000 m, and the polychaete *H. artifex* reaches a depth of 2000 m. The brittle starfish *O. lymani*, mentioned previously, penetrates to a depth of 3300 m; the glass sponge *Hyalonema boreale*, found in the shallow depths of the upper abyssal zone, also crosses the lower border to depths of about 3200 m. Similar distributions are found in the sea pen *Pennatula aculeata* (Figure 4-38) and the sponge *Euplectella suberea* (Figure 4-39). Even though these six species also occur outside the upper abyssal zone, this diverse assemblage appears to be recognizable as a separate zonal unit. On the basis of the large animals alone, further separation into subzones may be justified (Rowe and Menzies, 1969).

MESOABYSSAL ZONE (2540–3300 M)

The mesoabyssal zone starts 130 km from shore and is centered bathymetrically at the junction of the continental slope with the continental rise, extending an added 20 km seaward. Along the north section in this general area (3000–3200 m) the current velocity at the bottom is very low and the mesoabyssal zone is evidently located between tongues of the WBUC. However, along the south the current appears to be quite swift (Rowe and Menzies, 1969). This contact zone between the slope and rise is also characterized by detrital "fall-out" of the turtle grass *Thalassia* (Menzies et al., 1967; Menzies and Rowe, 1969). Likewise, suspended materials in the bottom water have been shown to accumulate here (Ewing and Thorndike, 1965; and Heezen et al., 1966). The accumulation appears to result from downslope movement and a slope-rise concentration by the WBUC (Rowe and Menzies, 1968).

Sediments along the whole area are a poorly compacted, light green to tan

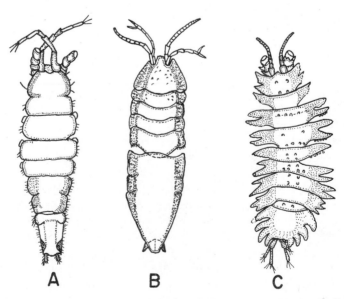

A   B   C

Figure 4-40. Isopod genera endemic to the mesoabyssal zone: (a) *Syneurycope*, (b) *Hydroniscus*, (c) *Rhacrura*.

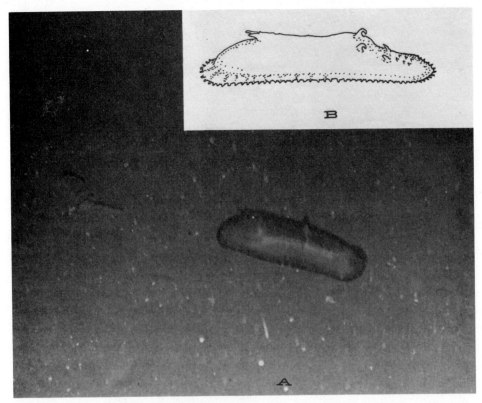

Figure 4-41. Mesoabyssal holothurian *Euphoronides cornuta:* (a) Underwater photograph at 3010 m, (b) pen and ink drawing of species from Thomson, 1874.

foraminiferal ooze. The mean particle size varies only slightly, but the percentage of sand changes somewhat (ca. 3.5–4.0% along the north, reaching its maximum of just over 15.0% to the south). Temperature in this area ranges from 2.3°C at 2600 m to about 2.2°C at 3300 m. Oxygen values were all just above or below 6.0 ml/liter.

Thirty-three species of isopods comprise the mesoabyssal zone; generic endemism is only 17%, consisting of the three genera *Rhacrura, Hydroniscus,* and *Syneurycope* (Figure 4-40). Species endemism is 71%. The mesoabyssal zone is distinct from the lower abyssal and the upper abyssal zones in 83% of the species (Table 4-1). In genera the mesoabyssal is more closely allied to the upper than to the lower abyssal. Isopod species diversity (species per genus) is highest in the mesoabyssal zone, precisely where Menzies and George (1967) located the "moderate depth species maximum."

Asellota dominate the mesoabyssal zone, comprising 88% of the species of the three major isopod groups (Figure 4-40). Between the upper abyssal and the mesoabyssal zones, two major groups, the Cirolanoidea and the Gnathioidea, drop out. It is quite apparent, therefore, that the diversity that exists is at a species level instead of being at a major group level as was the case with the fauna of the intertidal province.

Large epifaunal animals also appear to be restricted to a rather diverse assemblage that is characteristic of the mesoabyssal zone. Endemic to the zone are the sea cucumbers *Pseudostichopus villosus* and *Euphronides cornuta* (Figure 4-41). The soft

Figure 4-42. Abyssal glass sponge *Tentorium semisuberites;* underwater habitat illustration at 2700 m; height of specimen, 1.5 cm.

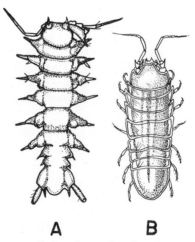

A   **B**

Figure 4-43. Isopod genera known only from lower abyssal (active zone) off the Carolinas: (a) *Echinothambema,* (b) *Nannoniscus.*

120

Figure 4-44. Lower abyssal (active) zone crinoid *Bathycrinus* sp., underwater photograph at 4000 m.

Figure 4-45. Lower abyssal (active) zone small sponge *Thenea delicata* Sollas: (a) Underwater photograph at 3800 m, (*b*) habitat sketch at 3800 m; diameter of specimen 1 cm.

Figure 4-46. Lower abyssal (active) zone stalked barnacle *Scalpellum regium,* underwater habitat sketch at 4600 m; length of body (including stalk) 8 cm, attached to clinker from ship.

coral *Anthomastus grandiflorus* (Figure 4-35) and other components of the region such as *O. lymani* have invaded from above, where they occur as separate populations at much shallower depths in the upper abyssal zone. Among these are the glass sponges *Hyalonema boreale* and *Euplectella suberea* (Figure 4-39), the hermit crab *Parapagurus pilosimanus,* and the sea pen *Pennatula aculeata* (Figure 4-38). Specimens of *Parapagurus pilosimanus* here, which are considerably smaller than those of the AZT, could be described as distinct subspecies on the basis of minor differences. It should be pointed out that although both *E. cornuta* and *A. grandiflorus* reach maximum densities in the mesoabyssal zone, they do extend up to 2000 m in very small numbers. The small sponge *Tentorium semisuberites* (Figure 4-42) makes its first appearance in the mesoabyssal zone (Rowe and Menzies, 1969).

LOWER ABYSSAL (ACTIVE) ZONE (3300–4100 M)

The lower abyssal zone commences around 3300 m and stops at 4100 m. It is 142 km from shore and 12 km wide. Essentially, it is located on the upper continental rise within a high-velocity part of the WBUC, ending near the lower continental rise. The sediments are reddish-tan and contain few foraminiferan tests. The per-

Figure 4-47. Lower abyssal (tranquil) zone, short-armed brittle starfish *Amphiophiura bullata*: (a) Underwater photography 5330 m, (b) habitat sketch at 5330 m, diameter of specimen 4 cm.

centage of sand increases slightly in the vicinity of the Transverse Canyon (Figure 4-1b).

Seventeen species of isopods are contained herein, with 42% of the genera distinct from the mesoabyssal zone and 15% distinct from the lower abyssal (tranquil) zone. Generic endemism amounts to 8% and is accounted for by the two genera *Nannoniscus* and *Echinothambema* (Figure 4-43). No genus makes its appearance here and instead several stop, thus typifying the end of the lower abyssal (active) zone. Those which stop are *Serolis*, *Ilyarachna*, *Eurycope*, and *Antennuloniscus*.

Of the nine species (50%) that are endemic, a greater percentage contributes to the mesoabyssal than to the lower abyssal (tranquil) zone. The species distinctiveness at the mesoabyssal boundary (83%) is equal to that at the lower abyssal (tranquil) zone boundary. Only two major isopod groups are present: the Asellota at 92%, and the Serolidae at 8% (Figure 4-11).

Large animals in this zone are *Bathycrinus* sp. (Figure 4-44), the sponge *Thenea* (Figure 4-45), and the barnacle *Scalpellum regium* (Figure 4-46).

LOWER ABYSSAL (TRANQUIL) ZONE (4100–4800 M)

Somewhere between 4100 and 4800 m on the lower continental rise is a separation of the lower abyssal (active) zone from a zone of low or negligible current velocity. This area has been identified by Heezen et al. (1966) as a tranquil zone (Figure 4-28). It appears to begin around 4500 m near the seaward margin or lower limit of the WBUC, continuing eastward as a "no-motion zone" until it reaches the Antarctic

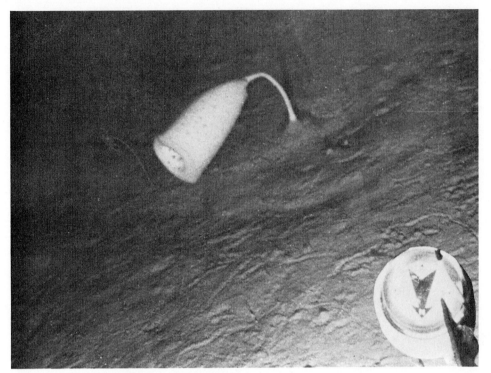

Figure 4-48. Lower abyssal red clay zone, crinoid *Bathycrinus* and stalked cup sponge. Underwater photograph, Hatteras Abyssal Plain, note current lineations. (From Heezen and Hollister, 1964.)

bottom current. Indications of this current are featured in the eastern margin of Figure 4-28. The tranquil zone is indicated by an absence of evidence of currents in submarine photographs. We have added some data to Heezen's diagram.

Sediments in this deep region are composed of fine silt and clay-sized particles. The middle of the area is cut by the extension of the Hatteras Canyon (Transverse Canyon of Rona et al., 1967), and there is accordingly a slight increase in the percentage of sand. Oxygen values are near 6.0 ml/liter, and the temperature is +2.3°C.

The genera between the lower abyssal (active) and the lower abyssal (tranquil) zones are not greatly distinctive, although the genus *Storthyngura* may be endemic to the latter. Fourteen species of isopods are in the tranquil zone. Twelve species found in the active zone are not present in the tranquil zone, and 10 species are endemic only to the latter. Because of this, the active and tranquil zones are 83% distinctive in their species.

The one large epifaunal animal characteristic of this zone is *Amphiophiura bullata* (Figure 4-47).

LOWER ABYSSAL RED CLAY ZONE (5070–5340 M)

The Hatteras abyssal plain, which is characterized by the northward flow of the Antarctic bottom water (Heezen et al., 1966), is mostly beyond the diagram of Figure 4-1*b* between 71°W and 67°W. The current flow is evident from current lineations as they appear in Figure 4-48 and from the bending of the crinoid *Bathycrinus* and the sponge (Figure 4-48), which are deflected to the northeast. An added characteristic is the presence of manganese nodules in varying abundance throughout this region (Figure 7-15a).

The eight isopod genera here are only 30% distinctive with the single endemic *Abyssianira*, whereas the species are 100% distinctive owing to the large number of species (14) that do not occur here. The eight endemic species are *Abyssianira dentifrons*, *Haploniscus* n. sp., *Desmosoma* sp. B, *Eugerda* sp. C, *Macrostylis* sp. A, *Ischnomesus* spp. A and C, and *Heteromesus* sp. B. The reduction in the number of species appears to be a significant feature of this zone.

Large animals occurring here include *Bathycrinus* sp. and *Umbelulla lindahli* of the lower abyssal (active) zone and the cup sponge probably *Hyalonema* sp. (Figure 4-48). The sessile brachiopod *Pelagodiscus* has been collected from manganese nodules, and the crinoid *Bathycrinus* occurs infrequently in bottom photographs.

Although we have distinguished three zones in the lower abyssal area which

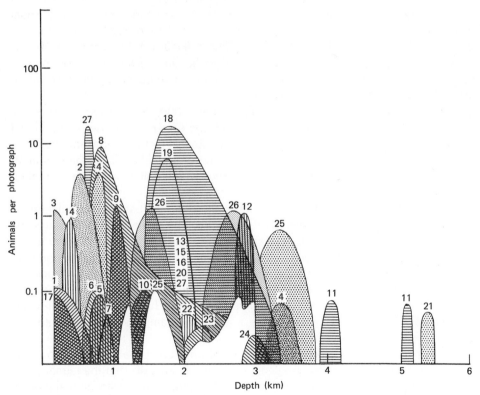

Figure 4-49. Quantitative distribution of epifaunal animals in relation to depth: 1, *Cancer borealis*; 2, *Munida valida*; 3, *Catapagurus sharreri*; 4, *Parapagurus pilosimanus*; 5, *Actinauge longicornis*; 6, *Actinoscyphia saginata*; 7, *Bolocera tuediae*; 8, Cerianthidae; 9, *Flabellum goodei*; 10, *Pennatula aculeata*; 11, *Umbellula lindahli*; 12, *Anthomastus grandiflorus*; 13, *Echinus affinus*; 14, *Cidaris abyssicola*; 15, *Phormosoma placenta*; 16, *Hygrosoma petersi*; 17, *Astropecten americanus*; 18, *Ophiomusium lymani*; 19, *Ophiacantha simulans*; 20, *Bathypectinura heros*; 21, *Amphiophiura bullata*; 22, *Pelopatides gigantea*; 23, *Pseudostichopus villosus*; 24, *Euphronides depressa*; 25, *Euplectella suberea*; 26, *Hyalonema boreale*; 27, *Hyalinoecia artifex*; 28, *Lopohelia* sp. (From Rowe and Menzies, 1969.)

correspond in depth range to features of water type and characteristic water current, we are not entirely satisfied with our divisions because of the limited data on which they are based. Only more collecting can determine whether we are correct in these zonal designations of the lower abyssal. We cannot rule out the possibility that a mosaic distribution obtains, similar to that suggested by Sanders (1963) and Sanders et al. (1965). There are so few data, mainly owing to the small number of large animals that occur in the lower abyssal zones, that conclusions based on such findings cannot be considered to be reliable. These large animals apparently are only rarely seen in underwater photographs.

EPIFAUNAL ZONATION

References have been made to large animals in the fauna which characterize provinces and zones as defined by the isopods. The intensive photographic survey carried out off the Carolinas (Rowe and Menzies, 1968, 1969; Menzies and Rowe, 1969) has provided quantitative information on the distribution of the major populations of large invertebrates seaward of the shelf break. A summary graph of individual epifaunal populations overlapping specific depths demonstrates where marked changes in the composition of the fauna occur (Figure 4-49). Depths with few overlapping populations may be considered to be sites of faunal discontinuities, whereas depths or depth intervals characterized by many overlapping populations are regions of limited epifaunal change.

Discontinuities are apparent at approximately 1000, 3000, 3800, 4200, 5000, and 5250 m, and these depths correlate with the upper boundaries of the upper abyssal zone, the mesoabyssal zone, the lower abyssal (active) zone, and the lower abyssal (tranquil) zone. Agreement between these zones based on large animals and isopod zones is closest for the higher zones where epifaunal species and their populations are numerous.

In deeper water (lower abyssal), however, the zones are based on populations of only a few species, and agreement is less precise. The large animals appear to be confined to specific areas which are separated by a large area of the sea bottom apparently devoid of any other large animals. This indicates that further zonation may yet be described when more data become available. But the latter kind of zonation, if it is not just an artifact of sampling, appears to result from differences in currents over the broad area and possibly from differences in the history of the sediments. Ecologically, the region of the lower rise and abyssal plain can be divided into several sections, as described earlier.

The remarkable agreement between faunal units composed of groups of large animals and zones of species and genera of isopods suggests that the faunal provinces and zones described herein will also be of significance to other animal groups. Similarly, it can be stated that the salient features that delimit these faunal groups are those characteristics of the environment which are of significance to life in the deep sea.

The recent work of Southward and Brattegard (1968) on Pogonophora off North Carolina, which was based to a considerable extent on collections made from R/V *Eastward* by us, permits a comparison of pogonophoran distribution with that for other animals. In general the pogonophorans show the three main faunal provinces. They are absent from the intertidal.

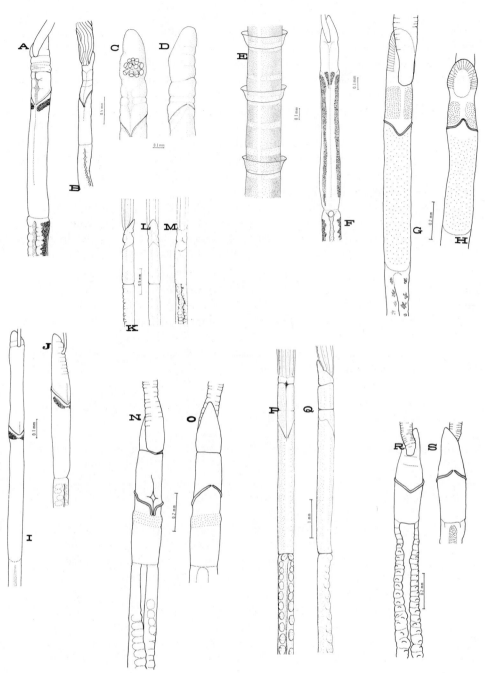

Figure 4-50. Pogonophora species off North Carolina in the Northwest Atlantic: (a) Outer shelf species, *Siboglinum holmei,* ventral view of forepart; (b)–(j) archibenthal species — *Polybrachia lepida,* anterior metameric region and (b) forepart, (c) ventral and (d) dorsal views; (e) *Diplobrachia* sp., anterior part of tube; (f) *Siboglinum candidum,* forepart; (g) and (h) *Siboglinum ekmani,* ventral and dorsal views of forepart; (i) and (j) *Siboglinum angustum,* ventral and latero-ventral views of forepart;(k)–(s) abyssal endemic species — (k)–(m) *Diplobrachia similis,* lateral, dorsal, and ventral views of forepart; (n) and (o) *Siboglinum fulgens,* ventral and dorsal view; (p) and (q) *Polybrachia eastwardae,* ventral and lateral views; (r) and (s) *Siboglinum pholidotum,* ventral and dorsal views. (After Southward and Brattegard, 1968.)

The shelf faunal province appears to contain only one or two species—
Siboglinum mergophorum Nielson and perhaps *Siboglinum holmei*. Both these
species are confined to depths beyond the topographic shelf, but both are found
within the water of the Gulf Stream in the outer shelf (160–450 m). Not one of these
extends into the archibenthal or the abyssal faunal provinces. The archibenthal
faunal province contains eight species; four are endemic to depths between 500 and
975 m and the remainder extend into the abyssal faunal province. This distribution
and affinity parallels that of the isopods and the larger epifauna, in that the AZT is
more allied to the abyssal faunal province than to the shelf. The upper abyssal zone
of the abyssal faunal province shows eight species; three also occur in the archibenthal and four extend into the mesoabyssal and the lower abyssal zones. Three
species are endemic to the upper abyssal zone.

The mesoabyssal zone has only four species of Pogonophora; none is endemic,
and three extend into the lower abyssal zones somewhat parallel to the distribution
of the large epifauna and in contrast to the isopod distribution. This distribution is
comparable to that of other species having a free or planktonic larval stage, such as
the decapods and the barnacles. Similarly, the species maximum among the pogonophorans is in the archibenthal and the upper abyssal and not in the mesoabyssal.
The assemblages of shelf, archibenthal, and abyssal Pogonophora appear in Figure
4-50. The Pogonophora, like the isopods and other species of animals, show a disjunct distribution off the Carolinas with allopatric populations widely separated
from one another. Certain species also show an emergence in the northern climate
and a submergence near the lower latitudes.

THE NORTHEASTERN ATLANTIC: BAY OF BISCAY

For the details of the distribution of communities in this region below intertidal
depths (43°N to 54°N), we refer to the book by Le Danois (1948), wherein the results
of many collections by European expeditions are summarized. Le Danois covered a
horizontal area far greater than that which we covered off North Carolina, and
because of this his summary transcends several faunal provinces and subprovinces at
the surface along the continental margin. His mixing of provinces and pooling of
bathymetric information resulted in a fairly confused picture of distribution with
depth. In general, however, he avoided such confusion and recognized the "boreal
outliers" or submergence of northern species as a general phenomenon throughout
the area between Northern Ireland and Spain.

The topography in this region is not at all simple, and the deep basins on the
continental shelf or border land are quite reminiscent of the topography off southern
California and Antarctica. The fauna in these "epicontinental" deep basins is a mixture of animals from the shelf of a given region and those from the north, as well as
those from deeper water from the south or west.

Zonation above the Abyssal Faunal Province

The fauna listed by Le Danois is a conglomerate of small infaunal species and large
epifaunal species. For the small infauna his lists are incomplete, but his lists are extensive for large animals, especially the molluscs. This feature has allowed certain
extremely interesting comparisons with the fauna off North Carolina.

Zone of Depth (m)   Facies at Zone (m)

1. Intertidal
2. Littoral or herbaceous (0–30)
3. Sublittoral (30–90)   I.   Inner shelf (40–50)
4. Neritic (90–200)   II. Outer shelf (120–200)
5. Atlantic Margin (200–1000)   III. Atlantic margin (200–1000)
                                                     (a) 200–500
                                                     (b) 500–1000
6. Semiabyssal (1000–2000) = Archibenthal   (c) 1000–2000 (semiabyssal)
7. Abyssal (2000–5000)   IV. Abyssal (2000–5000)
                                                     (a) 2000–3000
                                                     (b) 3000–5000
                                                     (c) 5000

The seven vertical zones just listed were recognized by Le Danois. He did not recognize provinces as we have done, but he did provide sufficient information to allow comparisons between our provinces and zones. Within and across various zones he recognized four principal animal facies, according to the methodology of C. G. Johannes Petersen.

The marine climate within the region from Northern Ireland to Spain is influenced by cold water from the north and warm water from the North African coast, as well as by invasions of Gulf Stream water from the Atlantic. The surface and shallow-water fauna reflect these complexities, being a transitional boreotemperate to subtropical fauna. The English Channel approximately marks the boundary between the Lusitanean region to the north (boreal elements) and the Mauretanian region from the Strait of Gibraltar to Cabo Blanco, Africa. But the majority of the area studied by Le Danois involves what Ekman termed a warm–temperate fauna. The Mediterranean Atlantic water, which is warm (13°C) and highly saline, sinks after leaving the Strait of Gibraltar, spreading over intermediate depths between 160 and 1200 m in the Northeastern Atlantic. Below this warm water is the cold bottom water, which has a minimum temperature of +2.5°C.

The abyssal fauna of the continental Eastern Atlantic margin can be considered as a unit that extends from 43°N latitude to Morocco (29°N–30°N latitude). It submerges to the south, staying substantially within the 9°C isotherm. Off Ireland its

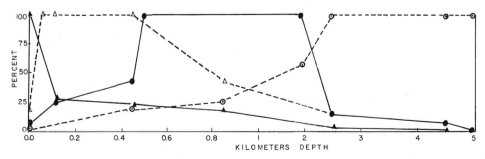

Figure 4-51. Percentage distribution of invertebrate species from one province to another in the Bay of Biscay as determined from data of Le Danois: solid triangle, intertidal; open triangle, shelf; solid circle, archibenthal; open circle, abyssal.

shallowest depth is 400 m, but a depth of 1000 to 1200 m is reached off Spain. It is precisely this kind of submergence that wreaks havoc with attempts to find a standard isobath for the localization of an abyssal fauna.

Le Danois gave particular attention to the submerged coral reefs (*Lophohelia*), which seem to be especially well developed along the continental margin of the Northeastern Atlantic. He incorrectly believed that these did not extend into the Arctic. Identical or very similar "deep-sea" coral reefs are being found along many continental margins, for example, at 400 m off North Carolina (Figure 4-4*b*) and off Argentina. They are probably worldwide features which were once continuous but have been obliterated in some places because of slumping and erosion. Le Danois devoted considerable attention to the extensive attached and clinging epifauna of these reefs. Utilizing his data from a restricted area, we are able to recognize provinces somewhat parallel to those off North Carolina but at different depths (Figure 4-51).

We equate his intertidal with our intertidal province and his shelf and Atlantic margin with our shelf faunal province. His semiabyssal zone (1000–2000 m) is partly equivalent to our AZT, which in the Carolinas extends only between 445 and 940 m. We have combined his semiabyssal zone and his lower continental margin into an archibenthal zone of transition (500–2000 m). We do this because 67% of the genera are in common between these two zones and because of the high proportion of abyssal elements in his continental margin fauna. His abyssal zone

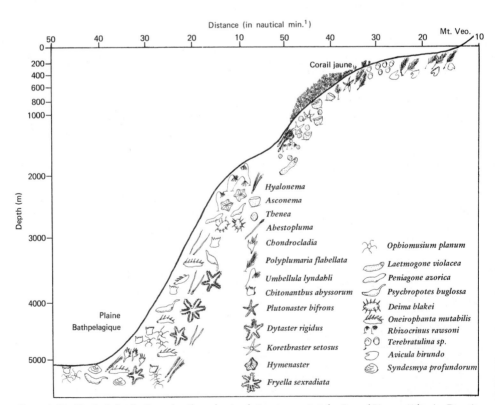

Figure 4-52. Vertical distribution of epifaunal animals in a transect at the Bay of Biscay. (After Le Danois, 1948.)

(2000–5000 m) is equivalent to our abyssal faunal province off Carolina (1000–5200 m).

The complexities of the littoral fauna (50–100 m) north to south result in four horizontal geographic units formed by the boreal fauna from the north and the subtropical fauna from the south. Similar horizontal boundaries are recognizable for deeper fauna on the continental margin (200–1000 m) and Le Danois suggested five horizontal geographic faunal units for this zone of depth.

Below 1000 m (archibenthal), Le Danois recognized only one horizontal geographic unit as a mixture of fauna from the north and from the south. This faunal depth range is most interesting in the high proportion of archibenthal types of animals, such as the decapods *Polycheles* and *Geryon*, the crinoid *Rhizocrinus*, the urchin *Dorocidaris*, and the mollusc *Pleurotoma*, to name a few of the better-known genera. The shelf echinoids *Spatangus rashii* and *Echinus acutus* disappear by 1000 m, marking the boundary of the shelf faunal province with that of the AZT or the semiabyssal zone of Le Danois. The sediments in this depth range are of terrigenous origin and the temperature is close to 9°C.

The fauna of the archibenthal zone has a high proportion of eye-bearing decapods, such as *Amalopennaeus elegans, Sergestes arcticus, Nephrops atlantica, Munida perarmata,* and *Parapagurus pilosimanus.* This shows its shallow-water affinities, which are also indicated by other shallow-water molluscan genera such as *Macoma, Lima, Octopus,* and *Schizotherus.* Deep-sea elements are recognizable with *Thenea muricata, Ophiomusium lymani, Aerosoma hystrix,* and *Mesothuria intestinalis.*

ABYSSAL FAUNAL PROVINCE

The abyssal faunal province commences at approximately 2000 m with the start of sediments of *Globigerina* ooze. The temperature ranges between +3.8 and 2.5°C; hence this area is always below the seasonal influences of the archibenthal.

Only 8% of the species known from the archibenthal are also in the abyss, but the proportion of genera is higher, reaching 15%. Similarly, 2% of the intertidal genera are known in the abyss but no shelf genus is known from the abyss at 5000 m.

The data of Le Danois are so clouded by his synthesis of northern and southern horizontal provinces that very little more can be done with them without a complete reworking of stations and species lists. For example, the species *Hyalinoecia tubicola* is listed as a shelf species off Ireland, whereas it is abyssal in the Bay of Biscay. It was treated in the tabulation as a shelf–abyssal species. Similar problems are encountered especially with the mollusc shells, which are not only subject to submergence like *Hyalinoecia,* but are also subject to artificial transport. We seriously doubt, for example, that *Hydroides norvegica* is an abyssal species.

To give an idea of the vertical distribution of epifaunal species (in the main) we have reproduced Le Danois' drawing (Figure 4-52) showing a section at 43°10′N and diagrams of the dominant organisms.

CHAPTER 5

Southeastern Pacific

The distribution of the fauna [of America's Pacific warm water region] is influenced to a high degree by the hydrographical conditions which are both peculiar and very variable. From the south comes the great cold Peruvian Current or Humboldt Current, whose main part turns west at the well-marked bend in the coastline at Point Aguja in the direction of the Galapagos Islands. [Ekman, 1953, p. 38.]

With these words Ekman described the impact of surface hydrography on the marine fauna of Peru. This fauna is temperate in marine climate, but is poorly known. Emphasis is provided by the collections of the R/V *Anton Bruun* [Menzies and George, 1972] in which 59 new species of isopods are described from collections containing 61 species in all. It is these isopods which form the basis for the following exposition on zonation at the Milne-Edwards Deep of the Peru–Chile Trench system. It will be seen that the Humboldt Current influences not only the surface fauna but the submerged subsurface fauna as well, even to the greatest depth of 6200 m.

The area sampled by the R/V *Anton Bruun* is outlined in Figure 5-1, where the intensity of trawl sampling is also given in an insert. The sampling intensity is shown along a vertical profile in Figure 5-2. The topography of the trench is also indicated in Figure 5-1, together with the derived faunal provinces and zones.

Several collections of animals were made on the topographic shelf within the region of the oxygen-minimum zone; these contained many animals and especially fishes, but no isopods were taken. The first sample bearing isopods at depths greater than the intertidal was between 5 and 20 m and the next isopod-bearing sample was from 907 m just below the oxygen-minimum zone. Data on isopods for the intertidal and the upper shelf are derived from Menzies (1962a). Beyond the topographic shelf at 220 m and starting at 900 m, 50 stations contained isopods (Figure 5-2, Tables 5-1, 5-2) representing 26 genera, 59 species, and 367 specimens. Data on large epifaunal animals are derived from nearly 1000 submarine photographs and from trawls (Figure 5-2). Biomass data are based on Campbell grab samples (Frankenberg and Menzies, 1968) and also on underwater photographs.

To date, only the Monoplacophora (Menzies, 1968), the Mysidacea (Băcescu, 1971), the Isopoda (Menzies and George, 1972), and the decapod Crustacea (Garth et al., 1971) have been reported on from these collections. It is these groups of fauna that we have used to characterize zones, previously defined on the basis of isopods.

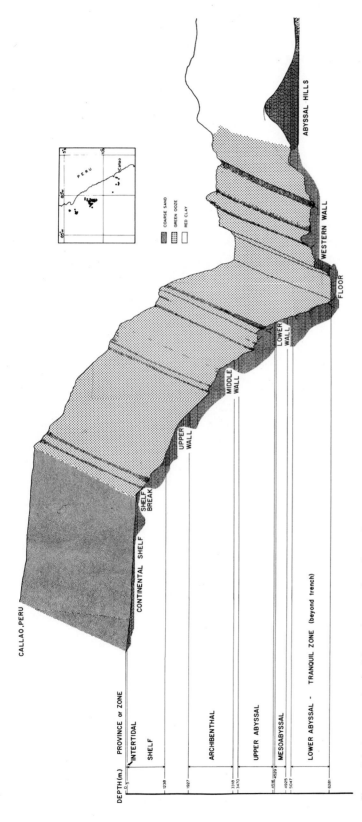

Figure 5-1. Area examined off Peru, density of trawl samples in insert. The profile is derived from continuous depth recordings and the sediments from observations by the senior author at sea, aboard the R/V *Anton Bruun*. The entire section is based on duplicate profiles to ascertain imaginary extent. The main point of the profile is the localization of the faunal zones with reference to topographic features. The lower abyssal red clay zone is indicated only by the red clay sediment area to the far right of the diagram.

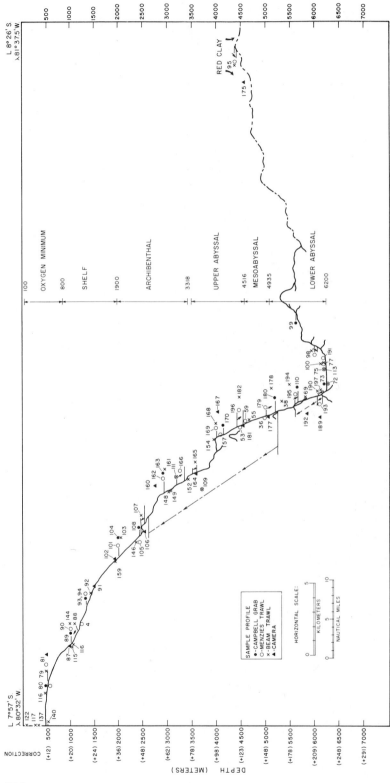

Figure 5-2. Sampling intensity and faunal zones of Milne-Edwards Deep, Peru–Chile Trench, R/V *Anton Bruun* Cruise 11.

134

GENERAL FEATURES: DISTINCTIVENESS OF PROVINCES AND ZONES

At the surface, the marine fauna of Peru may be characterized as subtropical or temperate (Menzies 1962a). The surface temperature ranges seasonally between 16 and 19°C. No major rivers enter the area today, and the estuarine fauna is reduced.

The topographic shelf edge reaches to about 30 nmi from shore to a depth of 220 m. Beyond this is a reeflike plateau at 490 m, followed by the eastern trench wall consisting of an upper wall (1927 m), a middle wall (3318 m), and a plateau at the top of the lower wall. The trench floor is flat (6200–6400 m, N–S). The western wall rises to about 3700 m, leading out onto the Pacific basin (Figure 5-1). The sill depth to the south is around 4750 m (Zeigler et al., 1957), but to the north it is less than 2000 m at the Pacific–Antarctic Ridge. The temperature structure suggests that the thick layer of bottom water of 1.8°C is derived mainly from the southwest.

This particular region of the Eastern Pacific is characterized by the superficial north-flowing cold Peru Current (*auct.* Humboldt Current) offshore, by intensive upwelling, and by a southward-flowing Peru undercurrent between 100 and 500 m in depth (Wyrtki 1966).

The major hydrographic feature of this region is the presence of the pronounced oxygen-minimum zone, which is at least 1000 m thick along the continental margin (Wyrkti 1966). Oxygen concentrations less than 1.0 ml/liter extend at least from 10 to 800 m (Figure 5-3). To the south in Chile the oxygen minimum also is a significant feature of coastal hydrography, but it is not as thick (wide in depth) as it is off Peru. Gallardo (1963) has sampled the shelf and upper slope off northern Chile and suggested, as did Frankenberg and Menzies (1968), that low oxygen in the water column has deleterious effects on the fauna.

Gallardo (*op. cit.*) divided the bottom fauna into three zones: the superior sublittoral, the inferior sublittoral, and the bathyal. The first is from the surface to 50 m and has hydrographic conditions favorable to life. It is typified by such molluscs as *Plagioctenium, Thais, Concholepas,* and *Acanthopleura,* and by crustaceans such as *Petrolisthes, Pachycheles,* and *Gaudichaudia.* The inferior sublittoral was referred to as "semiabiotic" and is in contact with the oxygen-minimum zone of the water column, reaching from 50 to 500 m. Here it is characterized by hydrogen sulfide in much of the surface sediment, which also contains remains of fish such as scales and bones. The average number of animals in this zone was 6.6 per square meter, and the wet weight was 0.17 g/m². Below about 500 m Gallardo found what he considered a typical bathyal (=archibenthal) fauna bathed by Antarctic intermediate water.

Data on surface sediments and Foraminifera are from Bandy and Rodolfo (1964) and from personal observations by the senior author. The sediments on the topographic shelf are coarse grained and greenish-black (Figure 5-1). The outer shelf sediments often have many fish bones and scales as well, and they give off the odor of hydrogen sulfide because of the low oxygen content of the shelf water. The sediments here are not devoid of benthic animals, although the standing stock of infauna is abruptly lowered (Figure 5-3) in the axis of the oxygen-minimum layer.

The percentage of total carbon shows significant peaks in quantity near 1200 and 3200 m, decreasing to less than 4.0% below 3260 m. The percentage of sand, consisting mainly of foraminiferan tests (beyond the shelf) is highest around 1200 m and decreases sharply below that depth; there is, however, a significant increase on

TABLE 5-1. DISTRIBUTION OF ISOPOD GENERA IN PERU–CHILE TRENCH: DEPTH IN METERS

Sampling Intervals

Isopod Genus	1 (0–2)	2 (5–20)	3 (21–40)	4 (41–100)	5 (101–300)	6 (301–500)	7 (501–900)	8 (907–1232)	9 (1927–1997)	10 (2335–2554)	11 (2945–2966)	12 (3086–3318)	13 (3470–3970)	14 (3994–4516)	15 (4529–4925)	16 (5047–5379)	17 (5586–5648)	18 (5750–5900)	19 (5968–6330)	20 (4200–4400 red clay)
Exosphaeroma	×																			
Cymodocella	×																			
Dynamenella	×																			
Amphoroidea	×																			
Dynamenopsis	×																			
Paradynamenopsis	×																			
Ligia	×																			
Neojaera	×																			
Chaetilia	×																			
Ianiropsis	×																			
Munna	×	×																		
Jaeropsis	×	×	×	×	|	|	|	×	|	|	×									
Cirolana		|	|	|	×	|	|	×												
Cleantis		×						×												
Gnathia								×												
Aega								×	×											
Desmosoma								×	|	×	|	×	×	|	×	×	×	×	×	
Munnopsis								×	×	×	|	|	|	|	×					
Munnopsoides								×	×	|										
Antarcturus									×	|	|	|	×	|	×	×	|	×	×	
Eurycope															×					
Storthyngura									×	|	|	|	|	|	×	|	|	×	×	
Ilyarachna										×	|	×	|	×						
Quantanthura													×							
Notoxenoides																				×

136

PROVINCES	INTERTIDAL	SHELF	ARCHIBENTHAL	ABYSSAL
Sugoniscus			×	×
Haploniscus		×	×	×
Mesosignum			×	×
Macrostylis			×	×
Austrogonium			×	
Acanthocope			×	
Haplomesus			×	×
Nannoniscus		×	×	×
Iolanthe			—	×
Ischnomesus			×	×
Ianirella			×	
Cyproniscus		×		
Zoromunna		×		×
Munnicope		×		
Identified faunal				

TABLE 5-2. BATHYMETRIC DISTRIBUTION OF ISOPOD SPECIES OFF PERU (SOUTH PACIFIC): DEPTH IN METERS

Sampling Intervals

Isopoda Genus	Species	1	2	3	4	5	6	7	8	9	10	11	12	13	14	15	16	17	18	19	20
		0–2	5–20	21–40	41–100	101–300	301–500	501–900	907–1238	1927–1997	2335–2554	2945–2966	3086–3318	3470–3970	3994–4516	4529–4925	5047–5379	5586–5648	5750–5900	5968–6330	4200–4400 (red clay)
Exosphaeroma	lanceolata	X																			
Cymodocella	foveolata	X																			
Dynamenella	tuberculata	X																			
Amphoroidea	typa	X																			
Dynamenopsis	bakeri	X																			
Paradynamenopsis	lundae	X																			
Ligia	sp.	X																			
Neojaera	elongatus	X																			
Chaetilia	paucidens	X																			
Ianiropsis	tridens	X																			
Munna	nanna		X																		
Jaeropsis	bidens		X	X	X	X															
Cirolana	robusta		X																		
Cleantis	chilensis		X																		
Cirolana	bathyalis								X												
Cirolana	deminuta								X												
Cirolana	ornamenta								X												
Gnathia	incana								X												
Aega	perulis									X											
Desmosoma	brevicauda								X												
Munnopsis	abyssalis								X	X	X										
Cirolana	natalis											X									
Munnopsoides	chilensis									X											
Munnopsoides	calidus															X					
Gnathia	lacunacapitalis								X												
Antarcturus	praecipius								X	X											
Eurycope	eltaniae													X		X					
Eurycope	latifrons																		X	X	
Storthyngura	octospinosalis									X						X					

Ilyarachna   *peruvica*
Eurycope   *cavusa*
Eurycope   *manifestus*
Eurycope   *profundum*
Desmosoma   *rotundus*
Desmosoma   *coalescum*
Desmosoma   *neomana*
Desmosoma   *acutus*
Desmosoma   *dolosus*
Desmosoma   *similipes*
Ilyarachna   *vemae*
Quantanthura   *globitelson*
Notoxenoides   *dentata*
Sugoniscus   *parasitus*
Haploniscus   *bruuni*
Haploniscus   *acutirostris*
Haploniscus   *concavus*
Haploniscus   *gratus*
Desmosoma   *funalis*
Storthyngura   *unicornalis*
Mesosignum   *truncatum*
Munnicope   *calyptra*
Macrostylis   *longifera*
Austrogonium   *abyssale*
Acanthocope   *orbus*
Haplomesus   *modestatenuis*
Haplomesus   *sp.*
Nannoniscus   *detrimentus*
Nannoniscus   *muscarius*
Nannoniscus   *ovatus*
Nannoniscus   *perunis*
Iolanthe   *neonotus*
Ischnomesus   *calcificus*
Ischnomesus   *simplex*
Ianirella   *latifrons*
Cyproniscus   *binoculis*
Zoromunna   *setifrons*
Ilyarachna   *defecta*
Mesosignum   *multidens*

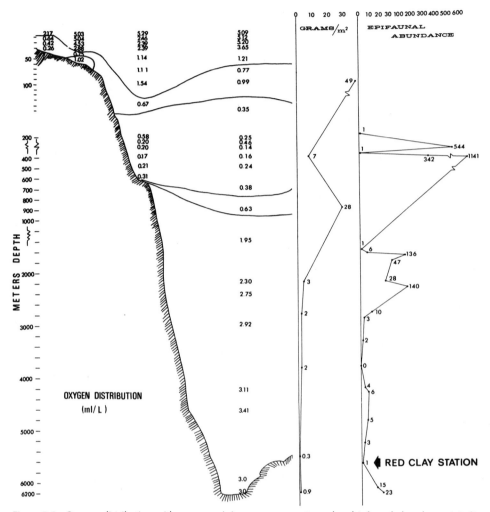

Figure 5-3. Oxygen distribution with oxygen-minimum zones contoured and infaunal abundance (g/m²) and epifaunal numerical abundance; Milne-Edwards Deep, Peru–Chile Trench.

the trench floor. The organic carbon is highest on the faunal shelf above 1000 m; below this depth it decreases but shows a relative increase at 2000, 2900, 3500, 4000, and 6200 m (Figure 5-4). Foraminifera and Radiolaria in the sediment vary sharply in abundance with depth, as well (Figure 5-5).

VERTICAL DISTRIBUTION AND ZONATION

Intertidal Faunal Province (0–2 m)

The intertidal faunal province off Peru commences at the upper level of the tide and ceases at about 2-m depth. As a whole the species of this province may be classified as temperate–subtropical. There are 13 isopod genera, and of these 78% are endemic in vertical profile (Tables 5-1, 5-3). Isopod genera endemic to the intertidal are

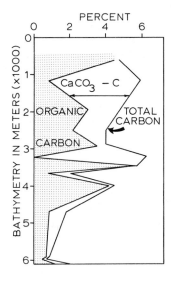

Figure 5-4. Percentages of organic carbon, calcium carbonate-carbon, and total carbon plotted against bathymetry (m), Peru–Chile Trench. (After Bandy and Rodolfo, 1964.)

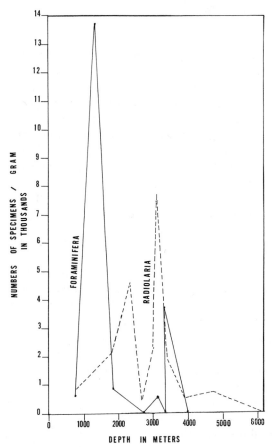

Figure 5-5. The Peru–Chile Trench: abundance of Foraminifera and Radiolaria (specimens/g of dry sediment) plotted against bathymetry. (After Bandy and Rodolfo, 1964.)

141

TABLE 5-3. DATA ON ISOPOD GENERIC ENDEMISM, DISTINCTIVENESS, AND COEFFICIENT OF DIFFERENCE, MILNE-EDWARDS DEEP

Sampling Interval	Depth (m)	Endemics	Starts	Stops	Total	(%) Endemism	Coefficient of Difference	Distinctiveness (%)	Faunal Provinces
1	0	10	3	0	13	78	77	78	Intertidal
2	5–20	1	0	1	4	25	50	50	Shelf
3	21–40	0	0	0	2	0	0	0	Shelf
4	41–100	0	0	0	2	0	0	0	Shelf
5	100–300	0	0	1	2	0	"50"	"50"	Shelf
6	301–500	0	0	0	1	0	0	0	Shelf
7	501–900	0	0	0	1	0	"80"	"80"	Shelf
8	907–1238	2	2	0	5	40	58	66	Shelf
9	1927–1997	1	3	0	7	14	14	25	Archibenthal
10	2335–2554	0	1	1	7	0	14	14	Archibenthal
11	2945–2986	0	0	1	6	0	29	37	Archibenthal
12	3086–3318	2	0	0	7	28	55	62	Archibenthal
13	3470–3970	3	3	0	11	27	27	33	Abyssal
14	3994–4516	1	0	1	9	10	41	50	Abyssal
15	4529–4925	4	1	2	12	33	50	50	Abyssal
16	5047–5379	0	0	0	6	0	0	0	Abyssal
17	5586–5648	0	0	0	6	0	14	14	Abyssal
18	5750–5900	1	0	0	7	14	14	25	Abyssal
19	5989–6281	1	0	6	7	14	100	100	Abyssal
20	4332–4423	0	0	0	1	0	—	—	Abyssal

Ligia (Oniscoidea), *Exosphaeroma, Cymodocella, Dynamenella, Dynamenopsis* (Sphaeromidae), the asellote *Ianiropsis,* and the valviferan *Chaetilia* (Figure 5-6). One intertidal genus, *Jaeropsis,* extends as deep as 200 m on the shelf. The species of the intertidal are more markedly set apart from the shelf fauna than are the genera; indeed 85% endemic and less than 15% of the species and 23% of the genera extend into the shelf (Figure 5-7). Qualitatively the intertidal zone has five major isopod groups. Species of the Sphaeromidae are dominant at 46%, followed by the Asellota (30%) and the Cirolanoidea, the Valvifera, and the Oniscoidea at 8% each (Figure 5-8).

Shelf Faunal Province (5–1238 m)

Beyond the intertidal, the topographic shelf drops slowly out to the shelf break (Figure 5-1) around 54 km from shore and at a depth of 220 m. Except for the nearshore beaches and rocks, the shelf sediments consist of coarse greenish sand often with hydrogen sulfide and many fish scales and bones in response to the oxygen minimum and the high organic production near shore. The water on the shelf changes seasonally and periodically according to special meteorological and hydrographical conditions. The Callao Painter or "El Nino" is the famous example of this. Near and beyond the shelf break, the Peru undercurrent flows southward between 100 and 500 m deep. Here the bottom drops abruptly to a submerged reeflike structure near 440-m depth and drops again to the upper wall of the eastern edge of the trench. The shelf water almost throughout is strongly influenced by the oxygen minimum. The temperature at the shelf break is 13°C, dropping rather rapidly to 4.7°C at the end of the faunal shelf province. It is no surprise to us that the topographic shelf and the faunal shelf do not coincide, because this has been observed by others elsewhere.

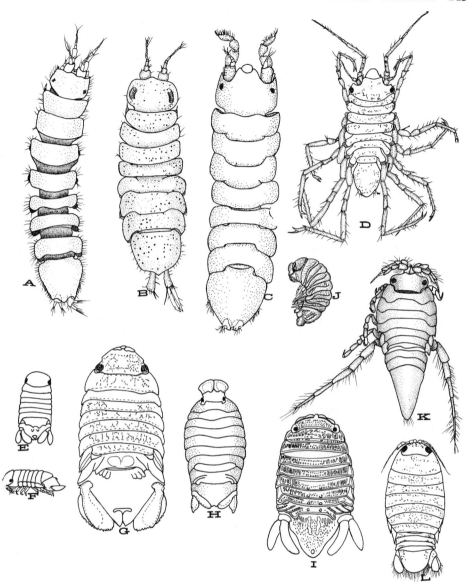

Figure 5-6. Assemblage of isopod genera found in the intertidal faunal province in Peru: (a) *Neojaera,* (b) *Ianiropsis,* (c) *Jaeropsis,* (d) *Munna,* (e)–(f) *Dynamenopsis,* (g) *Dynamenella,* (h) *Amphoroidea,* (i) *Sphaeroma,* (j) *Cymodocella,* (k) *Chaetilia,* (l) *Cirolana.* Scales and sources various.

Although isopods may be less than abundant from the shelf region, this is not the case with other kinds of organisms. Animals such as pennatulids, crabs, fishes, and ophiuroids (Figure 5-9) were plentiful between 450 and 470 m and are typically shelf animals elsewhere. The isopods collected at 907 m consisted of the eye-bearing genera *Cirolana, Gnathia,* and *Antarcturus* and two eyeless genera, *Munnopsis* and *Desmosoma.* The shelf genera of isopods are distinct from the archibenthal by 66% of the genera and by 91% of the species (Figures 5-7). The shelf contributes 24% of its species to the intertidal and only 1% to the archibenthal. No shelf species

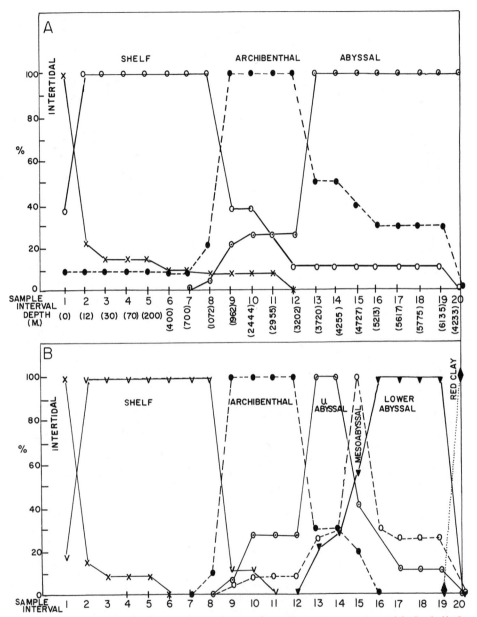

Figure 5-7. (a) Percentage distribution of isopod genera between provinces: ×, intertidal; ○, shelf, ●, archibenthal zone of transition; ⊙, abyssal; (b) percentage distribution of isopod species between faunal zones: ×, intertidal; ∨, shelf; ●, archibenthal; ⊙, upper abyssal; ○, mesoabyssal; ▼, lower abyssal; ◆, red clay.

extends deeper than the archibenthal, and at a species level the shelf faunal province is completely isolated from the abyssal and is almost completely isolated from the archibenthal as well.

Qualitatively, the shelf faunal province has the same major groups of isopods as the intertidal except for the semiterrestrial genus *Ligia* (Oniscoidea) and examples of the Sphaeromidae. The province is dominated by the Asellota at 50%, followed by

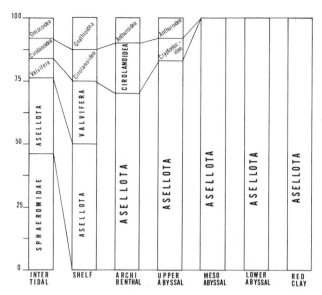

Figure 5-8. Relative abundance (percentages) of major isopod taxa in various zones.

Figure 5-9. Underwater photograph on the faunal shelf off Peru, *Anton Bruun* cruise 11, 460 m depth, showing numerous partly buried ophiuroids, pennatulids, and the crab *Munida*. This photo was taken in the core of the oxygen-minumum zone where the dissolved oxygen is between 0.17 and 0.21 ml/liter. Isopods were not collected here.

the Valvifera at 25% and the Gnathioidea and Cirolanoidea both at 12%; 20% of the genera and 88% of the species of isopods are endemic to the depth range of this province.

Because of the width and depth of the province, there were several sampling points involved; so few contained isopods, however, that our method fails us here to a certain extent. Thus there is an 80% generic distinctiveness between points 7 and 8 (500–900 m and 907–1238 m), but it is evident that this high value is an artifact induced by the oxygen minimum and the absence of isopods between points 5 and 8. There is little question in our minds that *Antarcturus, Gnathia,* and *Cirolana* would have surely been found in shallower depth north and south of the trench. Had this been so, the high distinctiveness would become reduced from 80% to 40%, and a substantially homogeneous shelf fauna would be recognizable (Tables 5-3, 5-4). The shelf off Peru, however, is probably more like the complex shelf elsewhere with an inner and an outer shelf fauna recognizable, as Gallardo (1963) suggested for Chile.

We have assigned the sampling interval between 907 and 1238 m to the shelf because of the presence of the species of *Cirolana, Gnathia,* and *Antarcturus,* which have prominent eyes and constitute more than 50% of the genera at that depth interval.

An assemblage of characteristic large animals captured from the outer shelf is presented in Figure 5-10.

Among these, the characteristic holothurian (Figure 5-10*d*) enters the AZT to a depth of 2454 m. Several fishes (Figure 5-11) collected between 300 and 900 m are

TABLE 5-4.   DATA ON ISOPOD SPECIES ENDEMISM, DISTINCTIVENESS AND COEFFICIENT OF DIFFERENCE, MILNE-EDWARDS DEEP

Sampling Interval	Depth (m)	Endemics	Starts	Stops	Total	(%) Endemism	Coefficient of Difference	Distinctiveness	Faunal Zone
1	0	11	2	0	13	85	85	86	Intertidal
2	5–20	1	0	1	3	33	66	0	Shelf (inner)
3	21–40	0	0	0	1	0	0	0	Shelf
4	41–100	0	0	0	1	0	0	0	Shelf
5	101–300	0	0	1	1	0	"100"	"100"	Shelf
6	301–500	0	0	0	0	0	0	0	Shelf
7	501–900	0	0	0	0	0	"100"	"100"	Shelf
8	907–1238	7	1	0	8	88	88	91	Shelf (outer)
9	1927–1997	3	1	0	5	60	60	70	Archibenthal
10	2335–2554	0	2	1	4	0	25	40	Archibenthal
11	2945–2966	1	0	0	4	25	40	50	Archibenthal
12	3086–3318	2	0	0	5	40	80	82	Archibenthal
13	3470–3970	7	5	0	15	46	47	58	Upper abyssal
14	3994–4516	1	1	1	10	10	66	69	Upper abyssal
15	4529–4925	12	4	4	24	50	66	66	Mesoabyssal
16	5047–5379	0	0	1	8	0	12	12	Lower abyssal (tranquil)
17	5586–5648	0	0	0	7	0	30	30	Lower abyssal (tranquil)
18	5750–5900	2	1	0	10	40	36	40	Lower abyssal (tranquil)
19	5989–6281	3	0	8	11	27	100	100	Lower abyssal (tranquil)
20	4332–4423	1	0	0	1	100	—	—	Lower abyssal (red clay)

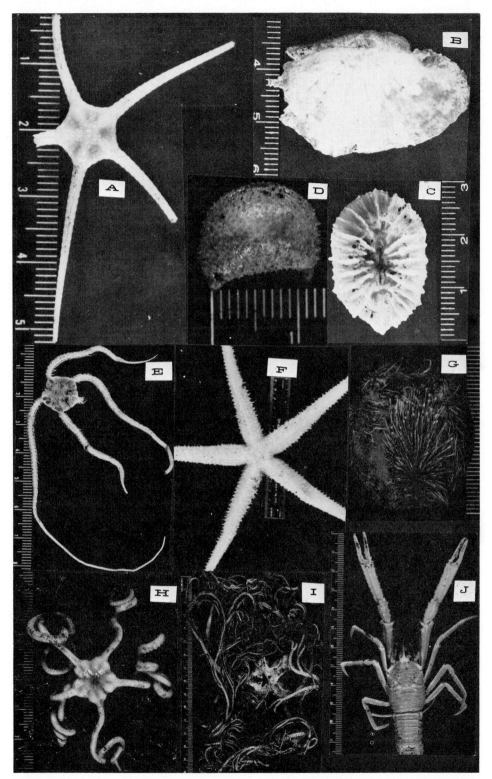

Figure 5-10. Assemblage of characteristic large epifaunal species trawled from the shelf faunal province off Peru (*Anton Bruun* cruise 11): (*a*) Ophiuroid; (*b*) and (*c*) dorsal and ventral view of coral, 1005–1124 m; (*d*) *Sphaerothuria,* 935 m; (*e*) Ophiuroid species, 935 m; (*f*) asteroid, 935 m; (*g*) echinoid, 935 m; (*h*) ophiuroid with arms coiled around alcyonarian, 935 m; (*i*) basket star ophiuroid, 935 m; (*j*) decapod (*Munida-?*) captured at 935 m.

147

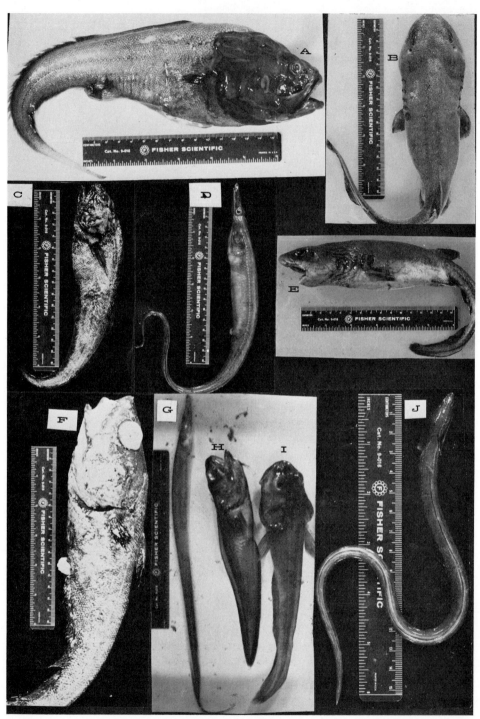

Figure 5-11. Assemblage of fishes trawled from the shelf faunal province between 330 and 935 m (*Anton Bruun* cruise 11) at: (a) 507 m, (b) 338 m, (c) 935 m, (d) and (e) 338 m, (f) 935 m, (g)–(i) 507 m, (j) 338 m.

148

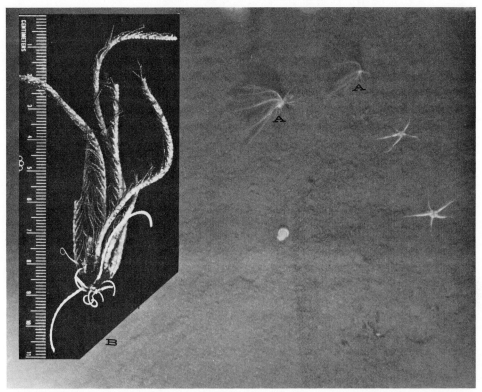

Figure 5-12. Large animals on the Peru shelf including an epifaunal leather-type ophiuroid and two specimens of a red *Antedon*-like crinoid: (*a*) From *Anton Bruun* cruise 11, at 455 m and (*b*) photograph of captured specimen from *Anton Bruun* cruise 11, 907–935 m.

characteristic members of the shelf faunal province. The other large species of
animals found on the shelf between 5 and 1236 m include a red *Antedon*-like
crinoid (Figure 5-12), the holothurian *Psolus* sp. (Figure 5-13), a purple pennatulid
(Figure 5-9), a buried *Asteronyx* (Figure 5-9, 5-14), and a brachyuran *Lophorochinia parabranchia* Garth (Figure 5-15). Six additional species of crabs were recorded by Garth from the shelf: *Trachycarcinus corallanus* Faxon, *Cymonomus
menziesi* Garth, *Trachycarcinus hystricosus* Garth, *Munida propinqua* Faxon,
Munidopsis hystrix Faxon, and *Munidopsis scabra* Faxon.

A total of 22 species of large epifauna are visible at the shelf depths in underwater photographs. Only five of these species enter the archibenthal as well. Both
the large epifaunal animals, as determined alone from the photographic evidence, as
well as the small infaunal isopods of the shelf are distinct from the archibenthal in
77% of the species (Table 5-4). Many of the large shelf animals are sestonophages,
and there is little doubt that by weight these dominate the shelf community.

Archibenthal Zone of Transition (1930–3320 m)

The archibenthal zone of transition is located entirely along the upper wall of the
Peru Trench, extending outward 21.6 km from the faunal shelf and thus starting
below the lower limit of the oxygen-minimum zone. The temperature ranges

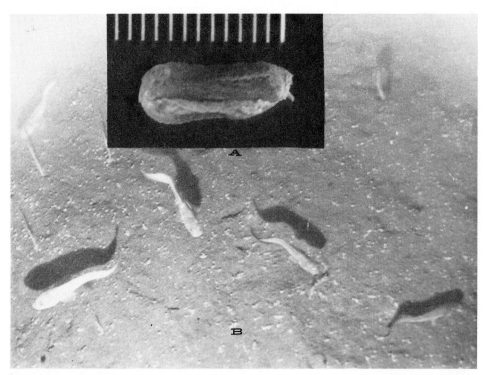

Figure 5-13. Large shelf animals, especially fishes: (a) Specimen of *Psolus* and several fish, 507 m, (b) innumerable specimens; *Anton Bruun* cruise 11, 519 m.

Figure 5-14. Large shelf animals, "x" indicates *Asteronyx*; *Anton Bruun* cruise 11, 1441 m.

Figure 5-15. Large shelf animal, spider crab *Lophorochinia parabranchia:* (a) Captured specimen (*Anton Bruun* cruise 11, at 320–338 m); (b) photograph of specimen on bottom at 500 m.

between 2.3 and 1.8°C from top to bottom, correlating reasonably well with the high percentage of ammonia-nitrogen in the sediments, ranging between 0.50 and 1.0% (Figure 5-16) in this zone (Bandy and Rodolfo 1964).

The AZT is known to contain 10 isopod genera (Figure 5-17) and 10 species, and of these, three genera and six species are endemic. It is distinguished from the shelf in 66% of the genera and 91% of the species (Tables 5-3, 5-4).

The AZT shows an increase in the Asellota to 70% and a loss of the Valvifera and the Gnathioidea. The Anthuroidea first appear, while the Cirolanoidea increase to 20%. In genera the AZT contributes 30% to the shelf and 50% to the upper abyssal, 40% to the mesoabyssal and 30% to the lower abyssal but none to the red clay zone. Generically the AZT appears to be more nearly allied to the abyssal faunal province than to the shelf.

Only one of the archibenthal species enters the shelf, but only to 1900 m, and none extends to a shallower depth. Not one of the species enters the lower abyssal or the red clay zone, but 30% enter the upper abyssal and 20% enter the meso-abyssal. Thus the archibenthal shows a greater affinity to the abyssal faunal province than the shelf province at a species level, as well.

EPIFAUNAL ARCHIBENTHAL ANIMALS

Characteristic large epifaunal species are a white *Antedon*-like crinoid (Figure 5-18a), an echinoid similar to *Phormosoma* (Figure 5-18b), a giant purple holoth-urian (Figure 5-18c), *Scotoplanes* sp. (Figure 5-18d), the ophiuroid *Bathypectinura* or one of its allies (Figure 5-18e), a long white pennatulid (Figure 5-18f), the quill worm

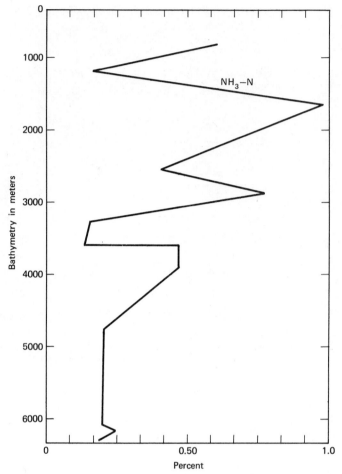

Figure 5-16. Distribution of percentage of nitrogen in sediments of the Peru–Chile Trench plotted against bathymetry. (After Bandy and Rodolfo, 1964, Figure 2b.)

Hyalinoecia (Figure 5-18g), an asteroid similar to *Fryella* (Figure 5-18h), and a large pycnogonid (Figure 5-18i). These are all large animals which are not present among the shelf assemblages. Several characteristic asteroids of the Archibenthal are shown in Figure 5-19. Garth and Haig recorded two species of *Ethusina* from the archibenthal, *E. robusta* (Miers), and *E. faxonii* Rathbun. Of these, the latter also extends downward into the upper abyssal. None was found in the faunal shelf.

Abyssal Faunal Province (3320–6400 m)

The abyssal faunal province starts on the eastern wall of the trench close to a submarine shelflike structure (Figure 5-1) below the middle wall and extends to the flat floor of the trench and slightly beyond. The water temperature is 1.8°C throughout. The oxygen is above 3.0 ml/liter (Figure 5-3). The start of this province is located only 400 m above the sill depth of 3700 m to the west, where the +1.8°C water enters this trench area. Accordingly, the start of the abyssal faunal province coincides approximately with the depth of origin of the bottom water from the sill to the west.

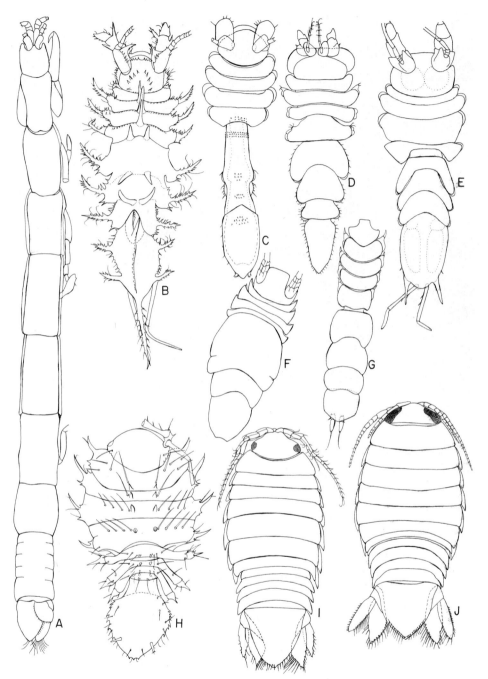

Figure 5-17. Assemblage of isopod genera known from the archibenthal zone off Peru: (a) *Quantanthura* (Anthuridae), (b) *Acanthocope*, (c) *Munnopsoides* (d) *Ilyarachna*, (e) *Munnopsis*, (f) *Eurycope*, (g) *Desmosoma*, (h) *Austrogonium*, (i) *Aega*, (j) *Cirolana*.

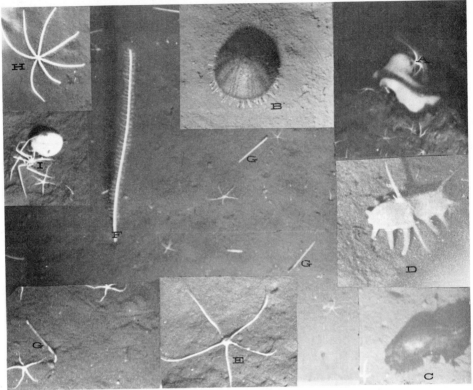

Figure 5-18. Examples of archibenthal large animals: (a) White *Antedon,* 1980 m; (b) echinoid similar to *Phormosoma,* 2961 m; (c) giant purple holothurian, 1980 m, (d) white *Scotoplanes* sp., 2525 m; (e) ophiuroid *Bathypectinura* sp., 2435 m; (f) long, white pennatulid, 1980 m; (g) quill worm, *Hyalinoecia* sp., 1980 m; (h) *Fryella* sp., 2519 m; (i) Pycnogonid, 2419 m.

Of the 19 isopod genera known from the abyssal faunal province 26% extend into the AZT; but only one genus (5%), *Desmosoma,* enters the outer shelf up to 1238 m. No abyssal genus enters the intertidal faunal province. Not one abyssal species is found to extend onto the shelf.

The Cirolanoidea do not enter the abyssal faunal province which now contains 90% Asellota, 5% Cryptoniscidae, and 5% Anthuroidea. Thus four major groups of isopods which are present in the intertidal, the shelf, or the archibenthal are absent from the abyssal faunal province. Certain eyeless genera of marine isopods, such as *Nannoniscus, Ianirella, Sugoniscus, Munnicope,* and *Notoxenoides* make their first appearance (Figure 5-20).

Large animals characterizing the abyssal faunal province include 33 species that are apparently endemic. The holothurian *Scotoplanes* sp. is the only large epifaunal species that enters the abyssal faunal province from the archibenthal. Possibly another species, an enteropneust, enters in the same manner; but this is questionable due to the different appearance of the two enteropneusts (Figure 5-21). The monoplacophoran *Neopilina* extends throughout the abyssal faunal province from 3318 to 6200 m, but not into the lower abyssal red clay zone. One species, *Neopilina* (N) *bruuni* (Figure 5-22c), is found in the upper abyssal zone; whereas two others, *Neopilina* (Vema) *bacescui* (Figure 5-22a) and *N.* (V.) *ewingi* (Figure 5-22b) Menzies

Figure 5-19. Assemblage of asteroid species collected from the archibenthal zone off Peru (*Anton Bruun* cruise 11): (a)–(c) Specimens of three different species, 2445 m; (d) Asteroid, 3117 m; (e) and (f) dorsal and ventral views of an asteroid species, 2554 m.

(1968) are found in the lower abyssal tranquil zone. The sea star *Eremicaster* (Figure 5-19e, 15-19f) also characterizes the abyssal faunal province.

A single brachyuran decapod, *Ethusina faxonii* (Figure 5-23d) was photographed in the abyssal, but only in the upper part at 3500 m.

UPPER ABYSSAL ZONE (3470–4516 M)

The upper abyssal zone commences at the start of the abyssal province near the middle trench wall and extends about 11 km from there, embracing a depth range of 1000 m to the sill depth of the southern edge around 4700 m. The lower boundary coincides with sediment grains of the lowest mean diameter (Bandy and Rodolfo,

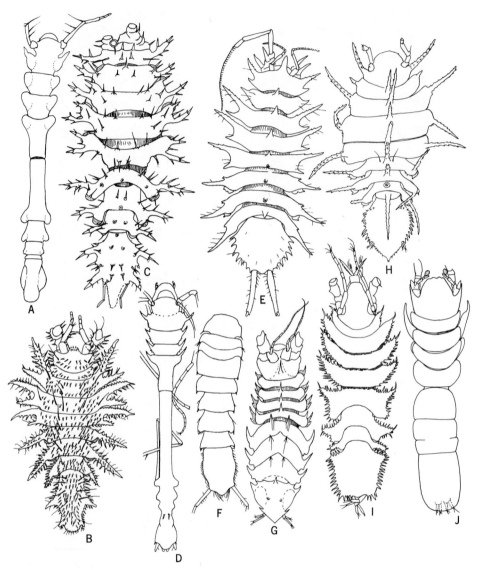

Figure 5-20. Assemblage of isopod genera (Asellota) endemic to the abyssal faunal province off Peru: (a) *Ischnomesus*, (b) *Mesosignum*, (c) *Ianirella*, (d) *Haplomesus*, (e) *Iolanthe*, (f) *Macrostylis*, (g) *Storthyngura*, (h) *Notoxenoides*, (i) *Sugoniscus*, (j) *Nannoniscus*. Scales and sources various.

1964). The temperature is uniformly 1.8°C. Tests for organic carbon of the sediments show a significant increase within this depth range (Figure 5-4).

Seventeen species of isopods are known from the upper abyssal, but only three species enter the archibenthal; the distinctiveness between these two adjacent zones, then, is 83% (Table 5-3). One species (6%) enters the shallowest depth of the archibenthal, but not one enters the shelf. Eight species enter the mesoabyssal, and only 24% go beyond this into the lower abyssal but not to the red clay zone. The species distinctiveness between the upper abyssal and the mesoabyssal zones is 69% (Table 5-4).

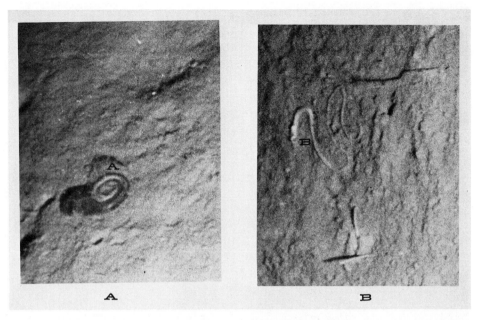

Figure 5-21. Examples of enteropneusts from (a) Archibenthal 2916 m, *Anton Bruun* cruise 11, Station 160, exposure 12; (b) lower abyssal trench floor, 6260 m, *Anton Bruun* cruise 11, Station 189, exposure 24. Note different kinds of trails in these two examples.

Asellota constitute 83% of the isopod fauna of the upper abyssal zone (Figure 5-8), and all species are blind. The endemic species of the upper abyssal zone are *Notoxenoides dentatus, Haploniscus acutirostris, Munnicope calyptra, Nannoniscus muscarius, Nannoniscus detrimentus, Ianirella latifrons*, and *Cyproniscus binoculis*.

Large animals of the upper abyssal zone include one species of *Scotoplanes* sp. (Figure 5-18d) which is also found in the archibenthal, as well as the crab *Ethusina faxonii* Rathbun (Figure 5-23d). The ophiuroid *Ophiomusium lymani* (Figure 5-24), a short black holothurian, and the pennatulid *Umbellula lindahli* (Figure 5-25) extend into the lower abyssal zone.

Endemic animals include a sponge (golf-tee sponge) (Figure 5-23g), the crab

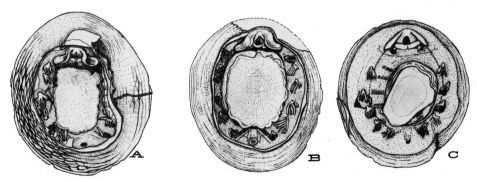

Figure 5-22. The three species of living fossils *Neopilina* from the Peru–Chile Trench: (a) *Neopilina (Vema) bacescui*, (b) *Neopilina (Vema) ewingi*, (c) *Neopilina (Neopilina) bruuni*. (After Menzies 1968.)

Figure 5-23. Upper abyssal characteristic epifaunal species from the Milne-Edwards Deep: (a) Stalked tunicate *Culeolus* sp., 4500 m; (*b*) *Hymenaster,* specimen from 2430 m; (*c*) bottom photograph of *Hymenaster* sp., 3900 m; (*d*) crab *Ethusina faxonii* specimen from 3489 m; (*e*) *Munnid* crab, bottom photograph, 3570 m; (*f*) specimen photograph of munid crab, 3489 m; (*g*) golf-tee sponge, bottom photograph, 3590 m.

Protobeebei miriabilis Boone, *Parapagurus abyssorum* Hendersen, *Ethusina faxonii,* a stalked tunicate *Culeolus* sp. (Figure 5-23a), and a species of *Hymenaster* (?) (Figure 5-23b). The species *Neopilina (N.) bruuni* (Figure 5-22c) is apparently restricted to the upper abyssal zone.

The upper abyssal zone shows a much higher relationship to the other zones of the abyssal province than it shows to the archibenthal.

MESOABYSSAL ZONE (4500–4925 M)

The mesoabyssal zone, in contrast to other zones, extends only 0.25 km outward from the upper abyssal and is less than 500 m high, occupying the upper part of the lower steep trench wall (Figure 5-1). Calcium carbonate and sand are present only as

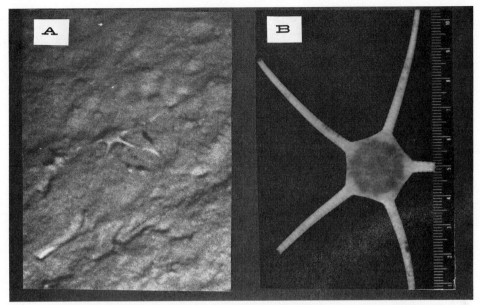

Figure 5-24. Ophiuroids common to upper and lower abyssal zones off Peru: (a) *Ophiomusium lymani*, in situ photograph, 3973 m; (b) O. *lymani*, animal captured at 3500 m.

traces in the sediments, and the organic carbon is low. The temperature is uniform at +1.8°C; the oxygen content is close to 4.0 ml/liter.

The mesoabyssal zone contains 24 species of isopods; all are blind and all belong to the Asellota. The mesoabyssal is distinct from the lower abyssal zone in 66% of the species and from the upper abyssal zone in 69% of the species. One species extends upward to the lower depth limit of the archibenthal. The fauna of this zone contributes 33% of its species to both the upper and the lower abyssal zones. Twelve of the 24 species are endemic to this zone. The mesoabyssal zone has twice as many species as the lower abyssal and more than both the shelf and the archibenthal combined, but only seven more species than the upper abyssal zone. This high species number persists in spite of the narrow range in depth of this zone, and even though fewer samples were taken here than from either the upper or lower abyssal zones (Figure 5-2).

The large animals of the mesoabyssal include species that are found also in other zones of the abyssal, such as a short black holothurian (upper abyssal and lower abyssal), *Ophiomusium lymani* (Figure 5-24; upper abyssal and lower abyssal) and a curious cluster sponge (mesoabyssal and lower abyssal). A dominance by weight of detritophages is indicated by the large animals. The mesoabyssal faunal zone is slightly more closely allied to the upper abyssal than to the lower abyssal. It is scarcely at all related to the archibenthal, and is totally isolated from the shelf faunal province. No crabs were found in the mesoabyssal zone or deeper.

LOWER ABYSSAL TRANQUIL ZONE (5000–6280 M)

The lower abyssal zone appears to start halfway up the lower steep wall of the trench from its floor extending around 60 km westward. The sediments on the eastern wall are green, whereas those beyond the western wall are red clay and mark the start of

Figure 5-25. Upper–lower abyssal sea pen *Umbellula lindahli:* (a) Habitat drawing at 4500 m, length of animal, 45 cm; bottom photographs from *Anton Bruun* cruise 11, Station 192, frame 2, 6100 m.

another zone. We did not locate the boundary between the red clay and green sediments. The trench floor shows a significant but slight increase in calcium carbonate, in organic carbon, in sand, and in ammonia-nitrogen, as well as an increase in radiolarians (Bandy and Rodolfo, 1964) and an increase in benthic biomass (Frankenberg and Menzies, 1968) of the infauna as well as the epifauna (Figure 5-3).

The lower abyssal tranquil zone contains 14 species of isopods; all belong to the Asellota and all are blind. The lower abyssal tranquil zone is 66% distinct from the mesoabyssal and 100% distinct from the red clay zone (Table 5-4). It contributes 56% of its species to the mesoabyssal and 28% to the upper abyssal. Not one species enters the archibenthal, and not one is known yet from the red clay zone. Only six of the 14 species are endemic to the tranquil zone. These are:

Ilyarachna vemae
Ilyarachna defecta
Zoromunna setifrons

Figure 5-26. Assemblage of in situ photographs of characteristic large epifaunal species from the lower abyssal zone off Peru (*Anton Bruun* cruise 11): (a) *Psychropotes longicauda*, 6260 m; (b) short-stalked *Umbellula* sp., 6260 m; (c) *Pheronema pilosum*, (d) sponge, *Euplectella*, 5027 m; (e) feeding bryozoan *Kinetoskias* sp., 6260 m; (f) cluster sponge (unidentified), 5027 m; (g) crinoid *Bathycrinus* sp., 6260 m; (h) holothurian *Pseudostichopus* sp., 5475 m.

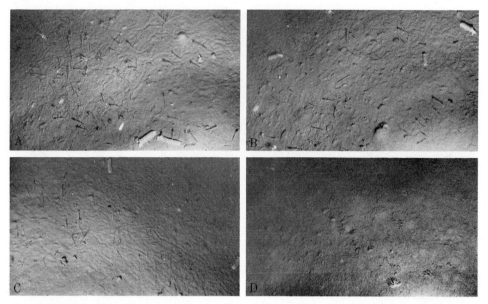

Figure 5-27. Peru–Chile Trench Floor showing large epibenthic fauna, *Eltanin* cruise, 3, 6200 m: (a)–(c) Holothurian *Peniagone* sp. seen with other animals, including the crinoid *Bathycrinus;* (d) the sponge *Cladorhiza.*

Iolanthe neonotus
Nannoniscus ovatus
Haploniscus gratus

The 17 large species of the lower abyssal tranquil zone include *Ophiomusium lymani* (cited previously), which also enters the archibenthal, and one that enters the mesoabyssal. On the other hand, four species enter the lower abyssal tranquil zone, bypassing the mesoabyssal. Typical lower abyssal tranquil endemics include the holothurians *Pseudostichopus* (Figure 5-26h), and *Psychropotes longicauda* (Figure 5-26a), and *Peniagone* sp. (Figure 5-27a–c). Three sponges typify the zone, *Cladorhiza* (Figure 5-27d), *Euplectella* (Figure 5-26d), and a curious "Chimney" sponge *Pheronema pilosum* (5-26c). Also found in this zone are the crinoid *Bathycrinus* sp. (Figure 5-27b) and the bryozoan *Kinetoskias* sp. (Figure 5-26e). There are two endemic species of *Neopilina* (Figure 5-22), *N. (Vema) bacescui* and *N. (V.) ewingi* (Menzies, 1968). Ophiuroids are represented by the eurybathial species *Ophiomusium lymani* (Figure 5-24).

LOWER ABYSSAL RED CLAY ZONE (4320–4430 M)

The western wall of the trench, extending onto the Pacific basin, is mainly red clay and contains a considerable amount of manganese nodules. It extends in depth from 4700 to 6000 m (Figure 5-1) and is generally impoverished both in epifauna and in isopods. The only species of isopod collected from this region is *Storthyngura unicornalis*. Doubtless there are other animals in the red clay environment, but these must be sparsely distributed. At a generic and species level among isopods, the fauna of the zone is 100% distinct from the trench floor fauna, but clearly we do not have enough data yet to characterize it as a zone, a province, or a subzone. The biomass is

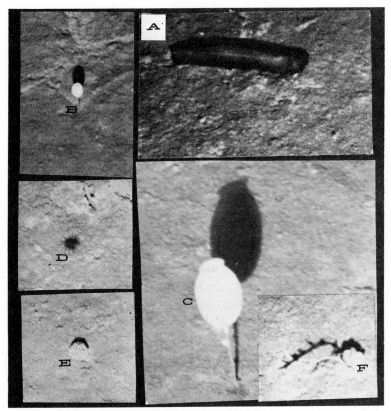

Figure 5-28. In situ photographs taken at 4567 m of assemblage of characteristic epifaunal species photographed from the only red clay station from the western trench wall of Milne-Edwards Deep: (a) Long black holothurian, (b) and (c) sponge on stalk; (d) burrowing sea anemone, (e) cluster sponge, (f) "reindeer horn."

probably close to other red clay environmental values, namely, less than 0.1 g/m² (Filatova and Levenstein 1961).

Large animals seen in underwater photographs characterizing the lower abyssal red clay zone (oligotrophic Pacific) include only eight species. One of these, which comes from the upper abyssal and apparently misses the mesoabyssal zone entirely, is a cluster sponge; the other is *Psychropotes* from the lower abyssal zone. The species characterizing the lower abyssal red clay zone include a long black holothurian (Figure 5-28a) (detritophagous), the isopod *Storthyngura* (detritophagous), the ophiuroid *Amphiophiura bullata,* a burrowing sea anemone (Figure 5-28d) (detritophagous), and a stalked sponge (Figure 5-28b and c). It seems likely that by weight this fauna, which is located in an oligotrophic situation, is dominated by detritophages and not by sestonophages, as suggested by Sokolova (1959).

Zonation of Benthic Foraminifera at the Peru–Chile Trench

The only published study on bathymetric zonation of the benthic Foraminifera of the Milne-Edwards Trench is that by Bandy and Rodolfo (1964). The majority of the samples were collected by one of us (Menzies, on the *Eltanin*) and were concentrated in the vicinity of the Milne-Edwards Trench, but there are many more samples from the south that were utilized by Bandy and Rodolfo (1964). Their work thus is a bathymetric synthesis of a wide geographic area (viz., 0–30°S) and hence suffers the usual mixing and loss of evident zones. Nevertheless, those authors found several foraminiferal zones associated with the zonal picture presented herein. Because foraminiferal shells are subject to transport from shallow to deep water, only upper limits of a fauna were determined. These have been related to our scheme of zonation.

SHELF (5–907 M)

Two species of Foraminifera, *Valvulineria inflata* and *Valvulineria inaequalis,* dominate this single shelf sample from 179 m.

Small calcareous species including *Epistominella pacifica,* four species of *Bolivinita,* and *Cassidulina delicata* had an upper depth limit between 300 and 500 m somewhere in the middle of the faunal shelf province.

ARCHIBENTHAL ZONE OF TRANSITION (1927–3318 M)

Bandy's Group 2A, dominated by *Cibicides wuellerstorfi,* has an upper limit between 878 and 2000 m; group 3, characterized by *Cyclammina cancellata* and *Uvigerina peregrina dirupta,* has an upper limit at 1171 m.

Group 4, represented by *Alveolophragmium subglobosum* (symmetrical form), *Eponides tumidulus,* and *Reophax nodulosus,* has an upper depth limit between 1863 and 1932 m at the start of the isopod archibenthal.

Bathymetric group 5 showed an upper limit near 2489 m, and *Hormosina ovicula, Alveolphragmium subglobosum* (asymmetrical form), and *Nonion pompilioides* were dominant.

The forams in group 4-5 are, as Bandy and Rodolfo (1964) pointed out, "eurybathials," the species *Epistominella levicula* being found also in the shallow water of the shelf off California. The genera are quite eurybathial, and several occur in shallow water as well as in deep water, such as:

Genus	Least Depth (m)
Nonion	796
Epistominella	796
Hormosina	1171
Alveolophragmium	1863
Reophax	878

The foregoing genera and others in the archibenthal thus extend from shelf depths as well as into the abyssal faunal province.

ABYSSAL FAUNAL PROVINCE (3470–6200 M)

Group 6 showed an upper limit between 3449 and 3247 m, only 200 m from our placement of an upper limit to the upper abyssal. Bandy and Rodolfo (1964) clearly stated that these index forms apparently are restricted to abyssal depths. The two groups involved are typified by *Planispirinoides bucculenta* and *Stilostomella antillea.* Bandy's group 7, in which *Recurvoides turbinatus* and *Bathysiphon* sp. are dominant (Figure 5-29), extends to the trench floor and appears to have an upper

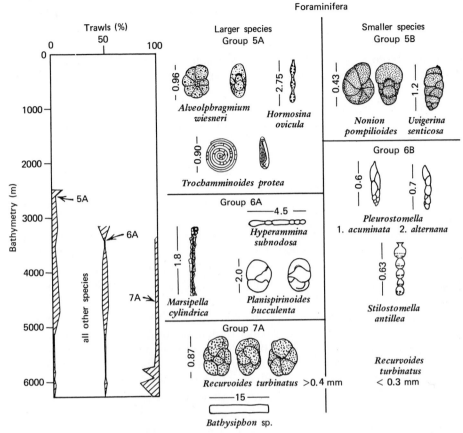

Figure 5-29. Frequency distribution of depth-index group: 5A, *Hormosina ovicula* group, 2489 m; 5B, *Nonion pompilioides* group, 2489 m; 6A, *Planispirinoides bucculenta* group 3149 m; 6B, *Stilostomella antillea* group, 3257 m; 7A, *Recurvoides turbinatus* group, 3404 m. Dimensions are given in millimeters. (After Bandy and Rodolfo, 1964.)

limit of between 3404 and 3495 m, exactly where we placed the upper limit of the abyssal faunal province as evidenced from isopods and large animals.

Zones within the abyssal faunal province are not identifiable using Foraminifera, due to the obvious mixing of shallow-water and deep-sea types and the broad geographic extent of the studies.

Zonation of Biomass at the Milne-Edwards Trench

From the standpoint of biomass or standing stock distribution, the Milne-Edwards Trench offers some special features which are useful in understanding factors influencing the quantitative distribution of deep-sea benthos. First, the trench floor is close to the continental margin; therefore, the principle elaborated by Zenkevitch of decreasing biomass with increasing distance from shore can be tested. Second, the trench is located in the vicinity of one of the world's most productive current systems, the Humbolt current, and should show this influence (providing surface productivity and benthic standing stock are related phenomena). The evidence obtained from Campbell grab samples (Frankenberg and Menzies, 1968) and from underwater photographs expressed as epifaunal densities (Figure 5-3) are illustrative. Although distance from shore is not great, the trench floor biomass is among the highest reported for that depth anywhere in the oceans.

Benthic biomass generally is highest in shallow water and lowest in the great depths. The highest standing stock of 49.46 g/m² was located at the shallowest depth of 126 m, which is just above or at the upper bathymetric limit of the oxygen-minimum zone. The low shallow-water value of 6.5 g/m² was located almost in the core of the oxygen minimum zone (Figure 5-3), and the next highest value of 28.05 g/m² occurs just below the oxygen-minimum zone. The irregularities of infaunal standing stock above 1000 m apparently are influenced by the oxygen-minimum zone more than by increase in depth. Below 5500 m the standing stock ranges between 0.33 and 2.07 g/m². This is in strong contrast to the oligotrophic red clay environment of similar depth in the Pacific, which has values between 0.03 and 0.05 g/m².

The shallow-water biomass of the Milne-Edwards Trench is 4 to 25 times less than that of the Antarctic as indicated from Ushakov (1963), but below 1000 m it is comparable to the Antarctic and to the North Pacific at similar depths. Thus Filatova and Levenstein (1961) reported 24.9 g/m² at 1500 m for the North Pacific, contrasting with our value of 28.05 g/m² at 1000 m.

The highest known and published standing stock for any trench floor is between 0.68 and 1.0 g/m² at the Kurile-Kamchatka Trench (6400–7500 m), and the lowest is that of the Sunda Trench at 0.076 to 0.043 g/m². At the floor of the Milne-Edwards Trench, the biomass is 0.8 g/m²; this value is among the highest known for that depth and is comparable to the Kurile–Kamchatka. Both trenches are located below highly productive surface currents, and both are close to land.

Both the trends in high and variable standing stock above 1000 m and the increase in standing stock on the trench floor are corroborated by the densities of large epifaunal animals as observed from the underwater photographs. The quantity of fishes is illustrative of unexpected irregularities. Fishes were not observed in photographs below 1940 m depth and the maximum number of an average of six fishes per photo frame occurred at 500 m in the axis of the oxygen minimum. The highest density of epifauna, as seen in photographs, was near the upper limit of the oxygen minimum zone where the holothurian *Psolus* sp. (Figure 5-13a) clearly dominated the benthos in numbers and weight as an example of oligomixity.

The regular and obvious increase in standing stock of the infauna as the trench floor is reached is also correlated with a comparable increase in epifaunal standing stock at depths between 5200 and 6200 m (Figure 5-3). These increases in standing stock are believed to be a special feature of this trench relating both to the influence of the productive Humboldt Current and the proximity of the trench to land. Most trenches are sediment traps being enriched more or less continuously by shallow water sediments. This has been illustrated for this trench by Bandy and Rodolfo (1964), who reported relative increases in organic carbon, calcium carbonate, and sand fractions on the trench floor. It is reasonable to believe that the relatively higher organic enrichment of the trench floor sediments should be reflected in an increase in benthic standing stock, and the data support this idea. A comparison of the oligotrophic red clay areas of shallower depth (Figure 5-3) and the trench floor epifauna 2 km deeper provides an obvious illustration of the differences between the two environments in standing stock. (See also Chapter 10.)

CHAPTER 6

The Arctic

The nature of the interaction of the faunas of the three oceans, the strong influence of the faunas of the Atlantic and Arctic Oceans upon each other, and the slight interaction between the faunas of the Arctic and the Pacific Oceans, are completely in keeping with the systematic interchange of water between the Arctic and its two neighbouring oceans. This, however, is only true, of course, for the position at the present time. The relationships differed to a considerable degree in the Quaternary Period, and even more during the Tertiary, not to mention the Mesozoic era. [Zenkevitch, 1963, pp. 27–28].

As the following discussion reveals, the quotation from Zenkevitch defines the basis for the general considerations of the Arctic fauna and its peculiarities. It is necessary to discuss these in some detail in order that the Arctic deep-sea fauna may be defined and better understood. At best, the basis for divisions of the Arctic deep-sea fauna into zones is poor, but it must reflect the origin of the Arctic fauna in general.

GENERAL FEATURES

The Arctic Ocean is roughly limited at the south by the Arctic Circle at 60°N latitude, although this is neither the hydrographic nor the faunal boundary isolating the High Arctic closer to the Pole. It was not until the historic voyage of the *Fram* (1896) that the presence of an ocean at the North Pole was discovered. At that time the area was thought to consist of a single ice-covered basin, but it since has been shown to contain five or six elliptical basins, 700 mi wide (east to west) and 1400 mi long, which are separated by submarine rises (Ostenso, 1963) (Figure 6-1). Going from the Pacific to the Atlantic along 165°W longitude, the basins encountered are described according to the following tabulation.

Basin	Symbol	Maximum Depth (m)	Ridge	Ocean
			Bering Strait (50 m)	Pacific
Canada	B-6	3500	Alpha Rise	Arctic
Siberia	B-5	3500	Lomonosov Ridge	Arctic
Eurasia	B-4	4500	Mid-Ocean Ridge	Arctic
Fram	B-3	3000	Nansen Sill	Arctic
Norway–Greenland	B-1-2	3000	Faero–Iceland–Greenland Ridge (440 m)	Atlantic

167

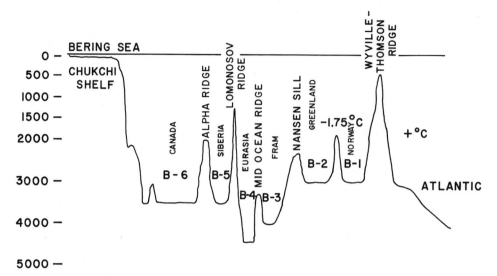

Figure 6-1. Section through Arctic basins and sills. Nomenclature for geographic names generally follows Ostenso (1963), with the following differences: Ostenso split the Canada basin into a Beaufort Deep, a Canada Deep, and a Hyperborean Basin and called the Siberia Basin the Markarov Deep. Nomenclature for Arctic Basin features is not yet stabilized.

The wide continental shelves of the Arctic constitute more than 30% of the Arctic Sea. Depressions in the shelves form the shallow epicontinental seas, such as the White, the Barents, and the Kara seas (Figure 6-3). The benthic fauna in these is a mixture of deep- and shallow-water components and reflects the origin of the water.

The shoal nature of the Bering Strait, only 50 m deep, prohibits any sizable exchange of water between the Arctic and the Pacific. The strongest hydrographic connection with the Arctic is through the entrance to the Atlantic at the Faero–Iceland–Greenland Ridge. The Norway current (Northern Drift), a warm current derived from the Gulf Stream, flows over this ridge into the Arctic. It is this current that keeps Iceland and much of the coast of Norway free of ice in the winter. The penetration into the Arctic of water of Atlantic origin is a dominant feature of Arctic hydrography. This water enters the Arctic as a high-salinity, warm inversion of positive temperature (Figure 6-2), only 100 m thick. It is restricted to depths less than 500 m (Mosby, 1963a).

The nearshore shelf waters off Murmansk and much of the water of the Canadian archipelago are influenced by summer melting; as a result, many isolated oligohaline conditions are created. This low-salinity cold water overlies the Atlantic penetration and has a distinct fauna. Below the water of Atlantic origin is the Arctic bottom water, which extends from 600 m to the bottom as an isothermal layer near minus 0.4°C. This layer is probably formed during the winter in the Norwegian Sea. The direction and velocity of bottom water movement are not known, but geostrophic currents would be expected. The High Arctic is characterized by surface temperatures below zero throughout the year. Temperatures at the continental margins reach as high as 5°C seasonally.

Today the cross section at the Bering Strait is only 2.5 km², whereas the opening into the Atlantic is 370 km². Around 8000 km³ of water (Kort, 1962) enters the Arctic

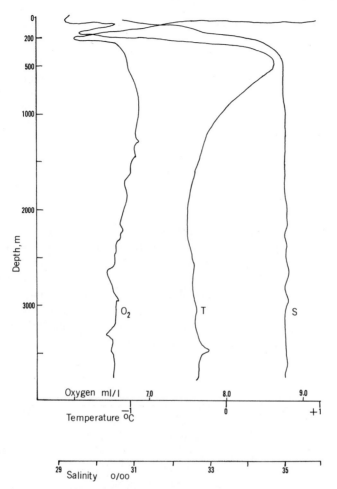

Figure 6-2. Representative oxygen, temperature, and salinity data available from Ice Station T-3. (After Kinney et al., 1970.)

through the Bering Strait. In contrast, 400,000 km³ water enters the Arctic from the Atlantic. A total of 436,300 km³ is carried out of the Arctic, with 6000 km³ in the form of floating ice (Table 6-1). In the geologic past, quite different relationships existed between the Atlantic and the Pacific, and these different features are reflected in the present composition of the fauna.

ZOOGEOGRAPHIC DIVISIONS OF THE ARCTIC

Zenkevitch divided the Arctic into three subregions: the abyssal, a shallow Lower Arctic, and a shallow High Arctic. He split the shallow High Arctic into two provinces, a marine province of high uniform salinity and a brackish water province of low and variable salinity (Figure 6-3). The penetration of the Atlantic surface waters is seen in the distribution of the Norwegian boreal fauna as far as Murmansk, where it mixes with the fauna of the Lower Arctic subregion.

TABLE 6-1.   COMPARISON BETWEEN BERING STRAITS AND WYVILLE
THOMSON RIDGE

Area	Cross-Sectional Area (km²)	Depth (m)	Inflow Annual Transported (km³)	Outflow (km³)
Bering Strait	2.5	70 (max)	8,000	8,000
Faero–Iceland–Greenland Ridge (Wyville Thomson Ridge)	370	440 (min)	400,000	436,300

See Zenkevitch (1963).

There are several special characteristics of the Arctic fauna which at times have perplexed and confused zoogeographers. We believe the picture is easily interpreted if one accepts the onset of glaciation as a geologically recent event occurring only 2 million years ago. The onset of the hypopsychral climate* in the Arctic appears to have been accompanied by a reduction of species in comparison with other regions of the world, as well as a reduction in certain major taxa (levels above species).

The Arctic polar fauna is distinguished from that of the Atlantic and Pacific mainly on the basis of species that are characteristically endemic to the Arctic, rather

* Climate below zero degrees centigrade.

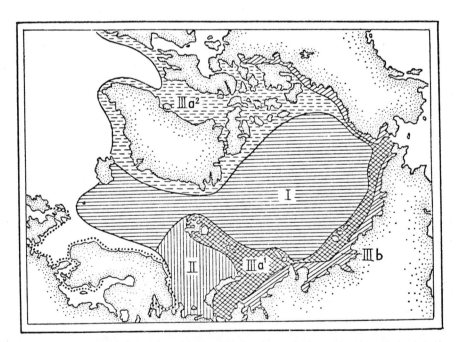

Figure 6-3. Zoogeographical zonation of the Arctic region (according to various investigators). *I*, Abyssal Arctic subregion; *II*, Lower Arctic, shallow subregion; *III*, High Arctic, shallow subregion; *IIIa*, shallow marine province; *IIIb*, shallow brackish-water province; *IIIa¹*, Siberian region, *IIIa²*, North American Greenland region. The propagation of the boreal littoral fauna northward and eastward is marked by a dotted line. (From Zenkevitch, 1963.)

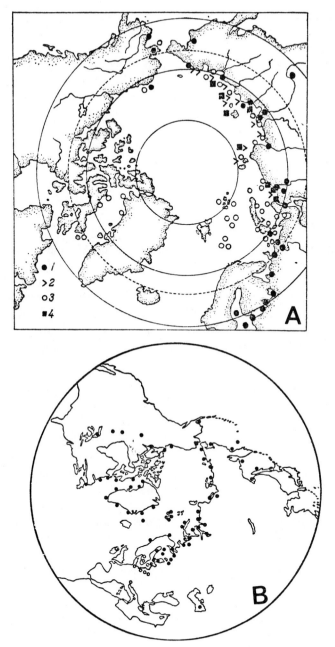

Figure 6-4. (*A*) Distribution of genus *Mesidothea* (Gurjanova, 1934)—1, *Mesidothea entomon*; 2, *M. sibirica*; 3, *M. sabini*; 4, *M. sabini v. robusta* (*B*) distribution of *Mysis oculata* (solid circles) and *Mysis relicta* (open circles). (From Zenkevitch, 1963.)

than on the basis of genera or higher taxa. This is because few genera in any group of organisms are endemic to the Arctic. A good example is the isopod genus *Mesidothea*, which is endemic to the Arctic in the deep sea but is broadly distributed in shallow water outside the Arctic in the North Sea, the Kamchatka Peninsula, and Aleutians of the Pacific. One species lives as a relict in the Caspian Sea (Figure 6-4a). The crustacean *Mysis relicta* shows a similar distribution (Figure 6-4b). Other species exhibiting circumpolar distribution with a +5°C temperature limit include the foraminiferan *Globigerina pachyderma* and the pteropod *Limacina helicina*. Ekman (1953) recorded several instances of species restricted to certain parts of the High Arctic, although they were not circumpolar.

FAUNAL CHARACTERISTICS OF THE SHALLOW-WATER ARCTIC

The shallow-water Arctic with direct influence of the Atlantic or to a lesser degree of the Pacific shows an existing fauna which is closely related to the fauna of both the Atlantic and the Pacific. The shallow Lower Arctic has an intertidal faunal province derived from the Atlantic Ocean through the influence of the Northern Drift Current and hence is Atlanto-boreal-Arctic in nature. The deeper waters of the large epicontinental seas have a mixed shallow-water and abyssal fauna affiliated with the Arctic bottom water.

Arctic Epicontinental Seas

THE BARENTS SEA

The Barents Sea, which has a maximum depth of around 400 m, is directly under the influence of the Atlantic water entering the Arctic. It has an intertidal faunal province south of the polar front which includes *Mytilus edulis*, *Littorina rudis*, *L. palliata*, *L. littorea*, *Acmaea testudinalis*, several species of *Gammarus*, and the isopod *Jaera marina*. These are all boreo-Atlantic and boreo-Pacific species.

The penetration of boreal species into the Barents Sea and the localization of a boundary between boreal (Low Arctic) and the High Arctic species has been investigated by Filatova, 1934, Derjugin, 1927, Vide Zenkevitch (1963). The boundary for plankton differs from that of the benthos (Figure 6-5). As depth increases and temperature decreases, the Arctic boundary for the shallow benthos moves farther southward (toward the Atlantic) than that of the plankton. Nearshore coastal waters are dominated by the fish *Cyprina islandica*, the echinoderm *Brissaster fragilis* and the brachiopod *Waldheimia cranium*. Offshore between 150 and 350 m, *Waldheimia* comprises 50% of the macroscopic population; as the polar front is approached and water temperature decreases, there is a change to an assemblage of sponges comprising 95 to 98% of the total biomass of 5 to 6 kg/m². The sponges involved are *Geodia baretti*, *Carriella cranium*, and *Thenea muricata*. Between 200 and 300 m, a *Brissaster* community replaces the sponge community in the warm submerged Atlantic water. Below 400 m in the colder waters, an arenacious foraminiferan *Rhabdammina* (Figure 6-6) predominates; it has a low biomass of 13.4 g/m². In the deeper parts of the Barents Sea the benthic biomass diminishes to a value of 5 to 8 g/m² and infauna predominate.

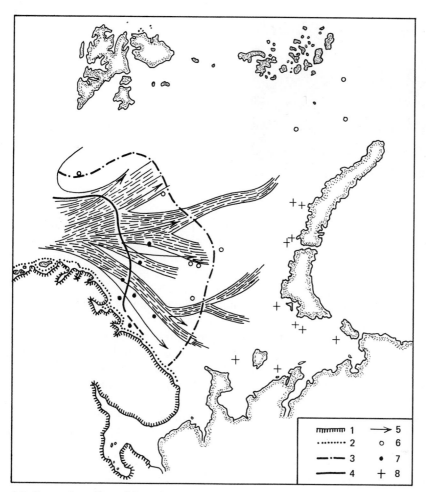

Figure 6-5. Penetration of boreal forms into Barents Sea: *1*, Littoral fauna; *2*, fauna of upper horizon of sublittoral; *3*, boreal pelagic fauna (Derjugin's boundary); *4*, boundary of boreal and Arctic faunas (Filatova); *5*, direction of migration. Places where cold-loving, bottom-living fish *Lycodes agnostus* (6) and *Lycodes vahli v. septentrionalis* (8) and the thermophilic *Lycodes seminudus* (7) (Knipovitch) are found. (After Zenkevitch, 1963.)

Except for the communities in the proximity of the polar front, the benthic biocoenoses to the south are similar to those found in Norway. *Lophohelia* and *Lithothamnion* "coral" banks are submerged in the Barents Sea toward Norway.

THE WHITE AND KARA SEAS

The annual temperature and salinity variations in relation to communities, biomass, and faunal distribution in the White Sea (maximum depth, 620 m) and other Arctic epicontinental seas of the Soviet Union have been presented by Zenkevitch (1963). The bottom of the Kara Sea is under the influence of Arctic temperatures below zero (below 15 m), whereas the sea surface is seasonally positive in temperature and of low salinity, and great variations in surface conditions occur from place to place.

Foraminifera typical of the deep sea of the Atlantic and Pacific (e.g., *Reophax curtus* and *Ammobaculites crassus*) occupy the brown muds of the central basin.

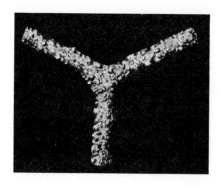

Figure 6-6. The foraminiferan *Rhabdammina abyssorum* M. Sars. (After Cushman, 1948.)

Here also, the Atlantic–Pacific abyssal sea pen *Umbelulla* is represented by *U. encrinus*. The benthic biomass of the brown muds of the Kara Sea are one-twentieth smaller than the biomass of the brown mud environments of the Barents.

POINT BARROW AND VICINITY

The communities and biomass of the fauna in the vicinity of Point Barrow remain to be described.

The collections of George and Nettie MacGinitie (MacGinitie, 1955) from Point Barrow, Alaska, constitute the first major American studies in the Arctic. But, MacGinitie, for several good reasons, did not apply quantitative techniques to his studies; nor was he interested in determining the dominant species of a "community."

Several taxonomic works appeared as a result of his collections on Foraminifera (Loeblich and Tappan, 1955), Polychaeta (Pettibone, 1951, 1954), Amphipoda (Shoemaker, 1955), and others.

Of the 428 species of animals studied by MacGinitie (*op. cit.*), who restricted his sampling to depths less than 66 m, the author reported cosmopolitan species at 48.6%, boreal immigrants from the Atlantic at 19.3%, from the Pacific only at 22.4%, and Arctic autochthons at 9%. His data for sponges and amphipods showed contrasting results: 37% of the amphipods were cosmopolitan, and none of the sponges were; 30% of the sponges were Arctic–Pacific, but only 7% of the amphipods were. Endemic sponges amounted to 30% and the amphipods to 25%.

PERCENTAGE DISTRIBUTION OF POINT BARROW MARINE FAUNA[*]

Total Number of Species	Pan-Arctic Pacific and Atlantic	% Distribution		
		Arctic–Atlantic	Arctic–Pacific	Arctic Only
428	48.6	19.3	22.4	9.6
Sponges (10)	0	40	30	30
Amphipods (100)	37	31	7	25

* Collections restricted in the main to less than 200 ft (ca. 66 m), maximum depth sampled was 714 ft (ca. 319 m).

CONTEMPORARY ARCTIC ZOOGEOGRAPHICAL RELATIONSHIPS

Evidence for the connection between the Pacific and the Arctic is especially strong also in the echinoderms.

The zoogeographic relationships of 121 species of echinoderms has been studied by Djakanow (1945). He found that 4% were cosmopolitan, 23% of boreo-Atlantic origin, 45% of Pacific origin, and 28% endemic. Close to one-half of the species of Pacific origin were identical to Pacific species. The genus *Leptasterias*, which has 27 species in the North Pacific, has only five in the Northeastern Atlantic, two in Greenland, three in Scandinavia, and only one Arctic circumpolar species. This kind of distribution is common in other groups of organisms and reflects the reduction of species in the Arctic and Atlantic since Mesozoic times, and especially since the onset of glaciation as well as the mio-Pliocene migrations from the Pacific.

SPECIAL CHARACTERISTICS OF THE ARCTIC FAUNA

The fauna of the Arctic has certain characteristics that substantially set it apart from the fauna of other oceans. These characteristics relate, as Zenkevitch believed (see chapter quote), to the history of the Arctic Ocean.

Cold Temperature Does Not Alone Inhibit Development of a Fauna

The existing Arctic fauna conforms closely to the special conditions of topography and currents of today. The absence of any special faunal features associated with low temperature *per se* was emphasized by Dunbar (1968). Cold water here does not inhibit the development of a flourishing fauna; this is also shown both by the diversified Antarctic fauna that lives at subzero temperature (Menzies, 1963) and by the profuse development of certain populations in the Arctic.

Youth of Ecosystem

Dunbar (*op. cit.*), whose interests were mainly in ecosystem evolution, pointed out that the Arctic is youthful and has not yet achieved stability.

His reason for believing that the Arctic ecosystem is young was based on only two major points. First is morphism, which may or may not be real, and second is a spotty distribution or anamolous distribution of species of plankton in the Arctic. Spotty distribution includes cases of amphiboreal distribution.

We account for amphiboreal distribution as a consequence of the temperate distribution of the former Pliocene trans-Arctic connection between the Pacific and the Atlantic (well known in molluscs, p. 184). Spotty distribution is accounted for by the oligohaline conditions themselves and by the selection pressure applied to expatriate boreal populations entering the Arctic from both oceans, rather than by any morphism *per se*.

Actually better arguments could be marshalled in support of Dunbar's contention: namely, a recent ecosystem as a consequence of sudden Pleistocene conditions in the Arctic and as a consequence of the present pronounced seasonal variation of the environment near shore, coupled with the range in Pleistocence temperature changes near shore during glacial and interglacial cycles.

It is highly probable that the Arctic fauna of today, as a hypopsychral one, has existed only in the last 2 million years (see p. 353, marine climate evolution), as a rather sudden transformation from a temperate Cenozoic fauna. The limited degree of generic distinctiveness of the North Polar fauna from the temperate fauna of the Pacific has been strongly emphasized by Ekman (1953). Part of the lack of generic and familial distinctiveness between the Arctic and the temperate zones is associated with Ekman's placement of the upper thermal limits of the Arctic fauna at a positive temperature of 5°C.

Morphism as a Special Characteristic of Arctic Fauna

A dominant feature of Arctic species appears to be genetic plasticity, which is reflected in morphologic variations as well as in ecologic isolation. Examples of species with wide-ranging morphologic variation were termed "morphs" by Dunbar (1968). As defined by Huxley (1953), "morphs" are cases of balanced genetic polymorphism and as such are difficult to treat according to the binomial system of nomenclature. There is no evidence to suggest that the number of "morphs" in the Arctic is truly higher than in any other faunal region of the earth, but the percentage in the Arctic in comparison with the total species complement is probably higher than elsewhere (mainly because the Arctic has a lower total number of species). The question of morphs in the Arctic is a subject that should be reexamined carefully by systematists. Many such specimens may in fact be valid subspecies rather than "morphs" due to the common association of distinctive morphological traits with unique ecologic preferences. Whether morph, race, subspecies, variety, or forma is the proper nomenclature is undecided in several instances.

It is true that closely related populations of the relatively few Arctic species have penetrated successfully into a wide variety of habitats. A good example of this resides in the genus *Mesidothea* (Figure 6-7) with four species, each with 0 to 4 subspecies.

1. *Mesidothea megalura megalura* (deep-sea Arctic).
2. *M. megalura polaris* (deep-sea Arctic).
3. *M. sabini sabini* (normal salinity, shallow).
4. *M. sabini robusta* Gurjanova (low salinity).
5. *M. sibirica* Birula (low salinity).
6. *M. sibirica* Birula, Menzies and Mohr (normal salinity-shelf).
7. *M. entomon vetterensis* Ekman (fresh water).
8. *M. entomon glacialis* Gurjanova (brackish water).
9. *M. entomon caspius* G. O. Sars (fresh water).
10. *M. entomon entomon* Linne. (brackish water).

Gurjanova (1938) cited three abyssal species of the amphipod genus *Onisimus* and eight in normal salinity, three in low salinity, and one in brackish water. She also cited an abyssal species of *Monoculodes* plus two shallow-water marine species, in addition to a low-salinity species and one brackish-water species. This diversity of species is repeated in the amphipod genera *Oediceros* (3 spp.), *Pseudalibrotus* (5–6 spp.), *Acanthostepheia* (3 spp.) and so on.

Among the copepods, Dunbar (1968) cited *Calanus finmarchicus* with four subspecies and *Thysanoessa* (3 spp.).

It is not known whether this genetic plasticity applies equally to other Arctic

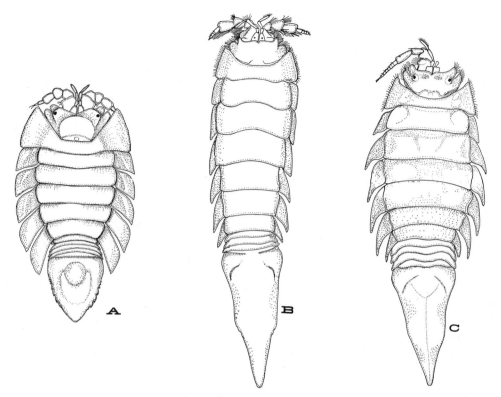

Figure 6-7. Three Arctic species of the valviferan isopod *Idotoega* (= *Mesidothea*) as typical examples of "morphism": (a) *Idotoega sibirica sibirica* (Birula), (b) *Idotoega sabini sabini* Kroyer forma *barentsi* Gurjanova, (c) *Idotoega entomon entomon* (Linnaeus). (From Menzies and Mohr, 1962.)

groups, but the taxonomic and probable genetic plasticity of High Arctic forms seems to be substantiated for the Crustacea and for the fishes.

The perplexing view about "morphs" which was presented by Dunbar (*op. cit.*) contrasts strongly with that presented by Zenkevitch (1963), following the works of Gurjanova (1938) and Gorbunov (1946). These Russian workers are inclined to believe that the numerous species and subspecies of various Crustacea and fishes in the Arctic represent an evolution of species from forms trapped in the Arctic at the onset of glaciation and earlier. The latter view is supported by the ecologic isolation of several species, some in fresh water, some in salt water, and some in the abyssal depths. Zenkevitch pointed out that many of the eurybiontic species (morphs) have moved far beyond the boundaries of the Arctic; indeed, species of the genus *Onisimus* occur in Greenland, the Skagarak, and the Kattegat, and along the Asian coast into the Bering and Okhotsk seas.

Biocoenoses of Oligomixity (Thienemann's Law)

The Arctic is unique among the oceans of the world in being characterized by high population densities of single species, both in the plankton and the benthos. Some plankton blooms of a species are seasonal and others (e.g., *Chaetoceros* and *Melosira arctica*), as reported by Ross (1954), are of long duration. For the benthos, Stuxberg (1882), Schorygin (1945), and Strelnikov (1929) listed *Mesidothea sibirica*,

Diastylis, Chionoectes opilio, Arca glacialis, Alcyonidium gelatinosum, Spiochae-toptrus typicus, Ctenodiscus crispatus, and *Ophiura (auct. Stegophiuria) nodosa* as examples of the Arctic oligomixic communities.

The oligomixitic community of *Rhabdammina* from the floor of the Barents Sea is an excellent example of single-species dominance. Most examples of exceptionally large numbers of individuals and few species appear to refer mainly to Low Arctic elements and not to High Arctic. Mohr (1959) also concluded that the High Arctic is limited in kinds and in individuals both in the plankton and the benthos.

Arctic Reductions in Quantity and Quality

The Arctic is unique in the kinds and number of reductions of biontic features affiliated apparently with its specialized history. There is a noticeable reduction in biomass, a pronounced reduction in the number of genera since the Tertiary, an absence of the intertidal in the High Arctic, a pronounced reduction of primary productivity, a low level of species endemism, especially in the High Arctic, with virtually no generic or familial endemism.

Biomass

The biomass of the High Arctic as determined from epifaunal organisms is much lower (Menzies, 1963, Ushakov, 1963) than that of the Antarctic at comparable depths. The floor of the High Arctic does not appear to have any appreciable epifaunal benthos. This is suggested not only by the absence of animals in bottom photos, but also by the absence of tracks made by animals. Certain holes visible in photographs of Arctic surface sediments are interpreted as being due to ice-rafted debris from the glacial ice and ice islands (Figure 6-8). The biomass of the Barents Sea at 400 m reaches a level of 5 to 9 g/m^2 and is even lower in the brown muds of the White and Kara seas. In contrast, the biomass of more shoal waters of the epicontinental seas of the Low Arctic may reach as high as 5000 to 6000 g/m^2.

Primary Production

Primary productivity appears to be low in the High Arctic because of ice cover and limited photosynthesis, even in the summer. Hopkins (1969) pointed out that the estimated average annual production of organic carbon in the High Arctic (central Arctic) is 1 to 3 × 10^6 metric tons, whereas the zooplankton average annual dry weight amounts to 1 to 2 × 10^6 metric tons; this places the zooplankton in a most unusual relationship to its food supply, necessitating 100% efficiency of food utilization (whereas 10% efficiency is more likely). This most interesting situation perhaps relates to the suggestion of Dunbar (1957) and English (1961) that leptopelic accumulations under the ice may indeed supplement the source of organic matter. Briefly the energetics of the biota of the shallow-water Arctic simply do not balance in a manner comparable to the hydrographic balance sheet.

Figure 6-8. Bottom photograph of Arctic Ocean basin at 2997 m at 83°49'N latitude and 165°05'W longitude, depicting apparent scarcity of epifauna and presence of holes, imprints, and trails. Photograph taken by Mr. Kenneth Hunkins from Drift Station "Alpha."

Reductions in Genera Since the Mesozoic

It may be accepted that the Arctic Tethys fauna was no different, or at least not much different, from the Tethys fauna of the rest of the world's oceans. Extinction of formerly cosmopolitan species and genera has been more marked in the Arctic than elsewhere. Additionally, because no new genera have since evolved in the Arctic, this ocean has been left with an impoverished generic composition. The genera of the Arctic furthermore (exclusive of fishes perhaps), belong to genera that today live in the Pacific and Atlantic. This situation applies to shallow-water fauna as well as to the deep-sea fauna at the generic taxon.

The impact of cooling on extinctions is especially evident in the decline of the species of cidaroids of the Mediterranean from 95 in the Upper Cretaceous to four in the Pliocene and only two in the Recent (Ekman, 1953).

Both in the deep-sea and in shallow water, the Arctic has shown fewer major taxonomic groups than the tropics above the level of genus. Many major taxa, especially the brown algae, are reduced in kinds and numbers in the Arctic. In the deep sea of the Arctic, several classes and orders known from the deep sea of the Atlantic and Pacific are yet to be found. A comparison of major abyssal taxa between Costa Rica and the Arctic at comparable depth illustrates this (Table 6-2).

Reductions in Species Complement

The distribution of planktonic species in the Pacific from south to north illustrates the impoverished nature of the Arctic planktonic fauna. Thus the Arctic basin shows

TABLE 6-2.   COMPARISON OF THE GENERAL DIVERSITY OF THE METAZOAN FAUNA OF THE EURASIA BASIN BELOW 2000 M AND THE ABYSSAL FAUNA OFF COSTA RICA*

Animal Group	Costa Rica	Eurasia Basin
1. Hydrozoa	×	○
2. Scyphozoa	×	○
3. Alcyonaria	×	○
4. Pennatularia	×	○
5. Coral	×	○
6. Actiniaria	×	○
7. Nematodes	×	○
8. Polychaeta	×	○
9. Hirudinea	×	○
10. Priapuloidea	×	○
11. Copepoda	×	○
12. Ostracoda	×	○
13. Cirripedia	×	○
14. Nebaliacea	×	○
15. Cumacea	×	×
16. Tanaidacea	×	○
17. Isopoda	×	×
18. Amphipoda	×	×
19. Decapoda	×	×
20. Pycnogonida	×	×
21. Monoplacophora	×	○
22. Solenogastres	×	○
23. Scaphopoda	×	×
24. Gastropoda	×	×
25. Bivalvia	×	×
26. Brachiopoda	×	○
27. Bryozoa	×	○
28. Holothuroidea	×	×
29. Crinoidea	×	×
30. Asteroidea	×	×
31. Ophiuroidea	×	×
32. Echinoidea	×	×
33. Pogonophora	×	×
34. Ascidiacea	×	○
35. Fishes	×	×
Total	35	15

* Data from Gorbunov (1946), Wolff (1961), and Menzies (1965).

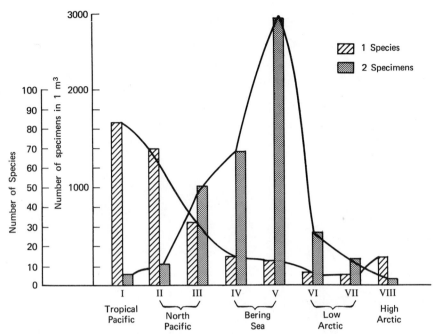

Figure 6-9. Plankton: Change in number of species (1) and number of specimens (2) per cubic meter from tropical part of Pacific Ocean through northern part of the Pacific, Bering Sea, Chukotsk Sea, and Arctic basin. (K. Brodsky, 1956, Vide Zenkevitch, 1963.)

both a great reduction in number of species and a reduction in numbers of specimens, especially in comparison with the plankton of the Bering Sea (Figure 6-9). We believe that this general picture repeats itself for the benthos.

Absence of High Arctic Intertidal Fauna

Although a small intertidal fauna develops in the Low Arctic under Atlantic water influence, there is none in the High Arctic. Ice cover, ice scour, and glacial grinding combine to make the High Arctic intertidal generally inhospitable to colonization by forms that are intertidal elsewhere in the oceans. A comparable situation occurs in the High Antarctic.

MacGinitie (op. cit.) emphasized that the absence of a tidal fauna at Point Barrow was due to ice cover and ice scour on the gravel beach. Additionally, the offshore benthic fauna is smothered periodically by mud carried out during storms. Moreover, simple ice formation, the grinding of ice against the shore, and the more abrasive action of glaciers are sufficient to destroy attempts at intertidal colonization (J. L. Mohr, personal communication).

Upside-Down Under-the-Ice "Benthic" Fauna

The Arctic, by virtue of its ice cover and the appearance of flourishing diatom growths from the ice in the summer months, has developed a unique "under-the-ice community" that was termed a specialized thigmotropic faunule by Mohr (1959). Barnard (1959) discussed briefly two large gammarid amphipod species, *Gammarus*

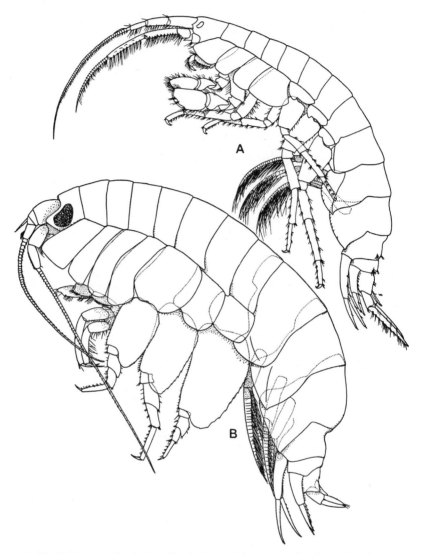

Figure 6-10. Upside-down under-the-ice "benthos": (a) *Gammarus wilkitzkii* Birula, (b) *Pseudalibrotus nanseni* Sars. (From Barnard 1959.)

wilkitzkii Birula and *Pseudalibrotus nanseni* Sars (Figure 6-10), living just below the polar ice pack and upside down as far as the true bottom is concerned.

HISTORY AND ORIGIN OF ARCTIC FAUNA

The Arctic fauna appears to have had a double origin from a cosmopolitan Tethys fauna of the Cenozoic, from the Pleistocene of the Atlantic mainly (Gurjanova, 1938), or from the Pacific mainly (Gorbunov, 1946), depending on the authority. The degree of isolation of the Arctic from the present Atlantic and Pacific below 2000 m suggests that it is an independent region, as was earlier claimed by Ekman (1953).

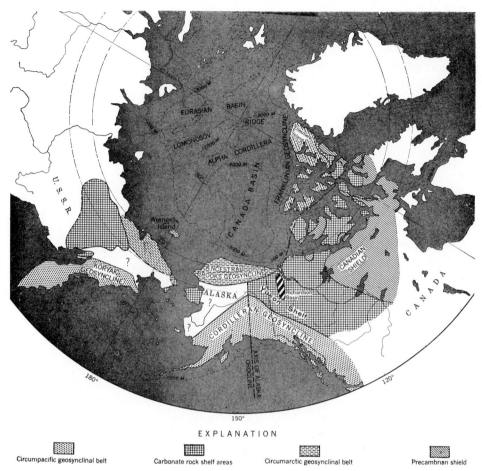

EXPLANATION

Circumpacific geosynclinal belt   Carbonate rock shelf areas   Circumarctic geosynclinal belt   Precambrian shield

Figure 6-11. Tectonic framework of early Paleozoic (pre-Late Devonian) sedimentation. Cordilleran geosynclinal rocks along the axis of the Alaska orocline may be displaced northward as a result of a long period of Pacific sea-floor movement against the continental margin in the Gulf of Alaska. (From Churkin, 1969.)

There is no agreement about the oceanic or continental origin of the Arctic Ocean, and the origin and configuration of an Arctic in past ages is clouded by conflicting opinion. Most recently Churkin (1969) rejected the Arctic landmass subsidence theory (Eardly, 1948) and wrote that the Arctic Ocean is a "very ancient ocean basin floored by oceanic crust and rimmed by an early Paleozoic geosynclinal belt." He identified not less than five ancient geosynclines (Figure 6-11). The present configuration of the basins results, in theory, from sea-floor spreading and the meeting of the Asian and North American continental blocks. The great thickness of the Mesozoic sediments strongly suggests that the Pacific and the Arctic had deep-sea connections at least throughout much of the Mesozoic. This seems to indicate that during that period of geologic time the Pacific and Arctic deep-sea faunas were rather completely mixed. Thus examples of High Arctic deep-sea fauna, in common with those of the Pacific, may be Mesozoic relicts, and those of Arctic–Atlantic affinities could be considered Cenozoic to Recent relicts.

The higher endemism of the abyssal species led Gurjanova (1938) to postulate an origin of the Arctic abyssal fauna from the Arctic shelf fauna during the Glacial times. Indeed, 30% of the abyssal isopod genera from the Arctic are not found on the Arctic shelf. In contrast, all shelf and archibenthal genera of isopods are found in the North Atlantic and/or the North Pacific. More than 50% of the abyssal genera are allied to cosmopolitan abyssal genera; 25% are found only in the North Atlantic, and 10% occur nowhere but in the North Pacific. These percentages are based on only 10 genera and hence are far from convincing.

It does appear, however, that the data fail to support Gurjanova's theory because there should be a greater affinity of the abyssal genera with the genera of the shelf than with cosmopolitan abyssal genera or genera found at abyssal depths of the Atlantic and the Pacific. There are many species from the shelf and from the plankton in common with species from the Pacific or the Atlantic. At abyssal depths the species of the Arctic are almost completely isolated from the abyssal fauna elsewhere.

This more solid geographic evidence is in agreement with Gurjanova's idea. The eurybathial abyssal species could easily have been derived from a glacial shelf fauna. It is likely that the endemic abyssal genera, which are allied evidently to abyssal genera in other oceans, achieved their submergence prior to the Pleistocene. Apparently their species and subspecies have evolved since then in abyssal depths and not as a result of a step-wise evolution from shelf genera and species.

Pacific Elements of the Pliocene

A late Cenozoic opening of the Pacific Arctic resulted in the dispersal of marine faunal elements from the Pacific into the Arctic and the North Atlantic. For example, the echinoids *Echinarachnius* and *Strongylocentrotus* (late Pliocene, England), were endemic to the North Pacific between the Miocene and Pliocene, and both appeared in the North Atlantic by the late Pliocene and Pleistocene but have subsequently vanished from the Atlantic and the Arctic.

Not less than 13 species of bivalves from the upper Pliocene of the North Pacific (central California) have been recorded by Soot-Ryen (*in* Durham and Allison, 1960) as ancient migrants into the Arctic. Many are amphiboreal relicts today.

1. *Chlamys islandica* (Müller).
2. *Mytilus edulis* Linn.
3. *Macoma calcarea* (Gmelin).
4. *Modiolus modiolus* (Linn).
5. *Nucula tenuis* (Montagu).
6. *Protothaca staminea* (Conrad).
7. *Saxidomus giganteus* (Deshayes).
8. *Siliqua patula* (Dixon).
9. *Spisula polynyma voyi* (Gabb).
10. *Thyasira gouldi* (Philippi).
11. *Yoldia limatula* Say.
12. *Zirfaea gabbi* Tryon.

Pleistocene Faunal Migrations: Atlantic Elements

The Pleistocene climate curve is shown in Figure 6-12, and sea-level changes appear in Table 14-4.

The presence of late Pleistocene fossiliferous strata 221 m above sea level in Norway indicates that Norway was High Arctic in climate up to 8000 to 5000 years B.P., during which time the climate improved (Spjeldnaes, 1964). This was followed by a cooling and another recent rapid improvement in climate.

The "warm-water" fauna during the postglacial climatic optimum migrated

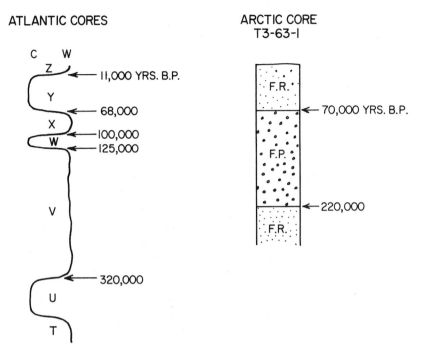

Figure 6-12. Comparison of the climatic curve of Ericson, based on variations in the abundance of *Globigerina menardii* observed in Atlantic cores to the "foram-rich" (FR) and "foram-poor" (FP) zones in core T3/63/1. The chronology for the Atlantic curve has been established by ^{14}C, ^{231}Pa/^{231}Th, and ^{230}Th methods of dating. (From Ku and Broecker, 1967.)

1500 km to the north and then back with a speed of several kilometers per year, according to Spjedlnaes; the retreating cold-water fauna was exposed to catastrophic selection, and certain populations became extinct. Others remained isolated from the retreating populations in fjords and in deeper water as relicts of the cold period. Included as relicts are the bryozoan *Kinetoskias arborescens* and possibly also *Umbellula*. This Pleistocene sequence of population, repopulation, extinction, and relict formation, accompanied by a depauperization of species, must have proceeded with much greater effect in the transition from the warmer Tertiary to the cold Pleistocene (see pp. 350–353 climate evolution).

Various workers have tried to indicate the shifts of isotherms in the Atlantic as a consequence of Glacial cooling of the Arctic and hence the Atlantic as well. The most recent and most successful attempt is the study on coccoliths by McIntyre (1967) who suggested that the northern boundary of the subtropical gyral (Gulf Stream) was shifted from its present position of 40°N latitude to 30°N latitude off North America. This amounts to a 600-mi southward shift of the subtropics in the Eastern North Atlantic (Figure 6-13). This shift placed the 12°C isotherm at the entrance to the Mediterranean during mid-Wisconsin times. McIntyre's reconstruction of isotherms is likely to be incorrect because Arctic species entered the Mediterranean during mid-Wisconsin times. It is thus more probable that the temperature at the entrance to the Mediterranean was then lower than 10°C and more likely near 5°C.

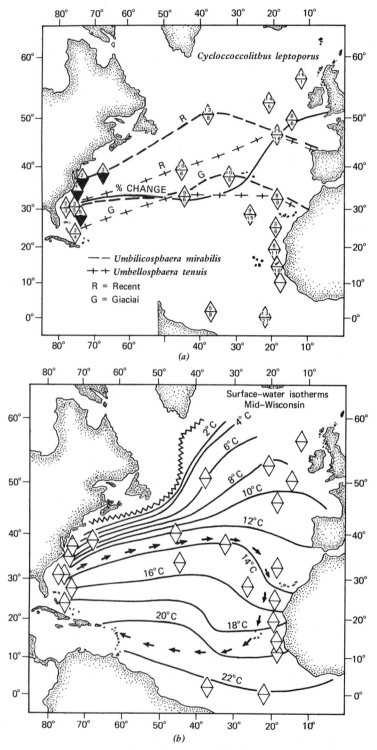

Figure 6-13. (a) Species population boundaries for Recent and mid-Wisconsin time and percent values for *Cycloccocolithus leptoporus* in diamonds with the upper Recent and the lower Glacial values; (b) paleoisotherm map of the mid-Wisconsin North Atlantic erected with the use of coccolithophorid temperature data. The dark arrows indicate presumed position of the subtropical gyral based on coccolith boundaries of subtropical forms. The jagged line represents pack ice (after Flint). (After McIntyre, 1967.)

Post-Pleistocene to Recent Faunal Changes

Ekman (1953) recorded cases of commercial exploitation of species entering the Arctic from the Atlantic as a consequence of historical warming of the Arctic. His citations regarding the codfish, the herring, the mackerel, and the haddock are good evidence of penetrations from the Atlantic. Faunal penetrations into the Arctic are logically associated with the Atlantic water entering the Arctic.

Pleistocene Temperature Changes in the Arctic

Arctic deep-sea sediment cores do not tell much about the past history of the High Arctic; they do, however, indicate that sedimentation has varied considerably and that foraminiferan-rich (FR) layers overlie and underlie sand layers with few foraminiferans (Figure 6-12). The foraminiferan-rich layers coincide with warm as well as cold periods of the Pleistocene. There has been little detectable change in planktonic fauna during the Pleistocene in the High Arctic, and only the remains of the two existing Arctic species, *Globigerina pachyderma* and *Limacina helicina,* are found in cores. Because of this the cores reveal one striking fact about Pleistocene High Arctic conditions. We know that boreal conditions were never achieved during interglacial times because the boreal planktonic species that would have been found in the sediments have not come to light. Thus, during the Pleistocene, the High Arctic was probably never warmer than 5°C (arcto-Boreal condition).

Ku and Broecker (1967) related the foraminiferal-rich and foraminiferal-poor (sand-rich) layers of the Arctic basin to changes in the rate of production of Foraminifera and not to climate change, concluding that the High Arctic has been covered by ice for the last 150,000 years. They suggested that the rate of Arctic basin sediment accumulation has been constant. This hypothesis is open to question.

VERTICAL FAUNAL PROVINCES AND ZONES OF THE ARCTIC

On the basis of the distribution of isopods and the larger epifauna, it is possible to divide the Arctic benthos into three main categories—an oligohaline shallow shelf fauna along Murmansk and the Canadian archipelago, a nearshore "shelf" faunal zone, an archibenthal zone of transition, and an abyssal faunal province, with one or more added zones.

Information about the Arctic deep-sea fauna includes the following regions studied by Soviet and Danish expeditions (Figure 6-14 & 15)

1. The northern and western parts of the Novo-Siberian shoal waters covered by 1932–1938 cruises of the *Sadko* and the *Sedov*. This region is 460 mi in latitude and 160 mi in longitude.
2. The central part of the Arctic between 70°55′N to 82°42′N and 87°03′ to 147°35′E (Gorbunov, 1946).
3. The deep stations of the *Ingolf* southeast of Jan Mayen Island between Iceland and Svalbad (Hansen, 1916) (Figure 6-15).

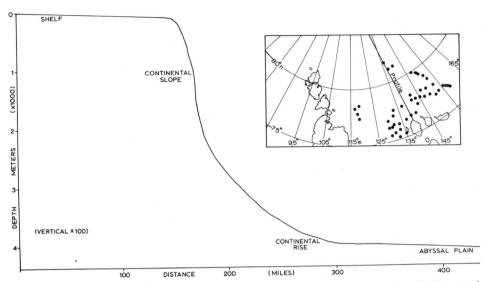

Figure 6-14. Area investigated in the Arctic Ocean; positions of sampling stations are plotted. The vertical profile, taken from the transect marked on the map, shows the extent of the shallow topographic shelf from the shore between 115°E and 165°E by 75°N to 82°N.

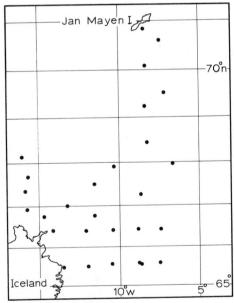

Figure 6-15. The Arctic area between Jan Mayen Island and Iceland showing the selected stations of the *Ingolf* at which isopod species were collected.

188

Archibenthal Zone of Transition (12–360 m)

In the Arctic we consider fauna occupying the topographic shelf to be typical of an archibenthal zone of transition, rather than a shelf fauna, as characterized in other regions. This decision is taken because few (if any) genera and few species are endemic to the Arctic shelf. The endemics that do occur are mostly affiliated with the abyssal zone and not with shallow water. For instance; all the endemic species of Arctic echinoderms recorded by Djakanow (1945) are found in abyssal depths.

Added evidence that the topographic Arctic shelf is in the main an archibenthal zone of transition is demonstrated by the high similarity between the major groups of orders of isopods on the topographic shelf. Thus the Gnathiodea (17–19%) and the Valvifera (25–31%) and the Asellota (50%) dominate the shallow waters. The Asellota increase significantly (to 76–80%) at the start of the abyssal faunal province (Figure 6-16). This break occurs at an approximate depth of 360 m. If the shelf is characterized as a zone, namely, the archibenthal zone of transition, it is nevertheless divisible into inner and outer subzones corresponding to the inner and outer zones of the shelf province elsewhere, but it is here distinguished on a species instead of a generic basis.

The inner zone occupies the shelf between 12 and 162 m and is in superficial Arctic waters under seasonal changes in salinity and temperature. The outer zone is located between 163 and 360 m just above the positive-temperature water of Atlantic origin. These zones are not strongly characterized, and 43% of the species found between 12 and 100 m also extend to a depth of 360 m.

Although all the shelf from 12 to 360 m has been assigned to the AZT because of the identical nature of the genera involved, the limited stands of the kelp *Macrocystis*

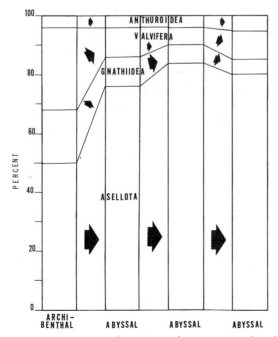

Figure 6-16. Relative abundance (percentages) of major isopod taxa in various faunal zones in the Arctic.

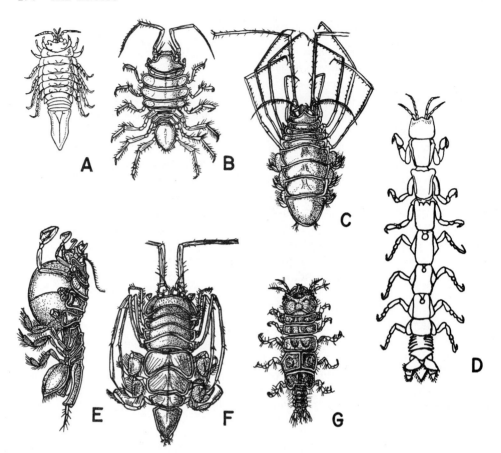

Figure 6-17. Assemblage of isopod genera from the archibenthal zone in the Arctic Ocean: (a) *Mesidothea*, (b) *Munna*, (c) *Eurycope*, (d) *Calathura*, (e) *Munnopsis*, (f) *Ilyarachna*, (g) *Gnathia*.

near shore, as well as the dominance of Low Arctic species (subpolar) very close to shore, argues otherwise. The question then is whether the AZT extends upward as high as 12 m or only to 100 m. Shallow-water as well as deep-water elements are on the Arctic topographic shelf to a depth of 360 m. Shallow-water elements are typified by eye-bearing species of the isopod genera *Munna, Gnathia, Pleuroprion, Mesidothea,* and *Synidotea;* deep-water elements are represented by the eyeless genera *Desmosoma, Ilyarachna, Eurycope,* and *Munnopsurus.* All these genera are also found in the Atlantic and the Pacific (Figure 6-17). They may be considered subpolar elements of the shallow-water Arctic seas. The isopod fauna of the topographic shelf contributes 36% of its genera also to the Arctic deep sea at least to depths as great as 2600 m. and the abyssal faunal province contributes 50% of its genera to the topographic shelf.

The Asellota occupy 50% of the fauna, and the Valvifera, the Gnathioidea, and the Flabellifera divide the remaining half. The absence of the Oniscoidea (intertidal elsewhere) and the great reduction in the Sphaeromidae are features unique to the Arctic.

Large species of other invertebrates characteristic of the archibenthal zone of the Arctic are:

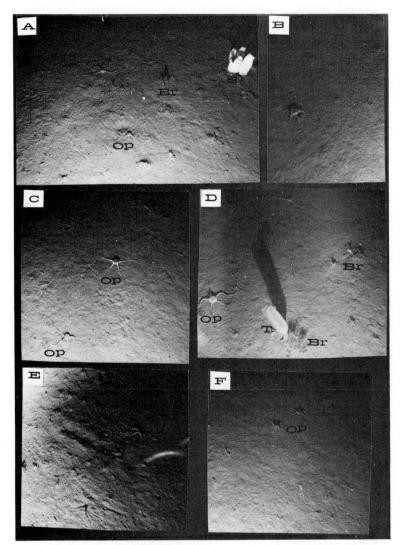

Figure 6-18. Arctic Ocean bottom photograph from the Chukchi Rise at 300 m (archibenthal zone): (a)–(f) Selected exposures showing epifaunal species, the ophiuroid *Stegophiura nodosa* (Op), sponge (S), bryozoan colonies (Br), and an unidentified stalked tunicate or sponge (T). Photograph taken by William Cromie from Drift Station "Charlie."

1. *Grantia miriabilis* (360–698 m).
2. *Trichasterina borealis* (162–1445 m).
3. *Umbellula encrinus* (162–1445 m).
4. *Nymphon robustum* (300–698 m).
5. *Siphonodentalium lobatum* (162–869 m).
6. *Poliometra prolixa* (162–698 m).
7. *Hymenaster pellucidus* (300–869 m).
8. *Ophioscolex glacialis* (300–698 m).

Photographs from the Arctic archibenthal show a sparse epifaunal assemblage (Figure 6-18) consisting of ophiuroids, sponges, and possibly also Bryozoa.

Abyssal Faunal Province (425–2600 m)

The start of the abyssal faunal province is recognized by the shift in major isopod taxa. The Asellota are now dominant at 76%, and the Anthuroidea, the Valvifera, and the Gnathioidea are correspondingly reduced. In addition, isopod genera such as the blind Atlantic genera *Katianira, Nanroniscus, Heteromesus, Haplomesus,* and *Nannoniscoides* make their first appearance (Figure 6-19). In species, the abyssal faunal province differs from the AZT in 33% of its genera and in 50% of its species (Table 6-3). The eye-bearing species that characterized the inner shelf are generally lacking. Ten genera are endemic bathymetrically to the abyssal faunal province. They are not endemic geographically.

We believe it is no mere coincidence that the start of the abyssal faunal province in the High Arctic coincides almost exactly with the end of the shelf water and the start of the submerged Atlantic water, and hence also coincides with the depth limit of any major seasonal changes in salinity and temperature.

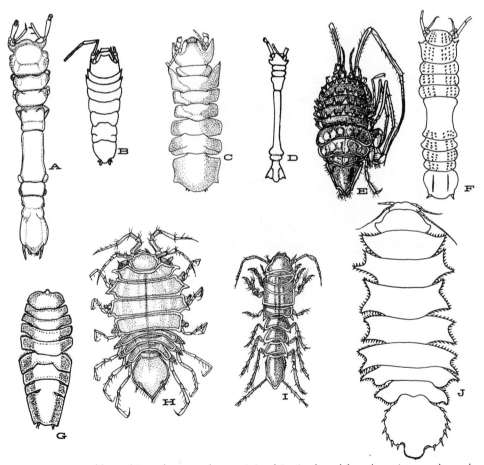

Figure 6-19. Assemblage of isopod genera characteristic of Arctic abyssal faunal province; scales and sources various: (a) *Stylomesus,* (b) *Nannoniscus,* (c) *Nannoniscoides* (d) *Haplomesus,* (e) *Ilyarachna,* (f) *Heteromesus,* (g) *Haploniscus,* (h) *Pleurogonium,* (i) *Macrostylis,* (j) *Katianira.*

TABLE 6-3. DATA ON DISTINCTIVENESS, COEFFICIENT OF DIFFERENCE, AND ENDEMISM OF ISOPOD GENERA AND SPECIES FROM ARCTIC

Series No.	Depth (m)	Endemics Gen.	Sp.	Starts Gen.	Sp.	Stops Gen.	Sp.	Total Gen.	Sp.	Endemism Gen.	Sp.	% Distinctiveness Gen.	Sp.	Coefficient of Difference	Probable Faunal Zones
1	12–100	1	9	10	8	—	—	11	17	9	53				Archibenthal
												9	62	56	
2	162–360	0	1	0	3	0	0	10	12	0	8				Archibenthal
												33	50	48	
3	425–570	1	4	4	6	1	3	15	21	13	19				Upper abyssal
												35	64	58	
4	610–869	3	15	2	2	5	8	18	31	16	48				Mesoabyssal
												44	76	68	
5	1000–1200	0	1	0	1	1	2	10	10	0	10				Lower abyssal
												10	50		
6	1400–1642	0	2	0	0	1	1	9	9	0	26				Lower abyssal
												11	33		
7	1800–2000	0	1	0	0	1	1	8	7	0	14				Lower abyssal
												12	50		
8	2360–2600	0	3	0	0	7	8	7	8	0	38				Lower abyssal

The abyssal faunal province may be divided into substantially distinctive bathymetric zones. We must emphasize that not one of the isopods of the deep floor of the Arctic basin has yet been reported on.

UPPER ABYSSAL ZONE (425–570 M)

Between 425 and 570 m is a zone that encompasses almost exactly the depth distribution of the Atlantic water in the Arctic (450–550 m). This region contains 21 species of isopods, and the species endemism is 19%. Distinguished from the archibenthal zone in 33% of the genera and 50% of the species, it is separated from the next deeper zone in having 64% of the species distinct. Genera that make their first appearance in this zone are *Heteromesus, Katianira, Haplomesus, Nannoniscoides,* and *Nannoniscus,* which are all in the Atlantic abyssal. None of these genera is represented in the archibenthal.

Between 30 and 50% of the species in the Atlantic water enter the AZT and 62% enter the next lower zone. Some species from this zone extend to the shallowest as well as greatest depth in the Arctic.

Large animals characteristic here are the sea star *Henricia scabrio,* the crinoid *Bathycrinus carpenteri,* the cirriped *Scalpellum striolatum,* the decapod *Sclerocrangon ferox,* and the fish *Lycodes pallidus.*

MESOABYSSAL ZONE (610–869 M)

The next deeper zone, which starts at the upper limit of the Arctic bottom water, contains 31 species; the species endemism is 48%, and the generic endemism is 16%. Starting at 610 m, the zone extends only 869 m. Many of its species are found in the Atlantic water (43%), whereas only 20 to 25% enter the archibenthal zone. The majority of the species belong to the blind genera *Haplomesus* and *Heteromesus* (Table 6-4). The blind genera *Stylomesus* and *Echinozone* are endemic vertically.

Large metazoan animals characteristic here are the sponges *Leucosolenia lacunosa* and *Abestopluma minuta,* the ostracod *Polycope* sp., the mollusc *Chrysallida sublustris,* and the echinoderm *Ophiopyren striatum.*

LOWER ABYSSAL ZONE (1000–2600 M)

The lowest or deepest zone, starting at 1000 m and extending to 2600 m and possibly deeper, has 17 species, and seven of these are endemic (41%). Five species (29%) and 33% of the genera of this abyssal fauna reach the archibenthal zone

TABLE 6-4.   BATHYMETRIC DISTRIBUTION OF ARCTIC ISOPOD GENERA: DEPTH IN METERS

Genus	Sampling Intervals							
	1	2	3	4	5	6	7	8
	12–100	162–360	425–570	610–869	1000–1200	1400–1642	1800–2000	2360–2600
Synidotea	X	X						
Munnopsis	X	X	X					
Munna	X	X	X	X				
Pleuroprion	X	X	X	—	X			
Munnopsurus	X	X	X	X	X			
Calathura	X	X	X	X	X			
Desmosoma	X	X	X	X	—	X		
Mesidothea	X	X	X	X	X	—	—	X
Eurycope	X	X	X	X	X	X	X	X
Gnathia	X	X	X	X	X	X	X	X
Ilyarachna		X	X	X	X	X	X	X
Pleurogonium			X					
Katianira			X	X				
Nannoniscoides			X	X				
Nannoniscus			X	X	—	—	—	X
Haplomesus			X	X	X	X	X	X
Stylomesus				X				
Heteromesus				X				
Echinozone				X				
Macrostylis				X	X	X	X	
Haploniscus				X	X	X	X	X
Faunal Zone	Archibenthal		Upper abyssal	Meso-abyssal	Lower abyssal			

to the shallow depth limit of 12 m. This is an example of polar emergence of abyssal genera and species.

Species found in this abyssal zone (Table 6-5) are *Haploniscus bicuspis, Haplomesus quadrispinosus, H. angustus, Macrostylis subinermis, M. longipes, Desmosoma plebejum, Ilyarachna hirticeps, I. derjugini, I. dubia, Eurycope hanseni, E. ratmanovi, Calathura brachiata, Mesidothea sabini, M. megalura, Gnathia stygia, Nannoniscus spinicornis,* and *Pleuroprion frigidum.* The genera *Gnathia, Pleuroprion, Mesidothea,* and *Calathura* are typically shallow-water genera, and most of their species occur in shallow water. Large animals inhabiting this zone are: *Cladorhiza arctica, Forcipina topsenti, Radiella sarsi, Thenea muricata;* the pogonophoran *Lamellisabella* sp.; the decapods *Bythocaris leucopis* and *Hymenodora glacialis;* the pycnogonid *Ascorhynchus abyssi;* the echinoderms *Bathycrinus carpenteri, Elpidia glacialis, Kolga hyalina, Pourtalesia jeffreysi;* and the fish *Rhodichthys regina.*

TABLE 6-5. BATHYMETRIC DISTRIBUTION OF ARCTIC ISOPOD SPECIES:* DEPTH IN METERS

Isopoda		Sampling Intervals							
		1	2	3	4	5	6	7	8
Genus	Species	12–100	162–360	425–570	610–869	1000–1200	1400–1642	1800–2000	2360–2600
Munna	fabricii	×							
Desmosoma	tenuimanum	×							
Desmosoma	polaris	×							
Desmosoma	zenkewitschi	×							
Eurycope	mutica	×							
Mesidothea	sibirica	×							
Gnathia	arctica	×							
Synidotea	bicuspida	×							
Synidotea	nodulosa	×							
Eurycope	cornuta	×	×	×					
Gnathia	elongata	×	×	—	×				
Mesidothea	sabini	×	×	×	×	×			
Munnopsurus	giganteus	×	—	×	×	×			
Pleuroprion	frigidum	×	—	×	—	×			
Gnathia	stygia	×	—	×	×	—	×	×	×
Ilyarachna	hirticeps	×	×	×	×	—	×	×	×
Mesidothea	robusta		×						
Munna	hanseni		×	×					
Munnopsis	typica		×	×					
Eurycope	producta		×	×	×				
Calathura	brachiata		×	×	×	×			
Nannoniscus	laticeps			×					
Nannoniscus	reticulatus			×					
Pleurogonium	intermedium			×					
Pleurogonium	spinosissimum			×					
Katianira	sadko			×	×				
Munna	acanthifera			×	×				
Nannoniscoides	reticulatus			×	×				
Eurycope	inermis			×	×				
Eurycope	hanseni			×	×	×	×	×	×
Haplomesus	quadrispinosus			×	×	—	×	×	×
Katianira	biloba				×				
Haplomesus	insignis				×				
Haplomesus	tenuispinis				×				
Stylomesus	gorbunovi				×				
Heteromesus	frigidus				×				
Heteromesus	longiremis				×				
Macrostylis	abyssicola				×				
Nannoniscoides	angulatus				×				
Desmosoma	globiceps				×				
Desmosoma	reticulata				×				
Ilyarachna	bergendali				×				
Echinozone	arctica				×				
Eurycope	neupokojevi				×				
Eurycope	laktionovi				×				
Nannoniscus	arcticus				×				
Haplomesus	angustus				×	—	×		
Haploniscus	bicuspis				×	—	×	×	×
Eurycope	ratmanovi					×			
Macrostylis	subinermis					×	×	×	
Macrostylis	longipes						×		
Desmosoma	plebejum						×		
Ilyarachna	dubia							×	
Ilyarachna	derjugini								×
Mesidothea	megalura								×
Nannoniscus	spinicornis								×

* Partly from Gorbunov (1946); partly from Hansen (1916).

Figure 6-20. Bottom photograph in the Arctic abyss: (a) Fish, probably *Lycodes* sp. latitude 78°28′N longitude 171°10′N, 2653 m; (b) deep-sea shrimp, latitude 81°44′N longitude 157°29′W, 3797 m; (c) bottom markings and a medusa, latitude 79°38′N. longitude 171°47′W, 2330 m. (After Ewing et al., 1969.)

Gorbunov (1946) listed 64 metazoan epifaunal and infaunal species from depths exceeding 1000 m as truly abyssal species in the central parts of the Arctic Ocean, and this number is the sum of all the abyssal species captured during six years of survey involving two research vessels. The relatively few samples from the Arctic provide only a clue to zonation; further confirmation can come only after intensive sampling of the Arctic abyss. In some outstanding photographs Ewing et al. (1969) have identified tracks and trails in association with the amphipod *Parathemisto abyssorum* Bock and have recorded the shrimp *Bythocaris leucopis,* showing eye-shine from 3797 m (81°44'N by 157°29'W), as well as an eye-bearing fish *Lycodes* from 2653 m (Figure 6-20a). The amphipod *Parathemisto abyssorum* Bock was recorded by Barnard (1959) as an inhabitant of shallow waters from the surface to 600-m depth; apparently it is capable of making excursions down to 3797 m.

Zonation is suggested within the abyssal faunal province of the Arctic, but it is not as sharply delineated as elsewhere. The absence of basin floor collections makes nomenclature for the zones impractical at this time. We have recognized three zones in this province: an upper zone between 425 and 570 m, a middle zone that may or may not be equivalent to the mesoabyssal of other oceans; and a lower zone that may or may not be equivalent to the lower abyssal elsewhere.

Green (1959) divided the Arctic into four major zones on the basis of assemblages of Foraminifera.

Zones	Depth (m)	Dominant Species
Shelf	433–510	*Cassidulina teretis*
Slope	619–1142	*Valvulineria arctica*
Apron	1532–2000	*V. horvathi*
Abyssal	2250–2760	*Eponides terner*

On the basis of isopods and other organisms, we split the abyssal from the archibenthal near 425 m and identified four faunal zones. The major coincidence between Green's analysis and ours is in the increase in Foraminifera species and genera in the middle abyssal zone, in exact agreement with the isopod data. Green did not investigate the inner shelf at all, and his data commence only at 433 m. Thus he missed entirely what we have identified as the archibenthal zone. We tend to relate his shelf zone as the start of the abyssal faunal province.

CHAPTER 7

The Antarctic

It is clear that the antarctic shelf has been a centre of development for marine animals during long geological periods. [Ekman, 1953, p. 229.]

The Antarctic continent occupies the "bottom" of the world (Figure 7-1) between 70 and 90°S latitude and is in open and deep contact with the three major oceans of the world—the Atlantic, the Pacific, and the Indian. A submerged ridge, the South Antilles Arc, connects the Palmer peninsula of Antarctica with Falkland Islands and Tierra del Fuego of South America. This ridge is continuous at 4000 m with a sill depth at 3000 m, connecting the South Antilles Basin with the Argentine basin. Accordingly, the water exchange to the Northwestern Atlantic is restricted to depths less than 3000 m. The open connection of the Antarctic with the three oceans is a strong contrast to the Arctic, which is so definitely isolated topographically and hydrographically. Other contrasts are notable as well; for example, no rivers are active today in Antarctica, and the oligohaline conditions that characterized the Arctic shelf are generally lacking. Instead a permanent ice cover on land and a permanent ice pack surround the continent. The continental shelf is generally quite narrow and the slopes are steep, meaning that deep water is close to the continent except at the epicontinental seas, such as the Ross and the Weddell seas (Figure 7-1). Australia is connected with the Antarctic continent at depths around 4000 m by the Macquarie Ridge (or East Pacific Rise). The Western Atlantic is in open connection to the Antarctic Ocean and permits entrance of Antarctic bottom water into the Western Atlantic. This is not the case with the Eastern Atlantic, which is effectively separated from the Antarctic bottom water by the Walvis Ridge, which connects the Mid-Atlantic Ridge with Africa, Deacon 1937, 1963.

GENERAL FEATURES

At the sea surface, near and below the permanent cover of the pack ice, cold water of low salinity moves eastward around the Antarctic continent in the Eastwind Drift. The temperature of this water is below 0°C, and the water sinks along the shelf and slope to form the Antarctic bottom water. At the boundary of this Antarctic divergence and to the north up to temperatures of 2°C, the Antarctic surface water flows westward in the Westwind Drift, and as a mass it also moves northward. This water sinks at the Antarctic convergence (Figure 7-2) to form the Antarctic intermedi-

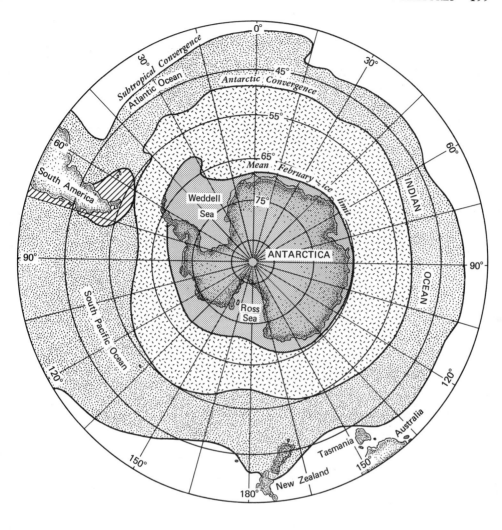

ANTARCTIC FAUNAL PROVINCES

High Antarctic   Low Antarctic   Antiboreal

Magellanic

Figure 7-1. Antarctic Ocean with zoogeographic divisions modified and simplified from the Antarctic Atlas. (From Hedgpeth, 1965b.)

ate water as a northward-flowing subsurface water mass. Its flow northward can be traced far above the equator. Below the Antarctic intermediate water north of the divergence is the deep water of the Atlantic, the Pacific, and the Indian oceans. This water rises as it approaches the Antarctic continent and feeds the surface water with nutrients to the south of the Antarctic convergence. Upwelling of nutrient-rich water accounts for the unusual productivity of the surface water and also probably for the unusual productivity of the Antarctic shelf benthos. The Antarctic surface water is in

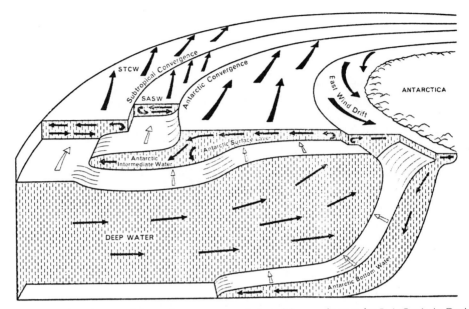

Figure 7-2. The structure of Antarctic water masses (adapted from a diagram by R. I. Currie in David 1965). (After Hedgpeth, 1969*b*.)

one of the largest surface currents on earth and appears to extend to a depth of 3000 m, where the velocity of this current is 75% of the surface velocity in the Westwind Drift. It has been measured between 30 and 50 cm/sec (Heezen and Holister, 1964).

FAUNAL REGIONS AND PROVINCES IN ANTARCTICA

The well-defined water types and water masses show corresponding faunal assemblages. The Antarctic region, which is better defined by plankton than by benthos, extends from the continental margin to the Antarctic convergence (55°S latitude). Ekman divided this region into two provinces. The first is a High Antarctic province between the continent and the limit of the pack ice at the Antarctic divergence. The seawater temperature is below zero and we identify it as a hypopsychral marine climate. The second is the Low Antarctic faunal province, bounded by temperatures between 0 and 2 to 4°C. This marine climate is subpolar and is generally confined to the water between the divergence and the Antarctic convergence. Northward between the Antarctic convergence and the ill-defined subtropical convergence at 40°S latitude is a province (Table 7-1) having a marine climate between 4 and 10°C. This marine climate has been termed "notal" by Soviet scientists and "sub-Antarctic" by other workers. We refer to it as antiboreal.

Other Provinces

Various workers have defined provinces and subprovinces within the Antarctic region (Figure 7-1) such as the Eastern and Western Antarctic, the Chilean, the Argentine, the Falkland, and the Kerguélen province. Just how these are defined depends to

TABLE 7-1.   HORIZONTAL FAUNAL PROVINCES IN THE SOUTHERN OCEAN AS
PROPOSED BY EKMAN (1953) SHOWING BOUNDARIES, TEMPERATURE, MARINE
CLIMATE, AND EQUIVALENTS

Faunal Province	Bounded by	Temperature (°C)	Marine Climate
High Antarctic	Antarctic continent to Antarctic divergence (limit pack ice)	0--1.8	Hypopsychral*
Low Antarctic	Antarctic divergence to Antarc-tic† convergence	0–2 or 4	Subpolar
Antiboreal	Antarctic convergence to sub-tropical convergence	4–10	Boreal‡

* Menzies (1963).
† Antitropical convergence of Ekman, see also Broch 1961.
‡ Notal of Soviet scientists.

a large extent on whether we consider the intertidal separately or the shelf fauna separately or both together as one. It is now generally conceded that the shelf fauna around the continent of Antarctica is a circumpolar one, being always under the influence of subzero temperatures and also within the eastward drift.

Insular Provinces

Within the boundaries of the subtropical convergence there has been a tendency to recognize subprovinces at each island or group of islands. This we do not accept. Instead we classify the islands and their fauna with respect to the type of water surrounding them at a definite depth. For example, only the continental islands are within the High Antarctic surface waters. Those within the low Antarctic include the South Orkneys, South Georgia, South Sandwich, the Bouvet Islands, and Heard Island. The Kerguélen Islands are on the boundary of the Antarctic Convergence and hence have a mixed fauna as explained by Ekman. The islands of Tristan, Gough, Prince Edward, Crozet, Marion, Macquarie, Campbell, Aukland, New Zealand (southern island), and the Antipodes are all within the antiboreal province at the surface of the water. At shelf depths, however, these islands are all within the influence of the Antarctic intermediate water, and by 3000-m depth they are all within the influence of the Antarctic bottom water or deep water and the hypopsychral climate. Various zoogeographers have remarked that the deep-sea fauna in Antarctica does not follow the surface boundaries, but there is a tendency to confuse the patterns of fauna by including deep-sea species in regional tabulations and ignoring water types and water masses as important boundaries.

The surface waters of Southern Australia, Northern New Zealand, and Africa are far removed from Antarctic surface currents and are outside the northern limits of the antiboreal (subtropical) convergence. Therefore, the surface fauna in these areas should show little if any affinity to Antarctica, and in fact it does not. Islands also belonging to this category include St. Paul and New Amsterdam, Kermadec, and Juan Fernandez. As depth increases and the Antarctic intermediate water is approached, the fauna resembles more closely that of the Low Antarctic. At the very bottom, the Antarctic bottom water determines the character of the fauna.

ANTARCTIC FAUNAL PROVINCES AND THEIR AFFINITIES

High Antarctic Faunal Province

The High Antarctic faunal province embraces all parts of the shelf which are covered by a hypopsychral marine climate. This fauna is generally abundant and diversified. Ekman recognized the generic endemism as a highly distinctive feature of the High Antarctic province, finding that 27% of 100 echinoderm genera and 65% of the fish genera were endemic to the high Antarctic. The recent data shown in the atlases of Antarctic fauna (Hedgpeth et al., 1969b) do not provide tabular information, but when analyzed they tend to suggest that the endemism for the High Antarctic genera is somewhat lower. The percentage of generic endemism in the Antarctic shelf fauna varies greatly, being 5% for the holothuroids and 32% for the ascidians. Each major taxon in Antarctica shows a different degree of generic endemism in the High Antarctic, and this is a major characteristic of the Antarctic fauna of the hypopsychral climate (Table 7-2).

SPECIES ENDEMISM

Ekman suggested that echinoderms showed a 73% endemism for the High Antarctic and fishes a 90% endemism. Actually again each major taxon shows a different degree of species endemism. For instance, Porifera and Mollusca exhibit low species endemism (between 21 and 25%) and Scleractinia the least (15%). Brachiopods and bryozoans are between 74 and 76% (Table 7-3). The average of 16 major taxa is 57% species endemism.

As a whole, 42% of the species of the High Antarctic are also found in the Low Antarctic. Only 20% of the High Antarctic species (in the 16 major taxa and 426 species) are also found in the antiboreal province (Figure 7-3), and 19% occur in the Magellanic region. Only 2% or less of the High Antarctic species are in common with Australia, New Zealand, Africa, or temperate eastern South America.

The situation in the High Antarctic for the isopod genus *Serolis* is about the same—27% is found also in the Low Antarctic, 11% in the Magellanic, in the Antiboreal, and temperate South America. South Africa and New Zealand–Australia have no High Antarctic species of *Serolis* in common.

Low Antarctic Faunal Province (Surface and Shelf)

The Low Antarctic has 251 species in the 16 taxa given by Hedgpeth et al. Species endemism is much lower than that of the High Antarctic, ranging between 0 and

TABLE 7-2. GENERIC ENDEMISM IN THE HIGH ANTARCTIC

	Number of Genera	Number of Endemics	% Endemism	Reference
Holothuroids	20	1	5	Pawson (1969)
Echinoids	—	—	25	Pawson (1969)
Sponges	116	10	8	Koltun (1969)
Ascidians	22	5–7	23–32	Kott (1969)
Fishes	37	—	65	Ekman (1953)

TABLE 7-3. PERCENTAGE OF SPECIES ENDEMISM OF VARIOUS TAXA, FROM
SOURCE CITED, IN THE HIGH ANTARCTIC (ANTARCTIC SHELF FAUNA)

Taxa	Number	Number of Endemics	% Endemism	Source	Remarks
Brachiopods	13	10	76	Foster (1969)	
Bryozoans	38*	28	74	Bullivant (1969)	
Stylasterina	19	15	79	Boschma and Lowe (1969)	
Holothuroids	38	22	58	Pawson (1969), Agatep (1967)	
Echinoids	24	16	66	Pawson (1969)	
Asteroids	82	50	60	Fell and Dawsey (1969)	
Crinoids	9	6	66	Dearborn and Rommel (1969)	
Scleractinia	6	1	16	Squires (1969)	
Ascidians	39	10	25	Kott (1969)	
Ascidians	—	92	82	Kott (1969)	As stated by Kott but not plotted
Pycnogonids	17	7	41	Hedgpeth (1969b)	
Ophiuroids	74	45	60	Fell, Holzinger, and Sherraden (1969)	
Molluscs	17	4	24	Dell (1969)	Common species only
Poriferans	42	9	21	Koltun (1969)	
Sipunculids and Echiura	14	7	50	Edmonds (1969)	
Cirripeds	17	11	65	Ross and Newman (1969)	Exclusive of pelagics

* Representative species only.

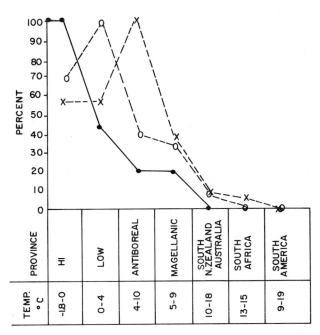

Figure 7-3. Percentage contribution of species from one surface marine climate or province to another in Antarctica: solid circles, high Antarctic; open circles, Low Antarctic, crosses, antiboreal.

TABLE 7-4. ENDEMIC LOW ANTARCTIC SPECIES (EXCLUSIVE OF DEEP SEA WHEN KNOWN)

	Number	Endemics	% Endemic
Brachiopods	7	0	0
Bryozoans	5	0	0
Stylasterina	3	0	0
Holothuroids	19	4	21
Echinoids	10	2	20
Asteroids	41	8	19
Crinoids	7	2	28
Scleractinia	10	0	0
Ascidians	28	0	0
Pycnogonids	14	2	13
Ophiuroids	44	10	23
Molluscs	14	0	0
Poriferans	29	0	0
Sipunculids and Echiura	11	3	27
Cirripeds	9	4	44
Total	251	35	195
Average			13

23% and averaging only 13%. This means that the balance of the species are either allied to the High Antarctic or to the antiboreal or to some other province (Table 7-4). No data are available on generic endemism but it is likely to be quite low. The closest species relationship is with the High Antarctic (67%) and next with the antiboreal at 37% or the Magellanic at 33%. Only 7% are common to New Zealand and Australia, 1% to Africa, and none to central eastern South America. In the genus *Serolis* 26% of the species enter the High Antarctic, 33% the Magellanic, 30% the antiboreal, 5% the area of Australia–New Zealand; none enters South Africa. Only 23% of species of *Serolis* reach temperate South America from the Low Antarctic.

Magellanic Province

We define a Magellanic province as the fauna within the continental shelf from southern Chile to the Falkland Islands. Some authors have pointed out the differences between the Chilean fauna and the Argentine shallow-water fauna but in all likelihood these differences are artificial and were noted more because of lack of information than as a result of real differences on the shelf. There is little doubt that the fauna of the intertidal differs considerably between the Argentine side and the Chilean side. The reasoning for this statement is based on the differences in origin of surface currents. The parts of southern Chile nearshore show fjord conditions and the main water current comes from the Antarctic as the cold Humboldt Current. On the Argentine side this is not the case; instead, the warm Brazilian current moves south from Brazil along the coast of Argentina and brings with it a variety of warm-water fauna. In contrast, the subsurface water at both places has its origin from Antarctica

TABLE 7-5. *SEROLIS* SPECIES: PERCENTAGE DISTRIBUTION IN IDENTIFIED
FAUNAL PROVINCES IN THE ANTARCTIC

Faunal Provinces	% in Each Province							
	I	II	III	IV	V	VI	VII	VIII
I. High Antarctic	100	39	27	11	11	11	0	0
II. Antarctic abyss	35	100	18	15	15	15	0	0
III. Low Antarctic	26	20	100	33	30	26	0	5
IV. Magellanic	15	22	54	100	13	13	0	0
V. Antiboreal	22	31	65	11	100	42	0	0
VI. Temperate	16	15	90	50	18	100	0	0
VII. South Africa	0	0	0	0	0	0	100	0
VIII. Australia–New Zealand	0	0	12	0	22	0	0	100

in the Antarctic intermediate water (Figure 7-2) and there is no a priori reason to
believe that the fauna in this water mass should be really distinct. The Magellanic
province of shallow water becomes then a considerable mixture of components from
the north and the south. It is of great interest to note that the fauna of the Magellanic
shelf shows its greatest similarity (54% of its species) to the fauna of the Low Antarctic, and hence with the fauna contained within the Antarctic intermediate water. On
the other hand, the Magellanic province contributes only 16% of its species to the
antiboreal island fauna.

Ekman pointed out that the Magellanic province shows a species endemism of
52%, which is close to that of the High Antarctic. Our Magellanic province is in reality a fauna living in subpolar thermal conditions on the shelf between 0 and 4° and
as such is Low Antarctic. It is not surprising, therefore, that the Magellanic shelf contributes 54% of its species to the Low Antarctic, and only 16% to the High Antarctic,
13% to the Antiboreal, 11% to the temperate, and none to South Africa or New
Zealand-Australia. The Magellanic, on the other hand, contributes 22% of its species
to the abyssal faunal province in the genus *Serolis* (Table 7-5).

Antiboreal Province

The antiboreal province has a species endemism of 30%, and contributes 55% of its
species to the High and Low Antarctic, 35% of its species to the Magellanic province,
and 5% or less to Australia-New Zealand, South Africa, and temperate South
America.

The conclusion seems fairly obvious that Africa, New Zealand, and Australia are
for all practical purposes isolated from the Antarctic region.

THE ABYSS AND THE ANTARCTIC REGION AS INDICATED
BY THE SPECIES OF THE GENUS *SEROLIS*

Because the abyss of the world ocean is fed by the Antarctic bottom water and by the
deep water of the Atlantic, the Pacific, and the Indian oceans, it is of interest to see
what part of the Antarctic region contributes species to the abyss of the oceans con-

cerned. Unfortunately, data are essentially lacking for the Indian and the Pacific oceans. Thus we have been able to use only data from the Atlantic with reference to the surface provinces we have recognized (Figure 7-1). The contributions from the various provinces to the abyssal faunal province are as follows:

Province	To the Abyssal (%)
High Antarctic	39 ⎫
Low Antarctic	20 ⎬ 81%
Magellanic	22 ⎭
Antiboreal	31
Temperate	15
Australia–New Zealand	0
South Africa	0

It seems especially obvious that the greatest percentage contribution of species to the deep sea in the Southern Ocean comes from the High Antarctic. In fact, the contributions that are made from the Low Antarctic and from elsewhere come from species that are also found, as a rule, in the High Antarctic. These are the eurythermal species of the Antarctic region.

The Antarctic region appears to have been a major source of deep-sea species. In rather strict contrast, the Arctic contributes no more than 10% of its species to the Atlantic and fewer yet to the Pacific deeps. It is likely that the distribution of water masses in the two polar oceans has played a major role in determining the percentage affinities of faunal provinces at a species level in the Recent era.

CONDITIONS IN THE PAST

The shelf fauna of Antarctica, with its low generic endemism and high species endemism, has been relict at least since the Miocene and has since gone its own evolutionary way. Apparently the extinction and evolutionary rates have varied among different groups of organisms since the Miocene and the increasing cooling, up to the maximum in the Pleistocene.

Prior to the Jurassic it is likely that Antarctica, New Zealand and Australia, and South America and India were joined as Gondwana Land. The faunas were then on the ancient epicontinental seas and were strikingly similar.

Little is known of the fossil marine fauna of Antarctica today, but those fragmentary bits of information are suggestive of past conditions. The sum of the data suggests that in several groups of organisms the marine fauna of Antarctica was even more diverse in the past than it is today. The Jurassic ammonites (Arkel, 1956) suggest a temperate warm-water fauna following the glacial conditions that prevailed in Pennsylvanian and Permian times. Antarctica was apparently tropical since the early Jurassic (King 1961), with the climate gradually cooling until the Miocene glacial conditions (p. 350). The general lack of tropical components of fauna in Antarctica today suggests a progressive loss of species and genera in those groups of marine organisms which now show a high diversity in the tropics, and finally an extinction of all tropical components from the Antarctic fauna as the climate cooled.

The echinoids of the Upper Cretaceous (Pawson, 1969) of Antarctica contained the Cidarinae as well as *Holaster* and the Eocene *Cassidulus*. These have since vanished from Antarctica. The Cenozoic fossils of the Brachiopoda of the Palmer Peninsula and South America are similar according to Foster (1969), as were the Cenozoic fossil faunas of Australia–New Zealand, and Antarctica. Today the number of species of brachiopods decreases as Antarctica is approached. The Cretaceous cirripedian fauna of Antarctica was more diverse than the contemporary fauna. The molluscan fauna shows a similar history—many families disappear between South America and the Palmer Peninsula (Figure 7-4), with the result that the Recent molluscan fauna of Antarctica is relatively impoverished.

Not all groups of organisms have reacted equally to the gradual decline of temperature in Antarctica. As noted earlier, certain groups show a decreasing diversity of genera and species such as the Mollusca, the Echinodermata, the Cirripedia, and the corals. Stomatopods are absent. In contrast, the Isopoda, the Ascideacea, the Porifera, and the Sipunculida all show an increasing diversity in Antarctica. Crablike decapods are represented by only one species, *Paralomis spectabilis*, which was found by Zarenkov (1968).

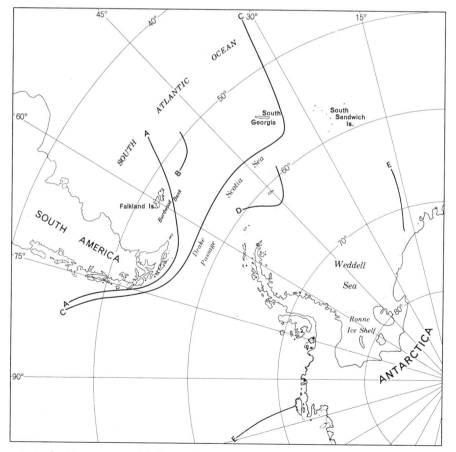

Figure 7-4. The disappearance of molluscan families from South America through the Scotia Ridge to the Antarctic continent. Southern limit of numbers of families A.13, B.1, C.5, D.1, E.5 involving 25 families. (After Dell, 1969.)

TABLE 7-6. NUMBER OF ISOPOD GENERA FROM THE SHELF FAUNAL PROVINCE OF VARIOUS REGIONS OF THE WORLD

Region	Number of Isopod Genera Known from Shelf	Source
Antarctica	55	Kussakin (1967), McMurdo collections
Peru	11	Menzies (1962a), Menzies and George (1972)
Puerto Rico	19	Menzies and Glynn (1968)
California	23	Menzies and Barnard (1959)
Carolina and Georgia	25	Menzies and Frankenburg (1966), *Eastward* collections
Arctic	16	Gurjanova (1933), Gorbunov (1946)

For the group Isopoda the picture is especially interesting. The increasing diversity of isopod genera in the Antarctic shelf is revealed when the total number of isopod genera known from the shelves in the Arctic, tropical, and temperate regions (Table 7-6) is compared with that of the Antarctic shelf.

The Antarctic shelf has the greatest number of isopod genera, more than double the number compared with any other shelf region. This high generic diversity in the group Isopoda appears to be characteristic of the Antarctic region even beyond the shelf depths in the archibenthal and abyssal faunal provinces.

Ekman's statement quoted at the beginning of this chapter now seems to be substantiated by the available recent data. The Antarctic shelf has been a center of development for marine organisms during long geologic periods. It has been a cold environment at least since the Miocene glaciations, and it is likely that most tropical elements have since vanished from Antarctica.

Impact of Pleistocene Glaciation on the Planktonic Flora in the Southern Ocean

The story of the impact of Pleistocene cooling on the fauna in the southern oceans is told best in the records from the abyssal sediments. Glacial marine sediments surround Antarctica near the continent out to the limit of the pack ice where these are mixed with diatom ooze and sponge spicules. Northward of the pack ice is a ring of diatom ooze whose outward boundary coincides almost exactly with the boundary of the Antarctic convergence at the surface (Figure 7-5a). North of this ring of diatom ooze is a band of calcareous foraminiferal ooze at depths less than 4500 m and red clays in the basins at greater depths.

Apparently the present ring or blanket of diatom ooze has accumulated mainly during the Pleistocene and Recent times, because the sediments below the diatom ooze consist of red clay (Figure 7-5b). The age of the red clay is identified as Pliocene by Hayes, mainly because the material lacks any evidence of glacial marine sediments. This implies that the sinking of Antarctic bottom water was not as strong during the Pliocene as it is now. Deep water was not upwelled then with sufficient vigor to supply enough nutrients to favor diatomite formation on the bottom at

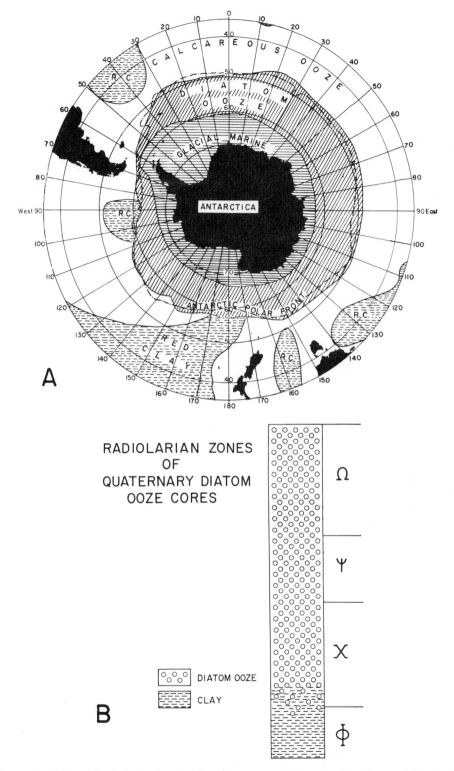

Figure 7-5. (a) Generalized distribution of sediment types around Antarctica [based on work by Hough (1956) and Lisitzin (1960, 1962), Goodell (1965), and *Eltanin* and *Vema* cores]; (b) typical diatom ooze core penetrating three Quaternary zones (after Hayes, 1967).

209

abyssal depths. Cores taken by Jouse et al. (1963) show southern and northern shifts in cold-water conditions around and off South Africa. Such shifts have also been noticed by Hayes and Opdyke (1967) and Donahue (1967).

BIOMASS OF THE ANTARCTIC SHELF

The biomass of the Antarctic at depths below the anchor ice is exceptionally high in comparison with other regions of the earth. At comparable depths, it is always higher than the biomass of the High Arctic. Between 200 and 400 m, silicious sponges often dominate the communities, but because of the wide variation in bottom conditions this dominance changes from one place on the shelf to another. Ophiuroids, for instance, contribute significantly to the shelf biomass in the Ross Sea, where one-third of the benthic biomass has been reported to consist of ophiuroids (Fell, 1961). Bottom photographs have indicated a density of ophiuroids reaching a maximum of one million per square meter.

Primary Production

The primary production of the Antarctic surface water reveals a gross productivity equal to 20% of the world ocean (1.5×10^{10} tons of carbon/year), according to El-Sayed (1968). The rate of sessile algae production is not known. Bunt (1964) made measurements of standing crop for the microalgae in the waters of McMurdo Sound and gave a recorded high value for *Phaeocystis* at 3×10^6 cells/liter at 20 m under the sea ice. The annual total must be high (El. Sayed and Mandelli, 1965).

VERTICAL DISTRIBUTION AND ZONATION IN ANTARCTICA

The bottom fauna along the margins of Antarctica is under the influence of the Antarctic bottom water, which flows downward from the continent into various basins of the Atlantic, Pacific, and Indian oceans. The main features of sediment zonation are shown in Figure 7-5. The shallow-water conditions described herein are based mainly on the work of Dearborn (1967) and Bullivant (1967) at the Ross Sea.

The topographic profile presented in Figure 7-6 is derived from various sources and has been modified to show the irregularity of the topography of the Antarctic shelf, with its epicontinental seas and abrupt continental slope. The edge of the continental shelf is located at about 400 m and the contact between the continental rise and the continental slope is located near 3000 m.

Data on isopods and photographic evidence are derived from the *Eltanin* stations shown in Figure 7-6. We have added the isopod data of Kussakin (1967) and the shallow-water collections taken by one of us (R. Y. G.) from McMurdo Sound in 1968.

Generic Distinctiveness and Faunal Provinces

Most species of isopods are not yet known from Antarctica, and thus we have had to be content with delimiting the provinces on the basis of generic distinctiveness as

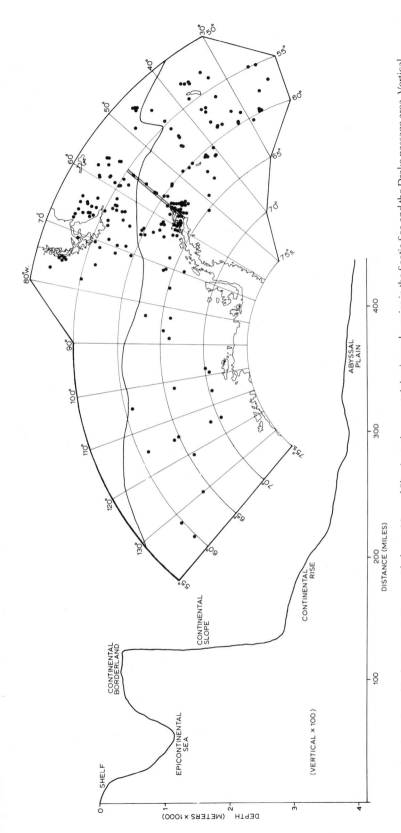

Figure 7-6. Area investigated in the Antarctic Ocean with the positions of *Eltanin* stations containing isopod species in the Scotia Sea and the Drake passage area. Vertical profile is drawn from the transect marked by double lines in the map. The Antarctic convergence is shown as a solid line crossing the profile near its northern limit.

211

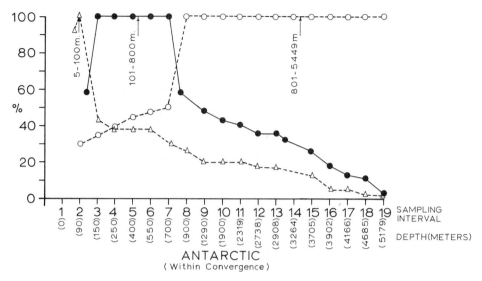

Figure 7-7. Percentage distribution of isopod genera in faunal provinces in the Antarctic Ocean: triangles, shelf; solid circles, archibenthal; open circles, abyssal.

revealed in Table 7-7 and Figure 7-7. We believe that high points of generic distinctiveness within the abyssal faunal province may indeed indicate added zones, and we have attempted to characterize these as best as possible.

A relatively homogeneous isopod generic assemblage extends from 5 to 100 m, where the first point of high generic distinctiveness of 62% separates the shelf province from an archibenthal zone of transition at 150 m. A total of 18 intervals of depth was examined between 5 and 5449 m for isopod generic distinctiveness (Table 7-7).

TABLE 7-7. DATA ON DISTINCTIVENESS AND ENDEMISM OF ISOPOD GENERA FROM THE HIGH ANTARCTIC

Series Number	Depth (m)	Number of Samples	Number of Endemics	Starts	Stops	Total	% Distinctiveness	% Endemism	Faunal Zone
1.	5–100	14+	31	24	0	55		56	Shelf
2.	101–200	14	2	6	2	32	62*	6	Archibenthal
3.	201–300	14	1	3	1	32	22	3	Archibenthal
4.	301–500	11	1	3	2	34	14	3	Archibenthal
5.	501–600	9	1	1	3	33	13	3	Archibenthal
6.	601–800	14	0	1	6	30	18	0	Archibenthal
7.	801–1000	8	0	1	4	25	20	0	Archibenthal
8.	1100–1481	8	1	3	1	25	29*	4	Upper abyssal
9.	1800–2000	5	0	1	2	24	18	0	Upper abyssal
10.	2078–2560	10	0	1	2	23	8	0	Upper abyssal
11.	2660–2816	7	0	2	0	23	19	0	Upper abyssal
12.	2817–3000	6	0	0	1	23	0	0	Upper abyssal
13.	3054–3475	15	5	1	3	28	18	16	Upper abyssal
14.	3510–3800	11	0	0	5	20	37*	0	Mesoabyssal
15.	3801–4005	12	1	0	4	16	33*	3	Lower abyssal
16.	4006–4328	6	0	0	1	11	25	0	Lower abyssal
17.	4484–4886	9	1	0	7	11	15	9	Lower abyssal
18.	4909–5449	3	0	0	3	3	—	0	Lower abyssal

* Significant points of distinctiveness indicating faunal break.

Only the shelf and the archibenthal are separated by a distinctiveness above 50%. Therefore in the Antarctic, like the Arctic, generic distinctiveness is low between all depths. Only detailed studies on the species will allow precise location of the zonal boundaries.

Intertidal Faunal Province

For the most part the intertidal is absent around the continent. The only Antarctic intertidal assemblage known south of Tierra del Fuego consists of an impoverished grouping of sessile diatoms, algae, and one limpet (*Patinigera polaris*) at the Palmer Station on the Palmer Peninsula (Hedgpeth, 1969a) (Figure 7-8). Here the tidal range is about 5 ft and the limpet achieves its maximum densities, not in the intertidal, but in the subtidal, where as many as 200 animals per square meter have been found at depths between 2 and 3 m. Similarly Zaneveld (1966) described zonation of submerged sessile algae. One algal belt, which he found below the extreme low tide to about 37 m, is divided into two belts–an upper belt or sublittoral fringe down to 10 m, and a sublittoral belt down to the limit of macroscopic sessile algae. The Palmer region is certainly atypical and is comparable to the impoverished intertidal at Murmansk in the Arctic.

We believe that the inner part of the shelf is composed of a mixture of typical shelf animals as well as a submerged "intertidal fauna." This is based on the large number of isopod genera, especially Sphaeromidae, which occur submerged on the Antarctic shelf and which, when found outside Antarctica in ice-free areas, are normally intertidal.

Shelf Faunal Province (1–100 m)

The temperature of subsurface water at McMurdo Sound (Bullivant, 1967) has a mean of $-1.86°C$ and varies less than $1.0°C$, and the subsurface salinity ranges between 34 and 35%. The shoreline and bottom out to 15 to 20 m is composed of cobble, gravel, and volcanic sand. Between 20 and 60 m, it is a silt and debris composed of invertebrate skeletons. Glacial rock and mud occur at various depths because of the irregular topography on the shelf (Dearborn, 1967).

In the Antarctic faunal shelf, the water column and the bottom (to an approximate depth of 33 m) contain aggregations of small ice crystals and large ice platelets throughout most of the year. Such individual platelets freeze to each other and accumulate to form the anchor-ice on the bottom in depths shallower than 33 m (Dayton et al., 1969). The rising anchor-ice has a significant effect on the distribution of the epifauna. In McMurdo Sound, repeated investigations by Dayton and his team of divers revealed the abrupt appearance of a dense association of sponges at 33-m depth, the lower limit of anchor-ice formation.

The possible effects of anchor ice on epibenthos in such shallow depths were pointed out by Bullivant (1967), who suggested that benthic animals could become trapped. In winter, ice that has accumulated along the shore may grind against the bottom to depths of 5 m or more. Grounded icebergs and floes may disturb the substrate at much greater depths, perhaps to the lower limit of anchor-ice formation. It appears that the zone of anchor-ice formation constitutes the upper shelf zone (1–33 m) having a limited fauna. The ice-free lower shelf fauna (40–100 m) is dominated by the sponge association (Figure 7-9).

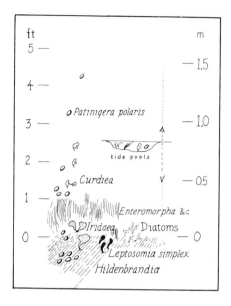

Figure 7-8. Diagram indicating intertidal zonation between tidemarks at Palmer Station. (After Hedgpeth, 1969a.)

Judging from the Ross Sea data (Dearborn, 1967; Bullivant, 1967), the irregular topography and sediments on the shelf have given rise to distinctive (recognizable) assemblages. Bullivant (*op. cit.*) suggested six assemblages or communities with various miscellaneous groupings. The dominant large-animal assemblage between 20 and 183 m is Bullivant's McMurdo Sound glass–sponge assemblage. The obvious isopod is the giant valviferan *Glyptonotus acutus* Richardson (Figure 7-10). There are 55 isopod genera on the Antarctic shelf (Table 7-8), and 31 are endemic to shelf depths. The Asellota constitute 56% of the shelf fauna, the Sphaeromidae 16%, and the Cirolanidae 11% of the genera (Figure 7-12).

Eight of the isopod genera are intertidal elsewhere in the world, including *Cassidinopsis, Euvalentinia, Cassidinidea, Sphaeroma, Ianiropsis, Exosphaeroma,* and *Cymodocella,* suggesting that the Antarctic shelf is composed of intertidal as well as shelf elements. The high proportion of Asellota is also different but comparable to that of the Arctic shelf. Many of the shelf genera enter the abyss (Figure 7-7) and continue in decreasing numbers down to 4685 m. The isopod genera that extend from the shelf (100 m) to depths greater than 2500 m include *Paramunnopsis* (3000 m), *Austrosignum* (2500 m), *Microarcturus* (3800 m), *Munna* (3800 m), *Gnathia* (3800 m), *Cirolana* (4005 m), *Serolis* (4866 m), *Antarcturus* (4005 m), *Janthopsis* (3475 m), *Eurycope* (4328 m), and *Desmosoma* (4880 m).

The most conspicuous large epibenthic fauna characteristic of the Antarctic shelf includes a large nemertean *Lineus corrugatus,* the ophiuroids *Ophiurolepis gelida* and *Ophiocantha antarctica,* the bivalve *Limatula hodgsoni* (Figure 7-11) and the asteroids *Odantaster validus* (Figure 7-11) and *Diplasterias brucei,* the gastropod *Neobuccinum eatoni,* the echinoid *Sterechinus neumayeri,* the crinoid *Promachocrinus kerguélensis,* the stylasterine coral *Errina carnea,* the scleractinian corals *Caryophyllia antarctica* and *Gardineria antarctica,* the sipunculid worm *Golfingia ohlini,* and the nototheniid fish *Trematomus bernachi.*

The extensive mats of sponges occurring at depths beyond the anchor ice formation are formed predominantly by the siliceous sponge species belonging to the hex-

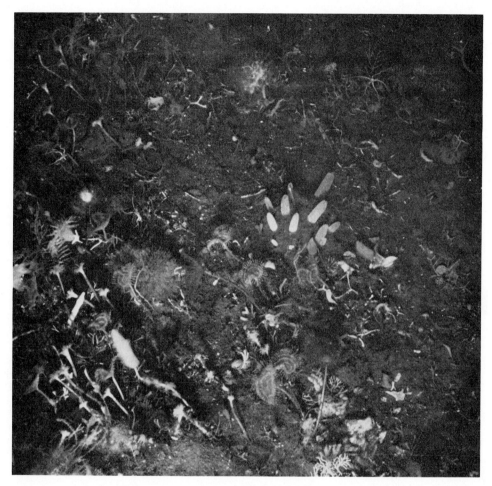

Figure 7-9. McMurdo Sound benthic assemblage at 260 m, showing bryozoans, gorgonaceans, sponges, echinoderms, and polychaete worms. (After Bullivant, 1967.)

actinellid genus *Rossella*. Several alcyonarian species of the genera *Primnoisis, Primnoella, Thouavella,* the deep-sea pennatulid genus *Umbellula,* and ectoprocts are commonly associated with the sponge mat. Generally the shelf bottom fauna comprises sessile species whose populations reach a peak density in areas of hard substrate formed by ice-rafted boulders and exposed rock. Organic terrestrial debris is absent from the physical environment, and brachyuran crabs are not common members of the fauna, as compared with other shelf environments.

Archibenthal Zone of Transition (150–900 m)

The archibenthal zone of transition appears to extend to the irregular topographic shelf boundary at 400 to 900 m where the second recognizable peak of isopod generic distinctiveness is located (Table 7-7). As expected, there is here a considerable mixing of shallow (43%) and deep (53%) generic components (Figure 7-7). Such mixing of generic components characterizes the AZT elsewhere. Of the 42

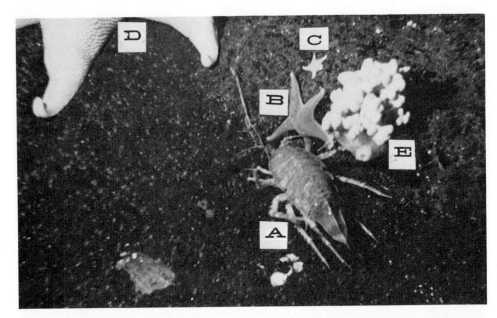

Figure 7-10. Antarctic bottom photograph taken at 47 m of shelf faunal province off Turtle Rock, Mc-Murdo Sound. The epifauna include: (a) the giant antarctic isopod *Glyptonotus acutus*, (b) the asteroid *Odantaster validus*, (c) *Odantaster meridionalis*, (d) the giant starfish *Macroptychaster accrescens*, (e) *Alcyonium paessleri*. (Photograph by Paul Dayton.)

genera present, 10 are endemic—*Austranthura, Accalathura, Pananthura, Leptanthura, Neoarcturus, Mormomunna, Pleurogonium, Cassidinella, Euneognathia,* and *Zenobianopsis*.

Nine shallow-water genera penetrate from the shelf and terminate within the archibenthal zone. Fifteen genera have species in the archibenthal besides having species in the shelf and in the abyss (Table 7-8).

Asellota constitute 51% of the fauna of the archibenthal zone. The isopod group Valvifera contributes 18%, the Anthuroidea 9%, the Cirolanoidea 7%, the Gnathiodea 5%, and the Sphaeromidae 7% (Figure 7-12). The representation of Sphaeromidae in the AZT is seen only in the Antarctic.

Bottom photographs taken in the Ross Sea and in the Drake Passage during the cruises of the *Eltanin* provide added data on the macrobenthic fauna in the archibenthal zone. This transitional zone appears to harbor both shallow-water and deep-sea faunal elements under different ecological situations. Obviously, the presence of shelf and abyssal communities within the recognized faunal boundaries characterizes the archibenthal zone. The trawl samples and bottom photographs of Bullivant (1967) demonstrate the high biomass and epifaunal abundance in the archibenthal zone. This includes a mixed assemblage of bryozoans, gorgonaceans, sponges, ophiuroids, tube worms and bivalves (Figure 7-9). Some sponges found here at 300 to 500 m are the same species occurring in the Shelf Faunal Province at 20 m. The giant isopod genus *Glyptonotus*, which is endemic to the Antarctic and a common large benthic animal in the shelf faunal province, seems to penetrate into the lower limit of the archibenthal zone.

TABLE 7-8. ANTARCTIC ISOPOD GENERA WITHIN THE ANTARCTIC CONVERGENCE: DEPTH IN METERS

Sampling Intervals

Isopod Genera	1	2	3	4	5	6	7	8	9	10	11	12	13	14	15	16	17	18	19
	5–50	51–100	101–200	201–300	301–500	501–600	601–800	801–1000	900–1481	1800–2000	2078–2560	2660–2816	2816–3000	3054–3475	3510–3800	3800–4005	4005–4328	4484–4886	4904–5449
Phycolimnoria	X																		
Anuropus	X																		
Cassidinopsis	X																		
Ourozeucktes	X																		
Euvalentinia	X																		
Plakarthrium	X																		
Anthura	X																		
Ectias	X																		
Notopais (E)	X																		
Notoxenus (E)	X																		
Astrurus (E)	X																		
Neasellus	X																		
Austroniscus (E)	X																		
Austronanus (E)	X																		
Austromunna (E)	X																		
Antennulosignum	X																		
Pleurosignum	X																		
Cassidinidea	X																		
Janiropsis	X																		
Austrofilius (E)	X																		
Cassidina	X																		
Notoasellus	X	X																	
Cymodocella	X	X	X																
Cymodoce	X	X																	
Antias		X	X	X															
Echinozone	X	X	X	X															
Pseudarachna	X	X	X			X													

217

TABLE 7-8. CONT.

Isopod Genera	1 (5–50)	2 (51–100)	3 (101–200)	4 (201–300)	5 (301–500)	6 (501–600)	7 (601–800)	8 (801–1000)	9 (900–1481)	10 (1800–2000)	11 (2078–2560)	12 (2660–2816)	13 (2816–3000)	14 (3054–3475)	15 (3510–3800)	16 (3800–4005)	17 (4005–4328)	18 (4484–4886)	19 (4904–5449)
Cilicoea	X	X	X	X	X	X													
Paramunna	X	X	X	X	X														
Iathrippa	X	X	X	X		X	X												
Aega	X	X	X	X		X	X												
Glyptonotus (E)	X	X	X	X	X			X											
Austrosignum	X	X	X	X															
Paramunnopsis (E)													X						
Microarcturus (E)	X	X	X	X	X	X	X		X						X				
Munna	X	X	X	X	X	X	X		X		X		X		X				
Gnathia	X	X	X	X	X	X	X								X				
Cirolana	X	X	X	X	X	X	X	X	X		X					X			
Antarcturus	X	X	X	X	X	X	X	X	X		X	X		X	X	X			
Serolis	X	X	X	X	X	X	X	X	X		X	X	X	X	X	X		X	
Exosphaeroma		X																	
Eisothistos		X																	
Arcturoides (E)		X																	
Haliacris		X																	
Coulmannia (E)		X																	
Iolella		X																	
Excorolana		X																	
Astacilla				X		X			X										
Rocinella			X				X	X											
Neojaera		X																	
Ianthopsis	X	X	X		X	X	X	X						X					
Eurycope		X	X		X	X	X					X		X	X	X	X		
Desmosoma				X		X	X					X		X					
Iaeropsis		X		X	X			X											
Austranthura (E)			X																
Pleurogonium			X																

Sampling Intervals

218

Euneognathia (E)
Accalathura
Edotea
Cassidinella
Stenetrium
Ilyarachna
Pananthura
Leptanthura
Haploniscus
Mormomunna (E)
Jaerella
Munnopsurus
Storthyngura
Neoarcturus (E)
Dolichiscus
Zenobianopsis (E)
Munnopsis
Nannoniscus
Stylomesus
Macrostylis
Munneurycope
Syneurycope
Heteromesus
Antennuloniscus
Ischnomesus
Ianirella
Microprotus
Acanthocope
Haplomesus
Iolanthe
Hydroniscus
Xostylus
Mesosignum
Exacanthaspidia

(E): endemic south of convergence.

Figure 7-11. Bottom photograph of shelf epifaunal abundance at 64 m near Hut Point, McMurdo Sound (lat. 77°50'S; long. 166°35'E). The dominant species is the lamellibranch *Limatula hodgsoni*. Also found are sponges, echinoids, asteroids (*Odantaster validus*), ophiuroids, and polychaetes. The sediment is gritty mud, matted with sponge spicules. (After Bullivant, 1967.)

Abyssal Faunal Province (950–5450 m)

Beyond the transitional zone at about 950 m is the abyssal faunal province, extending to 5450 m. This province embraces the upper continental slope, the continental rise, and the basin floor. The fauna from its upper limits is somewhat equivalent to Bullivant's (1967) deep ooze assemblage. The sediments are diatomaceous ooze with occasional ice-rafted rocks, extending to *Globigerina* ooze at 4500 m and finally to red clay at depths in excess of 5000 m. The bottom water is Antarctic bottom water, which spreads down and northward and has a temperature of less than 0°C within the convergence.

Menzies (1963) has described the fauna as hypopsychral based on the distribution of the Antarctic genus *Serolis*. Macrobenthic species are sparse according to Bullivant (1959). The province has 46 isopod genera and contributes 35 to 50% of

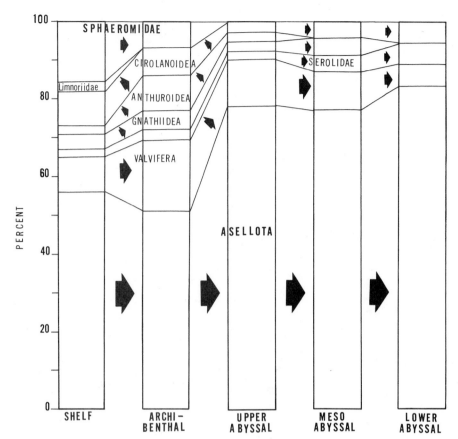

Figure 7-12. Relative abundance (percentages) of major isopod taxa in various faunal zones in the Antarctic.

these to the archibenthal and 30% to the shelf; 22 genera are endemic. Throughout this province there exists a similarity of 63 to 100% at the generic level between the sampling depth intervals, suggesting that the generic fauna below 1000 m shows little variation. Many isopod genera that are abyssal elsewhere make their first appearance. These include *Nannoniscus, Stylomesus, Macrostylis, Munneurycope, Syneurycope, Heteromesus, Antennuloniscus,* and *Ischnomesus* (Figure 7-13). It is a significant coincidence that Rankin et al. (1968) pointed out possible faunal change around 1000 m based on data from the Weddell Sea, and Bullivant (1967) placed the break between his deep-ooze assemblage and his shelf-edge barnacle assemblage somewhere between 521 and 1280 m.

UPPER ABYSSAL ZONE (950–3475 M)

Thirty-seven isopod genera are in this zone with 16% endemic. With the presence of so many typically deep-sea genera (see above), it is highly probable that the Upper Abyssal is distinctive. This can be demonstrated only after the species, sediments, and currents are known. The lower limit of the zone more or less coincides with the end of the continental slope and the start of the continental rise. The sediments are

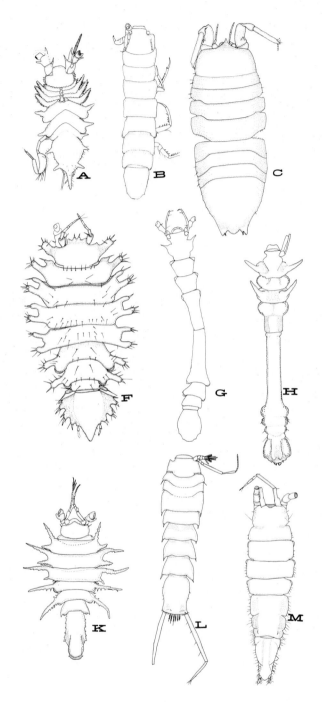

Figure 7-13. Assemblage of isopod genera characteristic of Antarctic abyssal faunal province (scales and sources various): (a) *Acanthocope*, (b) *Xostylus*, (c) *Antennuloniscus*, (d) *Acanthaspidia*, (e) *Exacanthaspidia*, (f) *Ianirella*, (g) *Ischnomesus*, (h) *Haplomesus*, (i) *Heteromesus*, (j) *Nannoniscus*, (k) *Mesosignum*, (l) *Macrostylis*, (m) *Syneurycope*, (n) *Stylomesus*, (o) *Hydroniscus*, (p) *Iolanthe*.

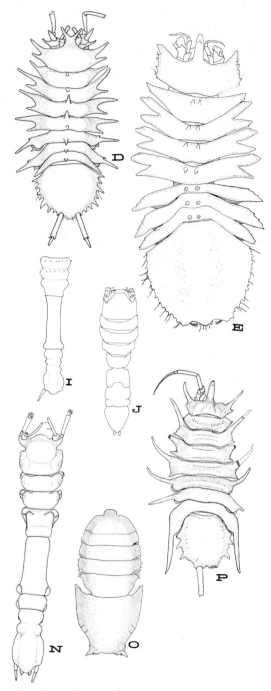

Figure 7-13. (Continued)

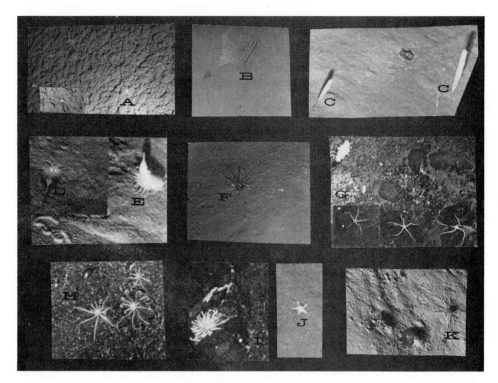

Figure 7-14. Assemblage of *in situ* photographs of epifaunal species from the Antarctic upper abyssal zone, Drake Passage, off Palmer Peninsula: (a) Ophiuroids, 1990 m; (b) feeding bryozoan, 3203 m; (c) sea pen, 3203 m; (d) isopod *Serolis* sp., 3020 m; (e) *Serolis* sp., 3020 m; (f) echinoid (unidentified), 1102 m; (g) ophiuroids, 2432 m; (h) and (i) crinoids, 2432 m; (j) asteroid, 3203 m; (k) *Echinus* sp., 3202 m. (All from *Eltanin*.)

diatom ooze. It is thus located, as elsewhere, on the steepest part of the continental margin.

Generic distinctiveness is 37% between 3475 and 3800 m, with this being the most distinct point within the Abyssal Province. It is probable that the break at 3510 m represents the end of the Upper Abyssal Zone and the start of the Meso-abyssal. Within the limits of this zone the following eight isopod genera make their first appearance and extend to greater depths as indicated below. These are: *Stylomesus* (3800 m), *Macrostylis* (3800 m), *Munneurycope* (5400 m), *Syneury-cope* (4886 m), *Heteromesus* (4886 m), *Antennuloniscus* (4005 m), *Ischnomesus* (4886 m), and *Iolanthe* (3800 m).

Unlike the Shelf Faunal Province, the Upper Abyssal lacks Sphaeromidae and Limnoriidae, and the four other major taxa, Cirolanoidea, Anthuroidea, Gnathioidea and Serolidae, constitute only 2% each of the total fauna. The percentage of asel-lotes has increased to 78%. The Valvifera, represented mainly by *Antarcturus*, con-stitute 11% of the generic fauna in this zone (Figure 7-12).

The large benthic invertebrates here include several echinoderm species represented by ophiuroids, crinoids, asteroids, and echinoids (Figure 7-14). The sponge complex and the gorgoneans apparently tend to diminish in the upper abyssal zone, but pennatulids (Figure 7-15) and bryozoans are sometimes pho-tographed. The gradual replacement of sessile by motile species and an abrupt drop in biomass and density as the depth increases are characteristic of this zone.

Figure 7-15. Assemblage of in situ photographs of epifaunal species from the lower abyssal zone, Drake Passage: (a) Unidentified ophiuroid on manganese nodules, 5092 m; (b) holothurian *Peniagone* sp., 4792 m; (c) pennatulid *Umbellula* sp. and *Peniagone* sp., 4404 m; (d) tunicate *Culeolus* sp., 5045 m; (e) stalked barnacle, 4442 m; (f) *Priapulus* sp., 4474 m; (g) holothurian *Scotoplanes* sp., 4451 m; (h) unidentified holothurian, 4000 m; (i) holothurian *Pseudostichopus* sp., 4404 m; (j) holothurian *Psychropotes longicauda*, 4653 m; (k) echinoid *Phormosoma* sp., 4237 m; (l) ophiuroid *Ophiomusium lymani* (?), 4237 m; (m) black stout holothurian, 4792 m.

MESOABYSSAL ZONE (3510–3800 M)

The mesoabyssal zone is located at the contact of the continental slope with the continental rise. The sediments of this zone are diatomaceous ooze and rock fragments from the continent.

The 37% distinctiveness in genera separates the mesoabyssal from the upper abyssal. This zone contains 22 genera including one endemic genus, *Hydroniscus*.

The Anthuroidea are now no longer present, leaving only the Cirolanoidea, the Gnathioidea, and the Serolidae each with one genus, and Valvifera and Asellota. The Valvifera decrease from four genera to two, and the Asellota from 29 to 17.

Large animals in the mesoabyssal zone off the Palmer Peninsula are the Holothurian *Scotoplanes* sp., *Umbellula* sp., a bryozoan, and an unidentified species of asteroid.

LOWER ABYSSAL ZONE (3800–5449 M)

The lower abyssal zone commences at the end of the mesoabyssal and extends out onto the abyssal plain. Because it embraces *Globigerina* oozes as well as the red clay of the basins, it should really be divided into an active (*Globigerina* ooze) and a tranquil (red clay) component as elsewhere.

It contains 18 isopod genera (Table 7-8), and two are endemic: *Xostylus* and *Exacanthaspidia*. The major groups present are Cirolanoidea, Serolidae, and Valvifera, each with one genus, and the Asellota now constituting 83% of the isopod genera.

Several species of holothurians seem to dominate the lower abyssal zone and these species belonging to the cosmopolitan genera *Psychropotes, Peniagone, Pseudostichopus, Scotoplanes* (Figure 7-15), *and Benthodytes* are mostly the same abyssal genera occurring elsewhere in other geographic regions. Other large epifaunal species include the ophiuroid *Ophiomusium lymani* (Figure 7-15l), the echinoid *Phormosoma* sp. (Figure 7-15k), the tunicate *Culeolus* sp. (Figure 7-15d), the pennatulid *Umbellula* sp. (Figure 7-15c), and a priapulid which is probably *Priapulus tuberculatospinosus*. This species was previously known only from shallow depths in the Antarctic, but is found to occur in the Antarctic abyss from the Lower Abyssal Zone at 4437 m.

CHAPTER 8

Fauna of Deep Trenches

~~~~~~
~~~~~~

The quantitative abundance of benthic fauna at all depths of the ultra-abyssal zone is equal to that of the abyssal of the same regions and not infrequently higher. The benthic fauna is most abundant in trenches of highly productive areas of the ocean. The greatest poverty of bottom animals is observed in the far from shore trenches of the oligotrophic areas of the tropical zone. [G. Belyaev, 1966, p. 242.]

As Belyaev noted, deep-sea trenches support a diverse benthic life, and most of the major invertebrate groups are represented. Belyaev was the first to recognize the extreme variation in trench biomass, and his compilation on the fauna of trenches is the most up to date, accurate, and significant.

A trench is a morphological feature of the earth's topography. Wiseman and Ovey (1953) defined a submarine trench as "a long, but narrow, depression of the deep-sea floor having relatively steep sides." To this definition should be added other characteristic features such as a flat floor (Figure 8-1), a high incidence of seismic activity (Ewing and Heezen, 1955), and negative-gravity anomalies (Worzel and Ewing, 1954). The flat floor seems to have resulted from sediment deposition from shallow water. For example, the floor of the Puerto Rico Trench has graded calcareous sand containing shallow-water Foraminifera and the alga *Halimeda* (Ericson et al., 1952). Coconut husks and bamboo were trawled from the great depths of the Philippine Trench (Bruun, 1956a, Bruun and Wolff, 1961).

The general features of a submarine trench are not necessarily depth related. Most trenches (Figure 8-2) exceed 6000 m in depth, but the Cedros and Hikurangi trenches have a maximum of less than 4500 m. Likewise, several broad ocean basins have a maximum depth in excess of 6000 m (Table 8-1).

There are about 23 trenches known to exist in the world oceans, 18 of them in the Pacific. The Romanche Trench, the South Sandwich Trench, and the Puerto Rico Trench are in the Atlantic. The Sunda Trench is the only trench of the Indian Ocean. The abyssal depths occupy approximately 83% of the whole area of the world oceans, but the trench depths cover only 1% of the entire ocean area. However, these enormous depths are far greater than the highest elevation on land, and even Mount Everest, the tallest peak of the Himalayas, could be totally submerged in a deep trench. Furthermore, the hydrostatic pressure prevailing at a depth of 10,000 m amounts to limits strikingly beyond imagination, with a square inch of surface bearing a pressure of 7 tons. A century ago, the *Challenger* Expedition (1872–1876) dredged up animals from the abyssal plain to a maximum of around 5000 m, and

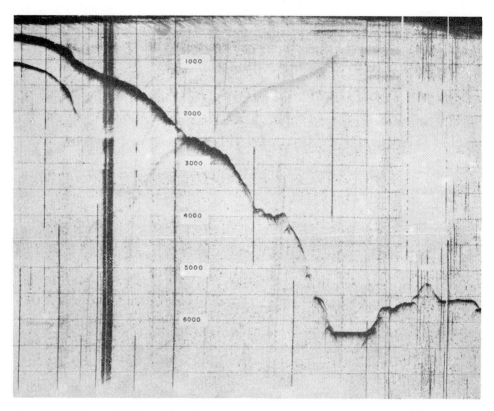

Figure 8-1. PESR depth profile of floor of Milne-Edwards Trench. R/V *Anton Bruun,* cruise 11, National Science Foundation. Note flat floor. (From Menzies and George, 1967.)

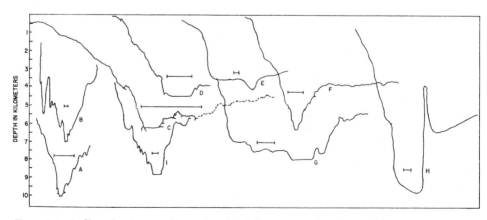

Figure 8-2. Profiles of various world trenches: (a) Philippine Trench (Fisher and Hess, 1963), (b) South Sandwich Trench (Heezen and Johnson, 1965), (c) Milne-Edwards Trench (original data from Menzies and Chin, 1966), (d) Cedros Trench (Uchupi and Emery, 1963), (e) Hikurangi Trench (Brodie and Hatherton, 1958), (f) mid-America Trench (Fisher, 1961), (g) Puerto Rico Trench (Ewing and Ewing, 1962), (h) Kermadec Trench (Brodie and Hatherton, 1958), (i) Kurile–Kamchatka Trench (Zenkevitch, 1969). Depth scale at left in kilometers, small horizontal scale on figure equals 20 km. (From Menzies and George, 1967.)

228

TABLE 8-1. MAXIMUM DEPTHS REPORTED FOR TRENCHES AND BASINS
(MENZIES AND GEORGE, 1967)

Trench	Maximum Depth (m)	Ocean Basin	Maximum Depth (m)
1. Mariana	10,915	1. Southwestern Pacific*	8581
2. Tonga	10,882	2. Great Pacific*	7521
3. Japan	10,554	3. Philippine*	7483
4. Kurile–Kamchatka	10,542 ± 100	4. East Caroline*	6920
5. Philippine	10,497 ± 100	5. New Hebrides*	6061
6. Kermadec	10,047	6. Tasman	5944
7. Idsu–Bonin (Ramapo Deep)	9,810	7. West Caroline	5798
8. Puerto Rico	9,200 ± 20	8. Guatemala	5682
9. New Hebrides	9,165 ± 20	9. West Marianas	5592
10. New Britain	9,140	10. New Britain	5419
11. North Solomons or Bougainville	9,103	11. South Fiji	5303
12. Yap (West Caroline)	8,597	12. Southeastern Pacific	5298
13. South Solomons	8,310 ± 20	13. Aleutian	5236
14. Palau	8,138	14. Kurile	5211
15. Peru–Chile (Ernst Haeckel Deep)	8,055 ± 10	15. Argentine	5072
16. Aleutian	7,679	16. North Fiji	4963
17. Nansei Shoto (Ryuku)	7,507	17. Cape	4950
18. Java (Sunda)	7,450	18. Coral Sea	4718
19. Mid-America (Guatemala)	6,662	19. Japan	4036
20. Guatemala	6,662	20. New Caledonia	4036
21. Cedros	4,350	21. East China	2719
22. Hikurangi	4,000 ± 100		

* Basins exceeding 6000 m.

much interest was expressed subsequently in testing the existence of animal life in the greatest depth of the trenches. In 1899 attempts were made from U.S. *Albatross* to sample the bottom of the Tonga Trench in a depth of 7632 m; disappointingly, however, the trawling failed. The possibility of life aroused great curiosity in the minds of oceanographers, and in the late 1940s expeditions were planned to explore the floor of these trenches.

The Danish *Galathea,* during the round-the-world expedition (1950–1952), succeeded in sampling the deepest depths in five trenches, namely, the Tonga, Kermadec, Philippine, Sunda, and Banda trenches. The recovery of living bacteria from the Philippine Trench at 10,000 m by ZoBell (1953) solved forever the question of the existence of life at the greatest depths of the ocean. Anton Bruun (1956), the scientific leader of the expedition, was the first to propose a new term, "hadal," for fauna living at trenches below 6000 m and "forming an ecological zone of their own." In 1957 Bruun published a first list of the names of all taxa known from hadal depths. Wolff (1956, 1960, 1962, 1970) produced a series of papers and popular articles about the concept of "hadal zone and fauna." He also listed the species belonging to the "hadal" community.

Russian investigations aboard the *Vityaz* contributed more to the knowledge of trench fauna from the chain of trenches in the Northwestern Pacific. Sixty trawls were made from 14 deep-sea trenches by Soviet expeditions to the Kurile–Kamchatka (1949–1953), Aleutian, Japan, Idsu–Bonin, Riu-Kin, Bougainville, Marianas, Tonga, Kermadec, New Hebrides, and Java trenches. Zeņkevitch (1959) identified the zone below 6000 m as "supraoceanic deeps." Instead of "hadal," the Soviet scientists used the term "ultra-abyssal," recognizing a "self-contained zone extending from 6000 m downward" (Zenkevitch et al., 1954, 1955; Birstein, 1957; Vinogradova, 1958, 1962; Belyaev, 1966; Zenkevitch, 1959). Vinogradova (1958) pointed out that this zone of supraoceanic depths was first delimited by Zenkevitch (1954, 1961). There are at least 300 presently identified species in deep trenches affiliated with the 30 classes of 140 families and 200 genera cited by Belyaev (1966), but probably a thousand species exist in all the world trenches. Although the species are mostly (75%) endemic to a given trench, endemism at the generic level is insignificant and endemism becomes negligible at the rank of family and other high taxa.

CHARACTERISTICS OF THE TRENCH FAUNA

The proponents of the concept of a hadal zone as a distinctive faunal unit proposed the following criteria to justify their claim.

1. Decrease in the total number of species in the hadal zone (>6000 m).
2. Change of faunal composition.
   (a) Dominance of groups such as Pogonophora, echiuroidean worms, holothurians, and isopods.
   (b) Remarkable reduction in representation of certain groups (e.g., fishes, tunicates, cirripeds, bryozoans, and sponges).
   (c) Absence of groups such as decapods, brachiopods, and turbellarians.
3. High percentage of species endemism.
4. Differences in morphological characteristics (e.g., blindness and lack of pigmentation, decreased armament and color, and lesser degree of calcification).
5. Gigantism due to the effect of pronounced hydrostatic pressure on metabolism.
6. Biocoenosis of oligomixity, one species dominating in density in an environment of low species number.

The first hadal criterion, *the decrease in number of species* with depth, appears to be a phenomenon associated with moderate depths elsewhere in the oceans, and not trenches alone. Vinogradova (1958) established a definite trend in the total number of species with depth based on her study of the total abyssal fauna and especially certain major taxa such as sponges and holothurians. Somewhere in the lower part of the continental slope, or sometimes at the contact zone between the slope and rise, a pronounced increase in species number occurs; beyond that depth, the species number tends to decline abruptly without any significant change at 6000 to 7000 m. Vinogradova's technique of pooling data from several sources for a single composite analysis may not provide a true picture of the accurate depth relationship with species number, although the general pattern is revealed. Moreover, she considered only those species known from depths below 2000 m (which, at that time, was assumed to be the upper limit of abyssal fauna).

If all the isopod species known from 79 stations off northern Chile (Menzies, 1962a) are somewhat representative of the shoal conditions off Lima, Peru, and if this information is included with the depth distribution of isopods in the Milne-Edwards Deep off Peru, a depth distribution of species numbers from the intertidal to the upper limits of the "hadal" is obtained. This distribution (Figure 8-3) shows an intertidal maximum; there is a decline in numbers at the shelf to 500 m, followed by a slight increase below the oxygen minimum zone and a uniform low species number to 2500 m. Below 2500 m two maxima are encountered—one at 3500 m and one at 4700 m—and a decline to 5500 m is followed by an increase between 6000 and 6200 m on the trench floor.

We accept the existence of what we call a moderate-depth species maximum, located generally in the mesoabyssal zone. As Vinogradova (1958) has pointed out (Figure 8-4), a species maximum is located at different isobaths for different groups. In all probability the moderate-depth species maximum will vary in depth depending on the group of organisms and their behavior (i.e., whether the species are obligatory benthonts) as well as on the physical environment in a particular locality. Obviously all depth levels below the moderate-depth species maximum must have fewer species. The decline in species number below the mesoabyssal applies to basins as well as to trenches and hence is not a criterion peculiar to trenches. It is probable that

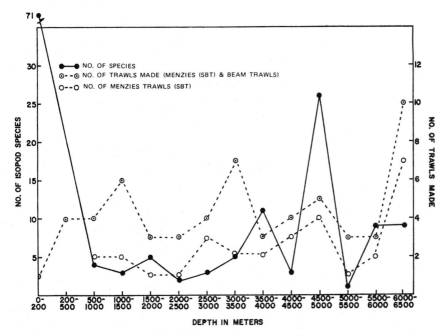

Figure 8-3. Vertical distribution of species of isopods by number of trawls (Menzies–(SBT) beam trawl combination) and Menzies trawls (550 μmesh diameter) at Milne-Edwards Trench, R/V *Anton Bruun*, cruise 11. (From Menzies and George, 1967.)

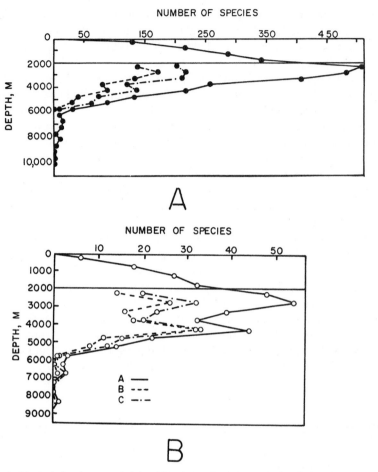

Figure 8-4. A. Vertical distribution of abyssal benthos (after Vinogradova, 1962, Figure 1); B. vertical distribution of abyssal species of sponges (after Vinogradova, 1962). (A) Total number of species, (B) number of species whose upper limit of habitat lies in this latitude, (C) number of species whose lower limit of habitat lies in this layer. Refers to species found at 2000 meters and deeper.

certain basins such as those in the Arctic will show fewer species than certain trenches.

The second hadal criterion, *change of faunal composition*, implies dominance of certain major taxa, recessive nature of some groups, and lack of certain groups.

Dominance of Various Groups

Wolff (1960) indicated that polychaetes, holothurians, and isopods are dominant groups of the "hadal" fauna. Belyaev (1966) pointed out that holothurians, bivalve molluscs, and polychaetes dominate both numerically and in biomass. Menzies (1964) suggested that Foraminifera are the dominant group in number, followed by polychaetes and bivalves at the Milne-Edwards Trench floor. Actually the "dominant" number of fauna is frequently merely a function of the type of collecting gear utilized. Thus the beam-trawl collections showed different numerical dominance than the small biology trawl (Menzies, 1964). The major abyssal groups are all well represented in the trenches at depths exceeding 6000 m.

The evidence suggests, and it seems altogether likely, that one trench may have a dominant group different from that of another. The wall of the Milne-Edwards Trench (Figure 8-5) has a different group dominant at each depth level. A summation of faunas with dominant groups selected for all trenches becomes rather meaningless in view of the great variation among trenches. The meaning of this variable group-depth dominance is obscure.

Insignificant Groups

Certain groups (e.g., fishes, cirripeds and bryozoans) appear infrequently in trench samples, and Wolff (1960) seemed to believe that this is a characteristic of hadal fauna. In the deep sea generally, fishes, cirripeds, and bryozoans are an insig-nificant (Menzies and Chin, 1966) component of the biomass everywhere (Figure 8-5); accordingly, this one feature is no special index of a hadal fauna. As pointed out by Menzies and Imbrie (1958), the cyclostomatous bryozoans are not even represented in the deep sea. The absence of Bryozoa of any type from trench collec-tions is hardly of consequence as a hadal characteristic.

Absence of Certain Groups

Reasoning from the "absence" of certain groups has inherent dangers. Perhaps groups remain absent only until they are once collected, and it is well known that the gear utilized (Menzies, 1964) plays a major role in determining what is collected. Wolff (1960) suggested that decapods, brachiopods, and turbellarians are absent from the "hadal" fauna, but these groups have been so rarely taken in deep-sea samples anywhere that this "absence" is not remarkable. Table 8-2 gives the known maximum depths reached by benthos including decapods and turbellarians. The common deep-sea brachiopod *Pelagodiscus* requires a substrate to attach to, such as manganese nodules, or rocks; if attachment surfaces are not present, *Pelagodiscus* could not be expected to occur. The maximum depth known for the decapod her-mit crab *Tylaspis anomala* is 4343 m, and another abyssal decapod *Protobeebei mi-riabilis* was recorded from 3570 m (Wolff, 1961). This demonstrates that decapods (Brachyura) are generally not known from the floor of the deep basins of the world oceans. Their absence from trench samples is not, on the other hand, a unique feature of trench fauna any more than their absence from basin floors. Indeed, a re-ported absence of turbellarians proved to have been an artifact of collecting, as Belyaev (1966) pointed out. Noodt (Menzies and Chin, 1966) reported numerous Turbellariomorpha to depths as great as 6200 m in the Milne-Edwards Trench floor.

Both Wolff and Belyaev suggested that trenches contain a high proportion of en-demic species (Table 8-3). Their reports indicated 58% (Wolff) and 68% (Belyaev) en-demism at the species level from depths greater than 6000 m. These figures do not prove a high proportion of endemism only in trenches, since other regions of the abyss of the world oceans show comparable or higher percentages. Gorbunov (1946) thought that the whole benthic fauna of the Arctic below 1000 m has a species en-demism of 60 to 70%. The Isopoda, excluding bathypelagic species, show an en-demism in the Arctic basin of 87% (Menzies, 1965).

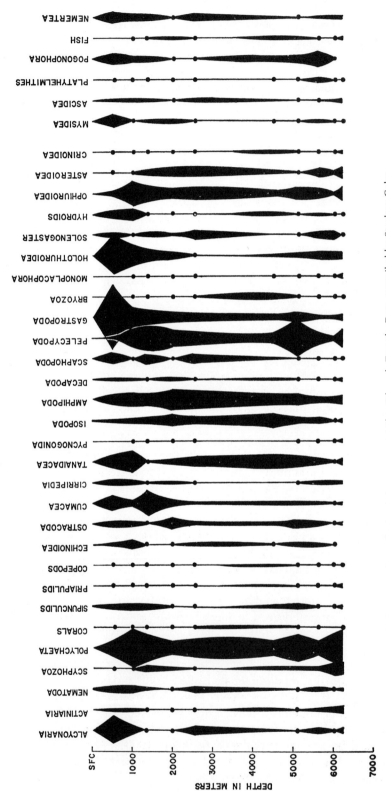

Figure 8-5. Vertical distribution of relative numbers of animals by groups at Milne-Edwards Trench. Data compiled by Stephen Coles.

234

TABLE 8-2.   KNOWN MAXIMUM DEPTHS REACHED BY BENTHOS*

Group	Genus and species	Depth (m)	Area
Foraminifera	*Sorosphaera abyssorum* Saidova	10,687	Kurile–Kamchatka Trench
Spongia	*Chondrocladia concrescens* (Schmidt)	8,660	
Hydrozoa	*Halisiphonia galatheae* Kramp	8,300	Kermadec Trench
Actiniaria	*Galatheanthemum hadale* Carlgren	10,210	Philippine Trench
Turbellaria	*Turbellariomorpha* sp.	6,200	Milne-Edwards Deep
Polychaeta	*Poecilochaetus vitjazi* Levenstein	10,687	Tonga Trench
Cirripedia	*Scalpellum (Arcoscalpellum) formosum* Hoek	6,860	Kurile–Kamchatka Trench
Cumacea	*Leucon* sp.	7,246	Aleutian Trench
Tanaidacea	*Leptognathia armata* Hansen	8,006	Bougainville Trench
Isopoda	*Macrostylis* sp.	10,710	Mariana Trench
Amphipoda	*Pardaliscoides longicaudatus* Dahl	10,000	Philippine Trench
Decapoda (Anomura)	*Tylaspis anomala* Henderson	4,343	Central South Pacific
Mysidacea	*Amphyops magna* Birstein and Tchindonova	7,230	Kurile–Kamchatka Trench
Pycnogonida	*Nymphon profundum* Hilton	6,860	Kurile–Kamchatka Trench
Gastropoda	*Aclis kermadecensis* Knudsen	8,230	Kermadec Trench
Bivalvia	*Phaseolus* n. sp. Filatova	10,687	Tonga Trench
Crinoidea	*Bathycrinus* sp.	9,735	Idsu–Bonin Trench
Holothuroidea	*Myriotrochus* sp.	10,687	Tonga Trench
Echinoidea	*Pourtalesia aurorae* Koehler	7,290	Banda Trench
Asteroidea	*Hymenaster* sp.	7,657	Mariana Trench
Ophiuroidea	*Ophiosphalma* sp.	7,230	Kurile–Kamchatka Trench
Pogonophora	*Zenkevitchiana longissima* Ivanov	9,500	Kurile–Kamchatka Trench
Pisces	*Careproctus (Pseudoliparis) amblystomopsis* Andriashev	7,587	Japan Trench

* Partly from Belyaev (1966).

A comparison of the fauna of two areas of the Argentine basin (Menzies, 1962b) makes this especially clear. Only three species are in common, hence the degrees of endemism in the two areas are 90 and 75%, respectively. Comparing the endemism of the Argentine basin samples with the endemism of the Cape basin we find only 10 species in common; hence the degrees of endemism are 74 and 84%, respectively. The depths are not hadal and trenches are not involved, and in the case of the intra-Argentine basin comparison, no special feature of topography separates the two areas of sampling! Thus an opposing statement could be made; namely, that trench endemism is not higher than endemism in basins or between basins. It does not seem true that only the hadal fauna is characterized by a high proportion of endemism.

Differences in Morphological Characteristics

The reason for differences in the morphological characteristics of a fauna, as a hadal faunal criterion, is obscure. Four characteristics have been implied by Wolff (1956)

TABLE 8-3. ENDEMISM AMONG HADAL SPECIES: DATA AFTER BELYAEV (1966)*

Group	Number of Hadal Species 6000 m (A)	Number of Hadal Species <6000 m (B)	Number of Species Exclusively at Depths 6000 m (A − B)	% Hadal Endemic Species
Foraminifera	128	73	55	33
Coelenterata-Spongia	26	3	23	89
Other Coelenterata	17	4	13	76.6
Polychaeta	42	20	22	52.4
Echiuroidea	8	3	5	62.5
Sipunculoidea	4	4	0	0
Cirripedia	3	2	1	33.4
Cumacea	9	0	9	100
Tanaidacea	19	4	15	79
Isopoda	68	18	50	73.6
Amphipoda	18	3	15	83.5
Solenogaster	3	3	0	0
Gastropoda	16	2	14	87.5
Bivalvia	39	6	33	84.7
Crinoidea	11	1	10	91
Holothuroidea	28	9	19	68
Asteroidea	14	6	8	57.2
Ophiuroidea	6	2	4	66.7
Pogonophora	26	4	22	84.6
Pisces	4	1	3	75
Total species	489*	168	321	68

* 489 includes about 300 identified species plus 189 unidentified species. This tabulation refers to trench depths and not to one trench or another.

as special peculiarities of the hadal Isopoda and Tanaidacea. These are size, armament with spines, eyes, and calcification.

Wolff suggested that the hadal isopods are preferably less spiny and have shorter spines than the related abyssal species. To the contrary this statement is not quite as correct as Wolff indicated. For example, the two trench species of *Mesosignum* — namely, *Mesosignum vitjazi* Birstein and *Mesosignum multidens* Menzies and Frankenberg — are the most spinous members of the genus *Mesosignum* (Figure 8-6). It cannot be concluded that trench species are preferably less spiny or that they have shorter spines.

It is of no special consequence that hadal Crustacea lack eyes (Birstein, 1963) because this is a characteristic of abyssal Crustacea as well. The distribution of eyes among the 56 benthic species of isopods from the Milne-Edwards Trench and isopod genera of Beaufort, N.C., is shown in Table 8-4. All the species lack eyes commencing at the depth of 2000 m in the North Atlantic, and, again, 100% lack eyes at a depth of only 500 m off Peru.

Deep-sea isopods in general are fragile and have a limited calcification. We could detect no difference between the degree of calcification as determined by inspection of a species of *Haploniscus* from the abyssal Atlantic and from the species

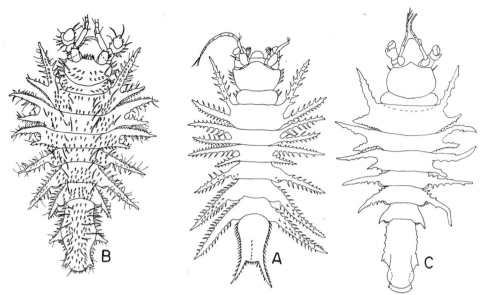

Figure 8-6. Species of *Mesosignum* from trenches and elsewhere: (a) *Mesosignum vitjazi* Birstein [after Birstein (1963)], Kurile–Kamchatka Trench, 7300 to 8439 m; (b) *Mesosignum multidens* [Menzies and Frankenberg (1968)], Milne-Edwards Trench, 6200 m; (c) *Mesosignum kohleri* Menzies [after Menzies (1962b)], Columbia abyssal plain, 4076 m [after Menzies and George (1967)].

in the Milne-Edwards Trench. If anything, the one species in the Milne-Edwards Trench is more heavily calcified than the others. Intertidal isopods (Asellota) show a considerable range in calcification but are similarly fragile.

The whitish color of hadal isopods is no different from the whitish color of abyssal isopods elsewhere as indicated by Zenkevitch et al. (1955) and by Zenkevitch and Birstein (1956).

Finally, in agreement with Belyaev (1966), we have been unable to detect any significant difference in morphological characteristics of spination, calcification, eyes, and color between abyssal isopods elsewhere and trench floor assemblages.

TABLE 8-4. PERCENTAGE OF ISOPODS WITH EYES IN RELATION TO DEPTH

Peru–Chile (South Pacific)			Off Beaufort (North Atlantic)		
Depth (m)	% with Eyes	Number of Species	Depth (m)	% with Eyes	Number of Species
Intertidal (Chile)	100	37	Intertidal	92	13
Shelf (5–200)	76	34	Shelf (0–200)	75	12
Trench wall (500–6000)	0	54	Slope (200–2050)	32	9
Trench floor (6000–6300)	0	9	Continental Rise (2000–3100)	0	8
			Abyssal plain (4500)	0	3

The absence of morphological criteria in distinguishing shallow-water Crustacea from deep-sea Crustacea was also emphasized by Dahl as early as 1954.

Gigantism

Gigantism due to the effect of pronounced hydrostatic pressure on metabolism merits special consideration because earlier workers on the hadal zone concept based their definition primarily on depth and correspondingly high hydrostatic pressure. It has been implied that hydrostatic pressure influences the physiological response of a deep-trench organism and that gigantism results as a consequence of metabolic readjustments. Because of practical difficulties in instrumentation and in the recovery of living animals from the deep sea, our knowledge of the influence of hydrostatic pressure on abyssal animals is so limited that only speculations may be made. Future experiments may yield some interesting results. Obviously, the subject of size of abyssal organisms needs some critical restudy. The Russian author Birstein (1957) used the species of the deep-sea isopod genus *Storthyngura* as the principal argument of abyssal gigantism. He found that the species from the Northwestern Pacific trenches are the largest in the genus. Within a decade, the species list of this genus increased to almost double what it had been at the time Birstein made his statistical analysis (George and Menzies, 1968a, 1968b). The largest species of the genus comes from the Antarctic shelf and slope depths, rather than from trenches, as Birstein believed. His curve of statistical reliability of the length—depth relationship in *Storthyngura* became invalid when a new analysis (Figure 8-7) was made with

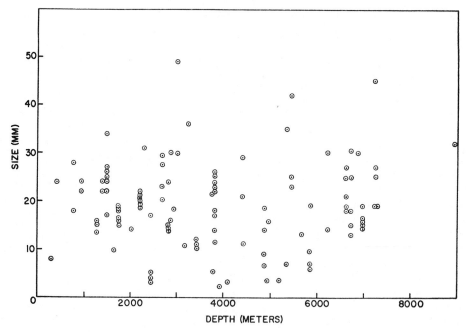

Figure 8-7. Relation of size to depth in 37 *Storthyngura* species, including the Antarctic slope data. Animals below 10 mm in size may or may not be adults because some species are small although adults. Animals above 10 mm length are probably all adults. (After Menzies and George, 1967.)

all known 37 species (Menzies and George, 1967). Since the largest isopods even in the genus *Storthyngura* come from an environment of modest hydrostatic pressure, we see no compelling reason to insist that a size effect was induced by hydrostatic pressure. We doubt whether further speculation is even justified until some physiological data on the influence of hydrostatic pressure on metabolism and growth are obtained.

It is of interest to note that the gigantic isopod *Bathynomus* (40 cm) does not occur in abyssal or trench depths but is instead restricted to shelf and upper slope depths in the Gulf of Mexico, the Sea of Japan, and the Bay of Bengal. The other giant isopod *Glyptonotus*, measuring 12 cm, is restricted within the Antarctic shelf depths. Perhaps even in other taxa, size of an animal may not have any definite relation to depth and consequent hydrostatic pressure.

The sixth hadal criterion proposed by Belyaev (1966) refers to an unique ecological situation, *biocoenosis of oligomixity*, wherein one species dominates in density in an environment having a low species number. We do not question that certain trench faunas are dominated in biomass by a single species of holothurian, as Belyaev pointed out. Belyaev reported 15,000 individuals of the holothurian *Elpidia* from 9000 m at the Kurile—Kamchatka Trench *Vityaz*, cruise 39, 1966). However, we seriously doubt whether this dominance is any more or less significant than the dominance of the holothurian *Psolus* sp. in the Peru–Chile Trench at the upper part of the trench wall near 1000-m depth, where 47% of the 4460 specimens collected at station 79 (5-ft beam trawl) belonged to one holothurian species of the genus *Psolus*.

If single-species dominance can be demonstrated for deep trenches in a manner comparable to that of *Artemia* in ultrahaline water (Thienemann's law), then indeed this peculiarity of hadal fauna would be an acceptable criterion. Bottom photographs and trawl results from many areas suggest that single-species dominance in density and number often occurs on the continental slope off the Carolinas and Peru, and even in the relatively shallow depths of the anoxic parts of the Gulf of Cariaco off Venezuela. It is unnecessary here to elaborate this interesting phenomenon of biocoenosis of oligomixity, but obviously the deep trench floor of the trenches that have been investigated are not, as a rule, unique in being characterized by single-species dominance. In the light of all these reconsiderations, we believe that the faunal composition of trenches is not strikingly different from the abyssal fauna elsewhere.

In general Wolff (1970) concurred with our conclusions but still adhered to a belief in abyssal gigantism, and continues to believe that hydrostatic pressure is responsible for certain characteristics of the fauna of deep trenches. Until more definitive collections are made from all trenches, shallow and deep, it is unlikely that continued analysis of the existing and admittedly limited information about the fauna of trenches will reveal a great deal more information.

ON THE ORIGIN OF TRENCH FAUNAS

It is commonly thought that the trench floor fauna as we call it has originated from abyssal depths and hence from the abyssal fauna. We agree, but it is not reasonable to presume that this fauna has come from the red clay environments to the ocean side of any trench. Instead, we suggest that it has come from the moderate-depth species

Figure 8-8. Habitat illustration of fauna of abyssal trench (lower) floor drafted from underwater photographs and collections from R/V *Vema*. Recognizable are the ophiuroid *Neopilina (Vema) ewingi*, foraminifera *Bathysiphon*, gastropod, and barnacles attached to worm tube. Drawing courtesy of Life magazine.

maximum, which likely has given rise to the fauna of the lower abyssal zones along the continental margins. If this is so, then each trench should have a greater affinity with the fauna on a nearby continental margin than with the fauna of adjacent red clay regions. At the Milne-Edwards Deep this is indeed the case, because the single red clay isopod species is not known from the trench floor, whereas 34% of the species from the moderate-depth species maximum are known from the trench floor, including *Neopilina (Vema) ewingi* (Figure 8-8). The presence of *Elpidia glacialis* on the floor of the Kurile–Kamchatka Trench is a good example of the affinities of that trench floor fauna with the Asiatic and Alaskan continental margins.

Seismic studies have shown that trenches are filling with sediments and sinking simultaneously. These seismic studies, as well as piston core samples, reveal repeated bedding of sediments with alternating ooze and sand layers. The sand layers are derived from turbidity currents or slumps, and it is therefore most probable that the trench environment is one of the least stable regions on earth, as far as the bottom fauna is concerned. The bottom fauna is being alternately buried and repopulated, and it is probable that this cycle has continued frequently since the Pleistocene glaciations.

The foregoing discussion about faunal repopulation is mainly theoretical because proof of repopulation has not yet been achieved through studies on benthic remains from deep-sea cores. We present it as a strong probability which can be subjected to testing.

The Peru–Chile Trench Fauna

It is difficult to assess the likely origin of the fauna of any trench or basin of the ocean, mainly because too many facts are lacking from which to build an easily evaluated hypothesis. This is especially true of the Peru–Chile Trench, where most of the species and several of the genera are new. The data on vertical distribution are, however, quite illuminating regarding some routes to the deep sea, and enough distributional information is available regarding *Storthyngura* and *Mesosignum* to allow some generalization that may or may not apply equally to other components of the fauna.

ORIGIN FROM SHALLOW WATER

The thought that the abyssal fauna has originated from a former shallow-water fauna is an old attractive one. But when we examine the affinities of the abyssal species with those of the intertidal and shelf provinces, we reach the inescapable conclusion that the abyssal province off Peru is one that is completely isolated today from the shallower provinces. Only the AZT shows connections with the abyssal province, and these connections are tenuous, containing only a few species. An origin of much of the abyssal fauna at this trench from a shallow-water fauna at this trench is clearly not evident.

ORIGIN FROM SUBMERGED CONTINENTAL MARGINS

Parker (1961) speculated that abyssal faunas could come about by simply sinking with the continental margin. This concept assumes that the sinking is a tranquil process and that the fauna that sinks is capable of abyssal existence. There is no biological evidence that this has happened.

ORIGIN FROM THE VAST RED CLAY ENVIRONMENT
OF THE PACIFIC BASIN

The red clay environments are so deficient in species of animals that an origin from them would appear to be unlikely today (geologically speaking). The reverse seems to be much more probable — namely, an origin of the red clay fauna from the abundant and diverse fauna in the abyss along the continental margin.

ORIGIN FROM GEOGRAPHICALLY DISTANT
CONTINENTAL MARGINS

In view of the isolation of the shallow-water fauna in Peru from the abyssal fauna at the same latitude, it follows that affinities of the abyssal fauna might be sought with some more distant fauna, shallow or deep.

The isopod genera that make up the abyssal fauna at the Milne-Edwards Deep consist of 15 genera, exclusive of the new or planktonic ones, and 87% of these are now known to occur along the continental margin of Antarctica. Only eight of the genera are in common with the fauna of the Arctic, yielding a 53% similarity, but all the Arctic genera in common are quite cosmopolitan.

The average least depth of occurrence for the 15 abyssal Peruvian genera is 3461 m, whereas the least depth recorded for 14 of these genera in Antarctica is only 1174 m (Figure 13-3). This is evidence of the operation of the phenomenon of polar emergence of abyssal genera, but it leads to the exciting possibility that 87% of the Peruvian abyssal genera of isopods could have come from the Antarctic continent simply through submergence along the continental margins.

A consideration of the total generic component in Peru, shallow and deep (and again exclusive of the eight cosmopolitan genera) shows that 78% of the 32 genera also are found in Antarctica and that 22% are endemic to the southern hemisphere. Again in sharp contrast, only 29% of the genera are in common with the Arctic, and again, too, these are cosmopolitan genera anyway. Accordingly, it seems reasonable to conclude that the Isopoda fauna of the Peruvian Trench is more nearly allied with the Antarctic fauna than with that of the Arctic.

The affinities and the geographical distribution of the abyssal genus *Storthyngura* on a worldwide basis are reasonably known (George and Menzies, 1968a, 1968b). The patterns of species distribution alone were not revealing regarding origins; however, when the species were combined into related groups and subgroups of similar morphology, a connection was found with most groups through Antarctica.

Two species of *Storthyngura* were in the abyssal faunal province at the trench, one from the lower abyssal red clay and one from the mesoabyssal of the trench wall, but both at similar depths. The species *S. unicornalis* belongs to subgroup B_2 of George and Menzies, which also contains *fragilis* [northern Pacific off Japan, *caribbea* (Caribbean), *challengeri* (Antarctica), and *gordoni* (off South Africa, Indian Ocean)]; thus the Antarctic connection exists with subgroup B_2.

The species *S. octospinosalis* belongs to subgroup C_3, together with *snanoi* (Caribbean), and *symmetrica* (off South Africa, South Atlantic), and as a subgroup it lacks an Antarctic representative. It is difficult to establish contemporary oceanic routes of connection between the three related species in this subgroup except by way of the southern hemisphere. Thus the connection to the Antarctic by these three species is a distant one by their membership in group C. Group C includes *S. elegans* in the Antarctic.

The species *Storthyngura triplispinosa* Menzies is found to the north and the south of the Milne-Edwards Deep, at 12° from the mid-America Trench (5680–5690 m) and at 2856 to 2596 m, as well as in the South Atlantic, and may be presumed to exist at the trench. This species belongs to subgroup E_1 together with *benti* (South Pacific) and *zenkevitchi* (South Atlantic). Birstein removed the Antarctic *robustissima* from subgroup E_1, but we believe that this is quite incorrect and, furthermore, that *robustissima, triplispinosa, benti,* and *zenkevitchi* form a natural unit of related species.

We concur with Birstein (1963) that the three North Pacific species *kurilica, herculea,* and *vitjazi* do form a natural unit, and indeed we placed them in subgroup E_2.

The species *Mesosignum multidens* Menzies and Frankenberg (1968) is distributed throughout the Milne-Edwards Trench on the trench floor. Its closest relative is *Mesosignum elegantulum* Birstein, from the North Pacific. The species *M. usheri* Menzies, although not known from the Milne-Edwards Trench, is found as far south as 35°S and also in the Caribbean. *Mesosignum* is now known as far south as 60°S at 3310 to 3530 m depth, and it is quite possible that an Antarctic origin can be established for the species of *Mesosignum*, comparable to that of *Storthyngura*. The genus, *Mesosignum,* like *Storthyngura*, is not known from the Arctic.

Thus both the generic composition and the distribution of species groups suggest an origin of the Peru–Chile Trench fauna from the Antarctic, rather than from the Arctic.

CHAPTER 9

Regional Comparison of Vertical Faunal Zonation

How are the environmental stability and homogeneity on the deep-ocean floor reflected in faunal vertical zonation? Known abyssal and hadal records of vertical distribution for 1144 species of deep-sea benthic invertebrates, reviewed by Vinogradova, showed rapid decrease of species from 2000 to 6000 meters, with a much slower reduction at greater depths. At about 3000 and 4500 meters, important changes occur in the taxonomic composition of the benthic fauna. Numerous species and higher taxa, broadly distributed on the slope and even shallower, disappear and are replaced by new species, genera, and families found only at greater depths. For these reasons, Vinogradova concluded that 3000 meters represents the true upper limit of the abyssal zone. We did not find such abrupt boundaries at bathyal and abyssal depths. [H. L. Sanders and R. R. Hessler, 1969, pp. 1420–1421].

Benthic living communities occur on and in the ocean floor to the greatest depths except in anoxic basins. Sanders and Hessler (1969), in summarizing the results of their investigations along a transect of the ocean floor between southern New England (Gayhead, Mass.) and Bermuda, concluded that abrupt faunal boundaries are not recognizable at bathyal (slope) and abyssal depths. They also named the topographic shelf–slope break as the true upper boundary of the deep-sea benthos in that part of the ocean, broadly delineating a continental shelf faunal assemblage of impoverished eurytopic (broad physical tolerances) fauna in depths less than 300 m and a deep-sea stenotopic (narrow physical tolerances) benthic fauna of high diversity in depths greater than 300 m. Our investigations along the Beaufort-Bermuda transect in the western North Atlantic Ocean revealed the presence of a zone of transition (archibenthal) between the shelf and the abyssal faunal province. Even in the continental slope environment, we are able to see pronounced changes in species composition and obvious peaks in species number that we call a "moderate-depth species maximum."

Vinogradova (1962a) recognized this trend of striking increase in species number for various groups as well as for the total fauna in her analysis of distribution of 1144 benthic species from various regions of the world oceans. Beyond the depths of the "moderate-depth species maximum," there is a definite decline in species number and also a corresponding decrease in diversity.

In our study of faunal zonation, we have purposely avoided pooling distributional data of species from different oceans or even from different regions of the

243

same ocean, as did Vinogradova. Instead we have restricted our analysis to a single transect in a given geographic region. Study of the vertical distribution of isopod species clearly brings out the trend of "moderate-depth species maximum" in the lower part of the continental slope in each of the regions investigated. It seems inescapable to conclude that the abyssal faunal province has three or more different faunal zones of varying species composition, density, biomass, and diversity. Here we make comparisons of the characteristics of the various vertical zones from one region to another in order to point out the major differences and similarities.

INTERTIDAL FAUNAL PROVINCE

The intertidal faunal province contains the fauna living between the tide marks from the highest spring tides to the low low water. By this definition, the fauna of the infralittoral fringe is a component of the intertidal. The following zones within the intertidal are recognized according to the universal scheme of Stephenson et al. (1949):

> Supralittoral fringe zone.
> Upper midlittoral zone.
> Lower midlittoral zone.
> Infralittoral fringe zone.

The tides are not uniform in their range, with latitude due to differences induced by land masses and topography as the "tidal wave" moves around the earth in response to the gravitational pull of sun and moon. The least average tidal ranges are at the poles and the tropics between 10°N and 10°S latitude. Maxima (averages) are at 60°N, 20°N, 10°S, and 50°S. In these regions the intertidal occupies the greatest vertical range, but at the most this range is reckoned in figures less than 15 m. It may involve a horizontal expanse of 1 km or more, depending on the local topography and the tidal range.

The intertidal is stratified latitudinally into the various marine climates such as the hypopsychral in the High Arctic and the High Antarctic, having temperatures less than 0°C, the subpolar, the boreal, the temperate, the subtropical, the tropical, and so on. Seasonal ranges in temperature are quite low at the poles and in the tropics but quite high in temperate climates, except along the eastern borders of the oceans, where upwelling moderates the temperature range seasonally.

Light is present usually throughout the intertidal province, and hence benthic algal productivity reaches its peak.

The primary productivity of the rocky intertidal zone is often great, especially in boreal regions. Here sessile algae fix carbon in amounts far exceeding that of the phytoplankton per unit area or volume. For instance, the Laminaria (brown kelp) of the central Californian intertidal produces daily 33 g of carbon/m² (Ryther, 1963).

The sediments of the intertidal are determined mainly by the geomorphology of a coastline, ranging from granite rocks in the boreal regions to coral in the tropics. The sands are variable in coarseness, depending on their source, on water currents, and on wave action. A muddy intertidal is usually restricted to estuarine conditions. The intertidal yields a wide variety of habitats for marine animals and plants.

An intertidal faunal province is lacking from the High Arctic and the High Antarctic. An impoverished intertidal fauna is present in the Low Arctic at Murmansk and on the Palmer Peninsula of the Low Antarctic. The order Oniscoidea, which has halophile genera and species, is lacking from High Polar regions, and the sand fauna is reduced or even absent from polar regions. The intertidal faunal province of the middle and the low latitudes is characterized by a high proportion of the Sphaeromidae and alga-eating Valvifera (p. 93, 145). The Sphaeromidae are represented in the Arctic by only one genus and in the Antarctic by six genera. In the polar regions the sphaeromids now live on the shelf instead of the intertidal, because both the Arctic and the Antarctic tend to have submerged intertidal components.

The absence of High Arctic and High Antarctic intertidal faunal provinces means that there is no relation between the intertidal species of these areas and the fauna on the shelf. However, at lower latitudes the intertidal shows some small relation to the shelf at a species and genus level.

Of the genera of the Peruvian intertidal faunal province, 20% enter the shelf and 10% enter the archibenthal faunal province. Not one genus enters the abyssal faunal province. Only 12% of the species of the intertidal enter the inner shelf, and not one species enters the archibenthal from the intertidal faunal Province.

Off North Carolina and Georgia, 50% of the genera of the intertidal enter the inner shelf, but not one enters the archibenthal or the abyssal; 32% of the intertidal species enter the inner shelf, but none penetrates as far as the archibenthal.

The number of species and genera in the intertidal in North Carolina and in Peru is about the same—there are 14 genera in both places, 14 species in Peru, and 15 species in the Carolinas.

One major conclusion emerges from this comparison: namely, that the intertidal faunal province is in the main isolated from the fauna of the shelf; it is almost completely isolated from the archibenthal, and totally isolated from the abyssal faunal province. This conclusion should have been obvious on an a priori basis because the organisms that are successful in the intertidal zone often have special morphological adaptations to life under conditions of wave stress, dessication, and the like. Plants have strong holdfasts, and animals often have stout clinging claws or the ability to burrow under rocks and in coarse sand. It is only in the intertidal that obligate herbivores live in abundance. Animals in the intertidal, except for the interstitial or burrowing species, very frequently have large eyes with black pigment; in many in-

TABLE 9-1. PROPORTION OF MAJOR ISOPODAN GROUPS IN THE ABYSSAL FAUNAL PROVINCE AT FOUR DIFFERENT LATITUDES IN %

Group	Arctic	Antarctic	10°S	35°N
Anthuroidea	4	2	17	0
Asellota	76	78	83	75
Sphaeromidae	0	0	0	0
Limnoriidae	0	0	0	0
Oniscoidea	0	0	0	0
Gnathioidea	10	2	0	10
Cirolanoidea	0	2	0	5
Valvifera	10	11	0	5
Serolidae	0	2	0	5
Total groups	4	6	2	5

stances, too, they have stouter and stronger skeletons than animals living at greater depths.

SHELF FAUNAL PROVINCE

The shelf faunal province extends roughly from the infralittoral fringe of the intertidal to the point at which there is a significant change in faunal composition. The average depth of the continental shelf for the oceans is 200 m, but this depth is seldom the boundary between the shelf fauna and the fauna of the AZT. For the regions examined, the topographic shelf showed considerable variation in depth:

Region	Depth of Shelf (m)	Faunal Boundary Depth (m)	Latitude
Arctic	100	<10	70°N
North Carolina	80	440	35°N
Peru	240	1240	10°S
Antarctic	400	100	70°S

The shelf of the Arctic extends out from shore as far as 440 nmi. The Antarctic shelf is usually narrow, however, having an average width of approximately 20 mi. In Peru the shelf extends 70 mi, whereas in North Carolina it extends only 60 mi in width.

Usually the sediments of the continental shelves are coarse grained, ranging from rocks to sand and shell interspersed with mud areas where depressions exist. The shelf water often experiences the effect of lowered salinity from runoff and is often quite turbid due to phytoplankton, detritus, or sediments in suspension. Light penetration is a function of water transparency and hence there are limits to the depth at which plants can live. Animals are often clustered in rather distinctive communities that may be classified on the basis of sediment type, rocks, sand, and mud. The diurnal or seasonal temperature range on the shelf is little or great depending on the geographical location, being least at the polar regions and in the tropics and greatest in boreal regions, where there is no upwelling to depress the annual temperature fluctuations.

It is generally recognized that the shelf province can be divided easily into two major components or zones: an inner shelf zone within the limits of light penetration for sessile algae, and an outer shelf zone where light is frequently absent and sessile plants are lacking.

The shelf fauna contributes no species or genera to the intertidal of the polar regions and few species (20%) to the intertidal of Peru or Carolina. Around 36 to 38% of the shelf genera of lower latitudes are found in the intertidal. Also in the low latitudes 28 to 35% of the shelf genera occur in the archibenthal. In the Antarctic 42% of the shelf genera are present in the archibenthal and 25% in the abyssal province; in the Arctic, the shelf genera seem to be identical with the archibenthal. The Arctic topographic shelf contributes most of its species to the archibenthal, and the same percentage enters the abyssal faunal province. In the lower latitudes of Peru and the Carolinas, shelf species are totally absent from the abyssal faunal province.

It appears then that the fauna of the shelf province is a closed system with regard to the abyss in the lower latitudes, whereas in polar latitudes it contributes a fairly high proportion of its genera and a lesser proportion of its species to the abyssal faunal province.

Valvifera are represented on the shelf in greater numbers than in the intertidal. The Sphaeromidae have shown a decline in numbers on the shelf everywhere except in Antarctica. The Antarctic shows the highest isopod group diversity.

The numbers of species, genera, and families in the shelf are higher than they are in the intertidal at the poles but are about the same in lower latitudes. The shelf has substantially as many families and genera of isopods as the archibenthal in all regions. Usually the number of species in the archibenthal shows a decline, and in Antarctica, where this decline is most apparent, the shelf has close to twice as many species as the archibenthal. The Antarctic shelf holds three to four times the number of families as the other regions and almost five times the number of species and genera.

ARCHIBENTHAL ZONE OF TRANSITION

The AZT commences at the lower limit of the shelf faunal province and extends deeper into the sea. It is in the aphotic zone along the continental margins and is hence devoid of living green plants. Sediments generally are transitional from terrigenous to pelagic. There are often strong correlations with water characteristics; thus off the Carolinas at 35°N it is between the Florida current and the western boundary undercurrent, whereas off Peru (10°S) it is just below the oxygen-minimum zone between the Peru undercurrent and the Pacific deep water or the Antarctic intermediate water. In the Arctic it is on the shelf and within the boundaries of the Atlantic warm-water inversion of positive temperatures. In the Antarctic it is located within the upwelling deep water, which here mixes strongly with the Antarctic bottom water. The temperature range of the archibenthal is greatest off the Carolinas and least at the poles. The sediments of the archibenthal in the Antarctic are glacial marine, containing increasing amounts of diatom ooze with depth; in the Arctic, on the other hand, the sediments change from glacial marine to the characteristic brown ooze between 100 and 360 m, and there is no sharp boundary line. Off the Carolinas the AZT ranges between 450 and 940 m. Here the sediments are sandy silt. In Peru the AZT extends from 1930 to 3330 m, with no particular correlation with sediments. The temperature range in the AZT is transitional between that of the shelf and the abyssal.

The AZT starts 60 nmi from shore off the Carolinas and extends another 7.5 nmi seaward, spanning 490 m in depth. The Peruvian AZT extends from 82.5 nmi from shore to 90 mi, also being 7.5 nmi wide, but here it is 1390 m in depth. In the High Arctic the AZT embraces the entire shelf, having a depth range of 360 m. The Antarctic AZT may extend 120 nmi beyond the shelf, and it is 800 m deep. Thus the archibenthal is farther from shore at the poles than elsewhere. It starts at shallower depths at the poles and extends deeper into the sea in low latitudes.

The antarctic archibenthal has more families, more genera, and more species of isopods than the archibenthal elsewhere.

In the lower latitudes the AZT fauna is more closely allied to the abyssal faunal province than to the shelf faunal province. Only in the Antarctic is the fauna of the AZT more nearly allied to the shelf than to the abyssal faunal province. In the Arctic we interpret the AZT to supplant the shelf fauna. The separation from the intertidal and the AZT is complete everywhere at the species level.

The archibenthal zone of transition may be looked upon not only as a transitional fauna between the shelf and the abyssal fauna, but also as transitional in depth, temperature range, and sediments.

For all practical purposes the AZT is equivalent to the mudline fauna of Murray (1895) from whence he derived the abyssal fauna. Although genera penetrate the abyss and the shelf from the AZT, the degree of species isolation of the AZT is great. Thus off the Carolinas the archibenthal species enter the shelf only by 10% and vanish by 85 m. Similarly, only 10% of the species penetrate the upper abyssal of the Carolinas, and none extends farther than 2062 m. In the Arctic the situation is markedly different, and 50% of the archibenthal species penetrate to the lower abyssal. We have no data on the Antarctic species, but we suspect that the situation there is similar to that of the Arctic.

In conclusion, we believe that the AZT is markedly separated at a species level from either the abyssal fauna or the shelf fauna in the lower latitudes, but that it is closely allied to the abyssal fauna and the shelf fauna at the poles, where it is indeed faunistically transitional.

ABYSSAL FAUNAL PROVINCE (AFP)

The abyssal faunal province embraces the balance of the sea floor. Its start is shallower at the poles than elsewhere, showing a tropical submergence where the region contacts the lower limit of the AZT.

Between the upper and lower limits of the AFP off Peru the temperature is constant at $+1.8°C$. Off the Carolinas it embraces a 2°C range between 4°C at the upper limit and 2°C at the lower limit in the Antarctic bottom water of the North Atlantic. There is no change in temperature in the AFP at the poles or in the Mediterranean.

The number of families, genera, and species of isopods increases in the abyssal faunal province in comparison with the AZT. The Antarctic AFP maintains its greater number of families, genera, and species in comparison with other regions.

Off the Carolinas the AFP commences with the start of the WBUC (Figure 4-7c) and the start of the pelagic foraminiferal oozes. In Peru the AFP starts with the deep water of the Pacific but the sediments show no particular change.

The asellote isopods range from 75 to 83% of the total isopod fauna in the AFP, and other groups of major taxa are absent or poorly represented in the AFP (Table 9-1).

Number of Species

The absolute number of species of isopods in the AFP increases, being higher than the number of species in shallower zones. This increase in species is in the main caused by the increase in asellote species. The Antarctic shows more species in the AFP than other regions (Table 9-2).

TABLE 9-2. NUMBER OF ISOPOD SPECIES IN THE SELECTED FOUR
GEOGRAPHIC AREAS

Region	Intertidal	Shelf	Archibenthal	Abyssal
Antarctic	0	120	70+	80+
Arctic	0	—	16	46
35°N, Atlantic	15	25	8	79
10°S, Pacific	13	11	13	37

Proportions

Forty percent of the Arctic abyssal genera are also found on the shelf and in the
archibenthal. In the Antarctic 30% are found on the shelf and 40% in the archibenthal. This high percentage of shallow-water abyssal generic representation appears to be due in the main to the phenomenon of emergence of abyssal genera.

In low latitudes the situation is quite different. Thus off the Carolinas (35°N) only
12% of the abyssal genera enter the archibenthal and 5% enter the shelf, whereas
none enters the intertidal. Off Peru (8°S) 20% are found in the archibenthal and only
5% on the shelf. Therefore in the high polar latitudes the alliance between the
archibenthal and the genera of the AFP is between 48 and 50%, whereas at lower latitudes the abyssal generic representation drops to 20 to 12%.

With species, the degree of isolation of the fauna of the AFP is much more
marked. In the Arctic 52% enter the archibenthal. Data are lacking for the Antarctic.
In the Carolinas only 5% of the abyssal species enter the archibenthal and none
enters the shelf. Off Peru 26% enter the archibenthal and none enters the shelf.

The species of the AFP of lower latitudes are thus totally independent of the shelf
and only a few enter the archibenthal; but in the Arctic as much as 34% of the
abyssal species enters the topographic shelf, or archibenthal. The emergence of
abyssal genera and species is most evident at the poles.

Abyssal Zones

We have divided the abyss into several zones. These are not necessarily comparable
to zones proposed by others (Chapter 2).

Upper abyssal (UAZ)
Mesoabyssal (MAZ)
Lower abyssal active (LAAZ)
Lower abyssal tranquil (LATZ)
Lower abyssal red clay (LARC)
Lower abyssal trench floor (LATF)

Each zone has certain distinctive ecologic features besides having a distinct
fauna.

The upper abyssal zone is characterized by a low species number and also
a relatively low number of genera. Off the Carolinas the UAZ ranges in depth between 940 and 2640 m. This depth is within a band of sediments 15 nmi wide con-

sisting of fine silt and clay and Foraminifera. The temperature range is 0.5°C (Figure 9-1). Off Peru the UAZ starts at 3320 m, extends to 4500 m, and has a width of only 3 nmi. The sediments are green radiolarian ooze. The temperature range is zero. In the Antarctic the UAZ ranges from 900 to 3480 m. Sediments are glacial marine with increasing amounts of diatoms with increasing depth. The water is Antarctic bottom water with a temperature range of 0°C.

The ocular index zero (OI_0), the depth where all species lack eyes, is located deeper at the poles and shallower at low latitudes. This polar depth depression of the OI_0 is due to the submergence of eye-bearing species at the poles.

The depth limits of the various zones and provinces (Table 9-3) which we have discussed appear in the composite Figure 9-1. This illustration is both fanciful and real. It is real from the standpoint of depth of zones and absolute temperatures, but it is a composite of Atlantic and Pacific data and in that sense is quite fanciful. It is useful mainly as an indication of the submergence of zones in the lower latitudes and the lack of correlation between the upper boundaries of a zone and bathymetry. We

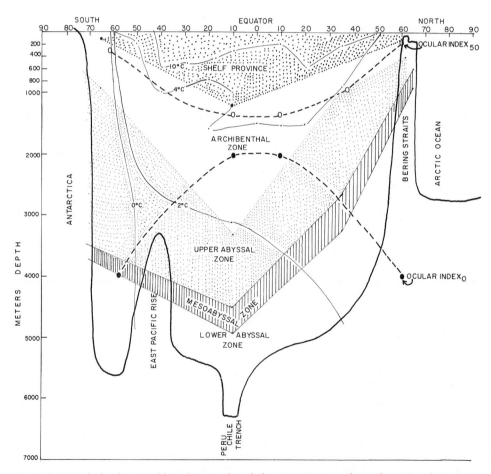

Figure 9-1 Depth distribution of faunal zones plotted along imaginary north–south section of an ocean; topography from Pacific section, temperature, and zones composites from all available data. Ocular indices also plotted.

TABLE 9-3. BATHYMETRIC RANGE OF FAUNAL PROVINCES AND ZONES: DEPTHS
IN METERS

Faunal Provinces and Zones	Off Carolinas (North Atlantic 30°N)	Off Peru (South Pacific) 12°S	Antarctic (within Convergence)	Arctic (High)
Intertidal	0–5	0–5	Not existing	Not existing
Shelf province	5–446	5–1238	1–100	<12(?)
		Oxygen-minimum layer		
Archibenthal zone (transition zone)	446–940	1927–3318	100–900	12–360
Abyssal province	940–5400+	3318–6280	900–5500	425–2800
Upper abyssal zone	940–2635	3318–4500	950–3475	425–570
Mesoabyssal zone	2640–3300	4500–4925	3500–3800	610–869
Lower abyssal active zone	3300–4400	5000–6280	3800–5400	1000–2600
Lower abyssal tranquil zone	4400–5080	—	—	—
		Eastern Pacific basin	No data	No data
Lower abyssal plain zone (red clay zone)	5080–5340	No data	—	—

present it as a hypothetical diagram that can be subjected to test and verification at other points in geography and in other oceans.

The UAZ always appears to have fewer species than the MAZ but more than the LAZ. The LAZ appears to have the least number of species in the LARC, although there is an increase in species on certain trench floors.

Each zone of the AFP appears to demonstrate tropical submergence, each being located at greater depths in the low latitudes than in the higher latitudes.

The proportion of asellote isopods is always high in the AFP and increases with each deeper zone and subzone. In the Carolinas (35°N), species belonging to the UAZ greatly diminish in numbers in the AZT and are not found in the shelf faunal province. Species from the MAZ and the LAZ all are confined entirely to the AFP, not even entering the AZT. A similar situation occurs off Peru (10°S), where species from the AFP enter the AZT at a 20% level but not the shelf faunal province. Those from the mesoabyssal enter the AZT at a level of only 10% and do not enter the shelf. Those from the LAZ are confined entirely to the AFP and do not even enter the AZT.

In contrast, in the high latitudes, as illustrated by the species of the Arctic, as much as 23% of the species from the LAZ enter the shelf, and in the Antarctic (as illustrated by the Serolidae) a similar situation obtains.

Apparently the abyssal faunal province in lower latitudes is more or less completely isolated in species from the intertidal, from the shelf, and even from the AZT. Furthermore, the AFP appears to show a decided tropical submergence, being deeper as the latitude decreases and shallower with increasing latitude.,

The correlation between sediment type and currents for the various zones of the AFP is rather good off the Carolinas and only minimal in Peru or the polar regions, where no such correlations are apparent.

SYNOPSIS OF THE MAJOR CORRELATIVE FEATURES OF ZONATION

1. An intertidal faunal province is lacking from the High Arctic and High Antarctic but is present in the lower latitudes. The polar absence of this province is associated with ice scour and ice cover.
2. The inner shelf fauna of the Antarctic region appears to consist of shelf components plus a submerged intertidal component.
3. The intertidal faunal province of the lower latitudes, although fully developed, is almost completely separated from the species of the submerged shelf faunal province.
4. The shelf faunal province of the high latitudes is closely allied in species and genera to the AZT.
5. The shelf faunal province of the low latitudes is a closed system which is independent of the species of the fauna of the AZT.
6. There is no correlation between the location in depth of the topographic shelf break and the shelf faunal province. Instead the SFP is lower in the lower latitudes and higher in the high latitudes.
7. Abyssal species enter the topographic shelf only at the poles.
8. The AZT is transitional in temperature range, in sediments (terrigenous to pelagic), and in fauna between the shelf faunal province and the abyssal faunal province. Usually the AZT is located just below the permanent thermocline. The presence of the ocular index 50 here suggests that it is also a region of morphologic transition.
9. A given isobath does not delimit the start of the AFP anywhere in the ocean because of the submergence of this province in the lower latitudes and its emergence at the poles. This contrasts strongly with all previous views regarding the depth of commencement of the abyssal fauna, which has been variously placed at 200, 1000, 2000, or 3600 m.
10. Similarly, a given isotherm does not describe the start of the abyssal faunal province. A temperature of 10°C (Bruun, 1956a) or 4°C (Madsen, 1961) has no particular meaning. The abyssal faunal province is confined to temperatures less than 2°C in the oceans but may well be at 14°C in the Mediterranean or the Red Sea. Generally the AFP is located in that part of the ocean depths where the range of temperature is less than 2°C. This may be the major thermal character of the AFP rather than any given isotherm.
11. As depth increases and succeedingly lower faunal units are reached, there is a continuing gradual decline of the major taxa present and also of the genera present along any given latitudinal depth profile.
12. Contrary to what can be inferred from Hessler and Sanders (1967), species number diminishes with depth from the high in the shelf and intertidal instead of increasing down to the abyssal. In the mesoabyssal zone there is a species maximum, but this is followed by a pronounced decrease in species in

the lower abyssal zone and its various subzones, with the true pelagic red clay environment having the lowest number of species.

13. Among the isopods (and in contrast to several other orders) there is a marked increase in families, genera, and species in the Antarctic throughout all the depth zones. In comparison, the isopod fauna of the Arctic is impoverished.

14. Endemism along a bathymetric scale shows that genera and species tend to have the highest endemism in the intertidal or shelf and in the mesoabyssal zone. The AZT generally has the lowest percentage of species and generic endemism.

CHAPTER 10

Abyssal Biomass

Except in a few cases it is not possible to assess the sizes of marine populations at different trophic levels, or the biological productivity on a worldwide basis. Quantitative data are absent or inadequate in many regions, while in others intercalibration of collecting and assay methods is necessary. [A statement made at 1958 meeting of Special Committee on Oceanic Research, Zenkevitch 1961, p. 329.]

As described in preceding chapters, benthic invertebrate assemblages are often zoned along continental margins. Zonation, however, is not the only characteristic of abyssal assemblages that has been studied. Any community of organisms can be described by a given set of quantitative characteristics. In general, these are the concentrations (density) and weight (biomass) per unit area, the number of different kinds that exist relative to their concentration (species diversity), and their distribution patterns on a smaller scale (dispersion or patchiness).

Abyssal biomass is defined no differently from biomass elsewhere in the oceans. It is the abundance of living matter per unit area over the seabed. It is often used as an indicator of productivity, and therefore some prefer the terms "standing crop" or "standing stock." These are valid indices of production only when rate of turnover is known. In practice, the measurement applies to the macroscopic living forms only and not to the microfauna or flora. The difficulties of quantitative sample procurement and the processes of biomass measurement have been discussed earlier (p. 66).

TOTAL AMOUNTS AND DISTRIBUTION IN BIOSPHERE

Marine biomass in the world ocean has been estimated at 10 billion tons (Table 10-1). More than half of this, namely, 6.66 billion tons, can be assigned to the benthos. Of this amount, 80% or 5.5 billion tons is distributed in the shoal waters of the continental shelves; only 56 million tons remains on the abyssal sea floor at depths greater than 3000 m. These depths cover three-fourths of the world oceans and have a low biomass comparable to that of the terrestrial deserts. The basic data for most of the world ocean and especially for the abyssal depths come from Soviet studies, and it was relatively recently (Kusnetzov, 1960) that a few abyssal biomass figures were known for the North Atlantic. Prior to that time, knowledge rested on the 65 quantitative samples collected by the Danish *Galathea* expedition of 1950 to 1952.

THE SHALLOW WATER BIOMASS

If data on the deep-sea biomass and its distribution can be considered sparse, then it may also be of some interest to realize that most of the data on the distribution of the shallow-water biomass are restricted to northern European seas and to the seas surrounding the Soviet Union. Very little data are published for the American shores, and of those which are accessible, most are restricted to the vicinity of Woods Hole, Massachusetts (Wigley and McIntyre, 1964), and southern California (Hartman and Barnard, 1958). In our discussion of the general trends of distribution of biomass, we have obtained our examples from European or Soviet sources. Even the simplest data are generally lacking from American shores in the intertidal and subtidal zones. Today these habitats are accessible almost everywhere along our shores. The advent of scuba equipment has made the subtidal not only accessible to but in some places overpopulated by divers. Perhaps the shovel and knife, the two instruments that might be used in such a study, lack the appeal of a 0.1 m² Petersen grab. Or perhaps simple, yet important data are not sufficiently interesting to sophisticated, modern science. Whatever the reason, this region can be considered nearly *terra incognita* as far as biomass estimates, even of the simplest kind, are concerned.

Intertidal biomass data are available for Peter the Great Bay in the Sea of Japan and are shown on Table 10-2. The eulittoral biomass of the boreal North Pacific is high, ranging from 86 g/m² (lowest value at the highest range of the tide) to 7038 g/m² at the zero tidal range in the *Pelvetia* beds. The majority of this consists of marine plants. The standing stock of the intertidal of the tropics is less by one or two orders of magnitude than that of the boreal regions, where it reaches its maximum. The High Arctic and Antarctic are both impoverished by contrast.

The intertidal biomass quite naturally shows considerable horizontal variation depending on the seabed involved, generally being least on sandy bottom and greatest in rocky areas. The temperate zones of the earth also have high eulittoral benthic biomass.

TABLE 10-1. QUANTITY OF BOTTOM FAUNA IN THE WORLD OCEAN* FROM VINOGRADOVA (1962)

Depth (m)	Area (millions of km²)	%	Approximate Mean Biomass (g/m²; tons/km²)	Total (million tons)	(%)
0–200	27.5	7.6	200	5500	82.6
200–3000	55.2	15.3	20	1104	16.6
>3000	278.3	77.1	0.2	56	0.8
Sum	361.0	100.0	220.2	6660	100.0

* After Zenkevitch et al. (1960).

TABLE 10-2. DISTRIBUTION OF BENTHIC BIOMASS, SEA OF JAPAN, PETER THE
GREAT BAY, FROM ZENKEVITCH (1963)

Depth (m)	Substratum or Dominant Species	Biomass (g/m²)
+1	*Gloiopeltis*	86
0	*Fucus*	250
0	*Fucus–Pelvetia*	4895
0	*Pelvetia*	7038
−0.1	*Tichocarpus*	2450
−0.5	*Corallina–Cystoseira*	6115
−0.7	*Zostera*	6570
15–35	(Sand)	522.9
30–35	(Sand)	343.4
51–58	(Silty sand)	398.5
55–64	(Silty sand)	182.4
80–200	(Firm sand)	212.6
177–238	(Fine sand and mud)	158
340	(Mud)	242
500–1000		36
1500		10
2500		2.2
3250		0.23

Subtidal biomass is influenced both by water depth and salinity characteristics,
as well as by the sediments and the food supply. Benthic biomass distribution in the
Elbe Estuary tends to show that the brackish parts have the least benthic biomass
(Caspers, 1948). In the marine part of the Elbe Estuary, it is as high as 6000 g/m², as
opposed to a high in the brackish part of only 37 g/m². The freshwater biomass is as
high as 143 g/m². This estuary feels the impact of dilution from surface to bottom,
and the foregoing results are therefore to be expected.

Offshore, the effects of depth begin to appear, but the influence of the increase
in salinity has a greater impact. This was demonstrated by Thorson (1957) for the
deep Scoresby Sound in East Greenland as follows:

Near Entrance		Midsound		Estuarine Part	
Depth (m)	Biomass (g/m)	Depth (m)	Biomass (g/m)	Depth (m)	Biomass (g/m)
10	360	10	240	10	35
100	200	100	65	100	20
150	110	150	30	150	15

These data for a sound or fjord are not much different from those elsewhere in
the oceans. They show biomass decreasing with increasing depth, except for some
anomalous situations. The polar environments constitute a prime exception in which
there is no measurable benthic biomass until the bottom layers of the ice sheet have
been passed. Beyond and below the ice sheet, the fauna is luxuriant throughout the

polar shelf. The epicontinental seas of the Antarctic show unusual increases in benthic biomass independent of depth *per se* (Ushakov, 1963).

VARIATION IN BIOMASS

Biomass (Tables 10-3 and 10-4) can be estimated in thousands of grams for each square meter in the intertidal zone. The continental shelf from 60 to 200 m depth has a biomass between 150 to 500 g/m² everywhere except in the Antarctic, where it is exceptionally high on the shelf and comparable to the intertidal values of the boreal North Pacific (Figure 10-1).

Between 400 and 1000 m, the biomass can be expressed in terms of tens of grams per square meter, whereas below 1000 m up to 4000 m, only 1 g/m² is present (Figure 10-2).

TABLE 10-3A.   BIOMASS VALUES (WET WEIGHT: G/M²) FROM REGIONAL PARTS OF THE WORLD OCEAN AS REPORTED BY VARIOUS WORKERS AT DEPTHS LESS THAN 1500 M

Depth (m)	Regions (see Table 10-3b)						
	I	II	III	IVa	IVb	V	VI
0	7038	420	*	*	*	0	0
10		360†	*	12	800	0	*
					1800		
18	533		4000			0	*
60	180		500				214
100		200†					
200	158	110†	260			1368	
300	150		170			483	
400							60
1500	24		20				

* Data not located.
† From Thorson (1934); Scoresby Sound, East Greenland.

TABLE 10-3B.   SOURCES OF DATA IN TABLE 10-3A

Region		Data from	Reference
I	Boreal Pacific	Peter the Great Bay	Zenkevitch (1963)
II	Boreal Atlantic	Massachusetts	Sanders et al. (1962)
III	Temperate Pacific	California	Hartman and Barnard (1958)
IVa	Tropical	Israel, Eastern Mediterranean	Gilat (1964)
IVb	Tropical	Florida, Atlantic, Caribbean	McNulty et al. (1962)
V	Antarctic polar	Indian Antarctic	Ushakov (1963)
VI	Arctic polar	Chukchi Sea, Arctic	Ushakov (1963)

TABLE 10-4A. BIOMASS VALUES (WET WEIGHT: G/M²) FROM REGIONAL PARTS OF THE WORLD OCEAN AS REPORTED BY VARIOUS WORKERS AT DEPTHS IN EXCESS OF 1500 M

Depth (m)	Regions (see Table 10-4b)				
	I	II	III	IV	V
1,500	24				
2,000	20	1.17		2.8	
3,000		1.87			1.8
3,500	5.6			1.4	
4,000		2.07		0.6	1.5
5,000	0.05				0.16
6,000	0.03	0.33	0.05		
6,200		0.85			
9,900	0.26				
10,000			0.0015		
10,600			0.001		
10,700			0.017		

TABLE 10-4B. SOURCES OF DATA IN TABLE 10-4A

Region	Reference
I Boreal Pacific	Filatova and Levenstein (1961)
II Temp. So. Pacific	Frankenberg and Menzies (1968)
III Tropics	Belyaev and Sokolova (1960)
IV Antarctic	Ushakov (1963)
V Temp. N. Atl.	Original data and LUBS samples off North Carolina

Measurements of abyssal biomass have not yet been taken with sufficient frequency to allow assessment of the significance of major variations in values that are or are not depth dependent. Thus the significance of an anomalous value is missed, even though it may be very important. Major attempts to date have been to describe the average condition, and of course this has importance.

Some anomalies are already understood. There is no measurable biomass on the floor of the Fossa de Cariaco off Venezuela at a depth of 500 m and greater, and the same is true also for certain fjords and the Black Sea, as well as for parts of the Red Sea. In these cases, it is perfectly obvious that the anomaly is due to the absence of oxygen and the presence of the toxic hydrogen sulfide. Off the shores of North Carolina at depths in excess of 70 m, there is a submerged reef at the edge of the continental shelf. Underwater photographs reveal an abundant epifauna, as well as fishes, and submarine trawls have recovered considerable benthic specimens suggesting an anomalously high biomass on the shelf (Menzies et al., 1966) (Figure 4-4a). The sands at shallower depths on the shelf and the sands below the reef are relatively impoverished (Figure 4-4g). This is a clear-cut example of anomalous

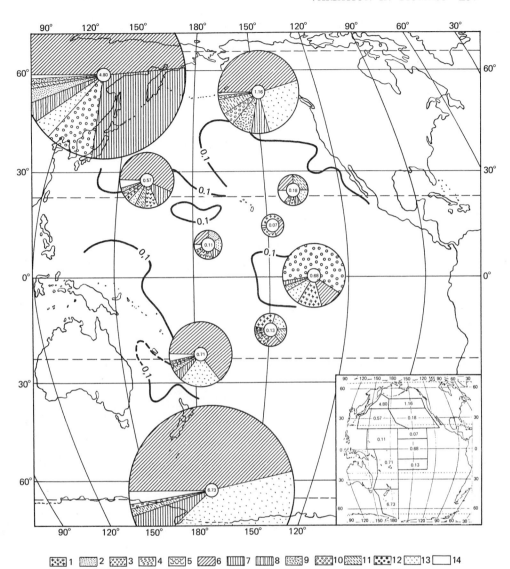

Figure 10-1. General distribution of average biomass and the composition of the bottom fauna by groups in the various regions of the Pacific Ocean. The figure in the center of each circle represents the average amount of biomass for the given region: 1, Soft rhizopods and related organisms of undefined systematic position; 2, sponge; 3, Coelenterata; 4, nematods; 5, Nemertinea; 6, polychaetes; 7, Echiuroidea; 8, Sipunculoidea; 9, gastropod mollusc; 10, bivalve mollusc; 11, Crustacea (Tanaidacea, Amphipoda, Isopoda, and Cumacea); 12, Bryozoa; 13, Echinodermata; 14, varia (rest). (After Zenkevitch, 1969.)

biomass as far as depth is concerned, and it has a special origin — namely, the presence on the reef of a solid substrate to which many epifaunal animals can attach (in comparison to the relatively impoverished sand above and below the reef).

There must be many similar anomalies that will be subject to reasonable explanation once environmental characteristics are known. For example, the 3803 g/m² at 960 m at 64°N and the 18,128 g/m² at 1278 m at 60°N latitude in the Atlantic

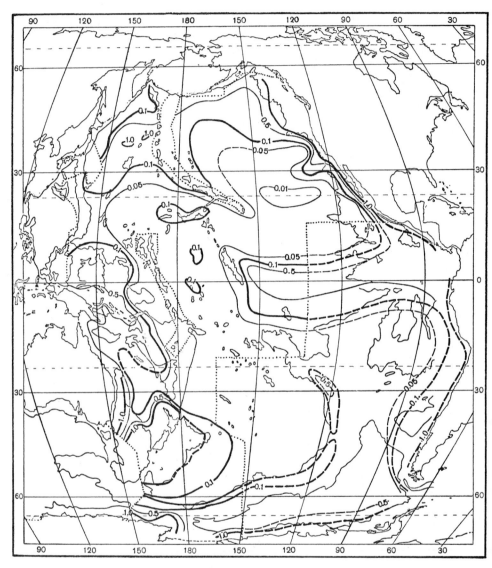

Figure 10-2. Distribution of general benthic biomass (g/m²) in the abyss of the Pacific Ocean: dots, boundaries of investigated areas; dashes, proposed distribution of general biomass in the noninvestigated areas of the Ocean. (After Zenkevitch, 1969.)

Ocean as recorded by Kusnetzov (1960) are orders of magnitude higher than others from the Atlantic at similar depths. The question remains how this can be explained. Until more quantitative sampling has been done in restricted areas, it is unlikely that the significance of such variations will be discovered.

Certain regularities in the distribution of biomass at abyssal depths have been described by Vinogradova (1962a). The Soviet data are based on the trawlograph and on the Okean grab samples (0.25 m²). In general, the biomass decreases with increasing depth and also with increasing distance from land (Figures 10-1 and 10-2). It is highest in the high latitudes and lowest in the low latitudes. There are some

TABLE 10-5. COMPARISON BETWEEN DISSOLVED OXYGEN CONCENTRATION, BIOMASS, AND RELATIVE ABUNDANCE OF LARGE EPIFAUNA AGAINST DEPTH OF MILNE-EDWARDS DEEP

| | Dissolved Oxygen (ml/liter) | | | | Biomass | Epifauna, Relative |
Depth (m)	to 30 m	to 83 m	to 638 m	to 6200 m	g/m²	Abundance
0	2.17	5.03	5.29			
10	0.64	5.04	2.46	5.12		
20	0.42	4.53	2.39	5.20		
30	0.26	2.88	2.39	3.65		
50	—	0.75	1.14	1.21		
	Bottom					
75		1.02	1.11	0.77		
100		—	1.54	0.99	49.46 (126)	
		Bottom				
150			0.67	0.35		
200			0.58	0.25		
250			0.20	0.46		
300			0.20	0.14		1 (305)
400			0.17	0.16		544 (450)
500			0.21	0.24	6.5 (519)	1 (498)
						342 (500)
						1141 (520)
600			0.31	0.26		
700			—	0.38		
			Bottom			
800				0.63		
1000				1.16	28.05 (995)	
1499				1.95		
						1 (1810)
1996				2.30		6 (1840)
						136 (1940–1980)
						47 (1980)
2491				2.75	3.44 (2384)	28 (2435)
						140 (2525)
2984				2.92	1.89 (2936)	10 (2975–2985)
						3 (3125)
						0 (3198)
						2 (3580)
3968				3.11	2.07 (4067)	0 (4074)
4462				3.41		4 (4410)
						6 (4560)
5750				ca. 3.00	0.33	5 (5010)
						3 (5475)
	Red clay tranquil zone					1 (5890)
6100				ca. 3.00		15 (6100)
6200				ca. 3.00	0.85	23 (6200)

* Actual depth in meters given in parenthesis.

261

regular indications of an increase in bottom biomass associated with the equatorial countercurrent (Zenkevitch, 1969) (Figure 10.2).

Beyond the continental slope and rise on the red clay deposits, the biomass is generally 0.01 to 0.05 g/m², or at least always less than 0.1 g/m². At depths in excess of 6000 m, (Table 10-4) there is considerable variation in benthic biomass. At least two trenches have anomalously high biomass for the depths (e.g., the Kurile–Kamchatka and the Milne-Edwards trenches). Values on the trench floor are orders of magnitude higher than the adjacent red-clay-filled basins at shallower depths.

RELATION BETWEEN NUTRIENT RECYCLING AND THE BIOMASS OF THE ABYSS

Many writers have believed that there is a direct relation between the surface productivity and the benthic biomass. This is indeed the case when the biomass is considered at depths less than the thermocline. Above the deep thermocline of the oceans, more than 90% of the organic matter produced in the sunlit layers is remineralized and hence is not available to the benthos for food. With increasing depth, the remaining 10% is similarly lost at an unknown rate to the water column. This does not leave much for the benthos to feed upon. Belyaev (1966) believed the relation of biomass to depth to be entirely dependent on the amount of nutrient or food matter transferred to the depths from coastal shallows of the continents.

Vinogradova (1962b), in contrast, believed that the rich plankton of the neritic zone is important in supplying food to the abyssal fauna. Menzies (1964) related several mechanisms for food supply to the abyssal fauna, which were all associated with transport from shallow water to the abyssal depths.

The biomass of the red clay environments, which ranges from 0.01 to 0.05 g/m², likely represents the maximum biomass possible when the fauna is supplied with surface organic food directly from plankton production. In contrast, the richer fauna near the coastlines has a standing crop that is two to three orders of magnitude greater at comparable depths than the red clay biotope. This reflects both a greater neritic surface productivity and a transport of nutrients from benthic sediments into the deeps from the continents.

RELATION TO THE MODERATE–DEPTH SPECIES MAXIMUM

The moderate-depth species maximum reported by Menzies and George (1967) is around 3200 to 3800 m in the eastern South Pacific. Estimating from Vinogradova's data, the biomass should be between 1.6 and 5.6 g/m²; our data suggest a biomass between 1.5 and 1.8 g/m², Table 10-4. In contrast, the biomass of the intertidal zone is in the order of hundreds of grams to thousands of grams per square meter. Because the intertidal species diversity is not similarly hundreds to thousands of times greater than that of the species maximum, it is reasonable to conclude that standing stock and species diversity are not directly related. These data imply that the species at the moderate-depth species maximum are much smaller in size than their shallow-water counterparts.

CHAPTER 11

The Structure of the Abyssal Benthic Community

Ecological communities . . . do exist, but what are linked in them by biotic factors are not the faunistic units, the species, but the ecological units, the life forms. [Thorson, 1957a, p. 470.]

This statement by Thorson reflects his insight into factors governing the development of level-bottom communities. If Thorson was correct, we expect that before too much longer a life-form diversity index may replace the species diversity index as a measure of community diversity as well as environmental stability. Meanwhile the problems associated with patchiness in the distribution of species — or life forms — remains highly significant.

PATCHINESS

Patchiness or dispersion refers to the spatial patterns or arrangements to which a group of animals conforms at a given time. The activities of describing and accounting for the spatial arrangements have been the subject of considerable quantitative work. The word distribution has been used in different ways. In one instance it describes a numerical relationship and in another the habitable range. Because of this, and also because of the statistical implications that may be attached to this usage, "distribution" on a small scale has given way to the term "dispersion."

Although a variety of methods have been devised to measure, assess, or describe dispersion — Grieg-Smith (1964) and Pielou (1969) offer good reviews — few have been carried into the marine environment. The horizontal patchiness in zooplankton has been studied using a Hardy–Longhurst Plankton Recorder (Wiebe, 1970), but no one has approached this problem in deep-sea benthic communities.

Scuba diving and submersibles have recently been used to great advantage in the investigation of patchiness (Caddy, 1970; Emery et al., 1970; Barham, 1968). The expense involved and the impossibility of covering great distances with such equipment probably preclude its use extensively.

There are basically and most simply three types of spatial patterns: overdispersion, in which the organisms are further apart or more evenly spaced than would be

263

expected due to chance alone; underdispersion (often called contagious distribution), in which the organisms are clumped in patches surrounded by areas of unexpectedly sparse densities; and random pattern, in which the location of each individual does not appear to affect the location of another individual. However, Clapham (1936) used "over" and "under" dispersion in the opposite sense. It has been suggested (Grieg-Smith, 1964) that the terms be replaced by "contagious" and "regular."

The first and second forms generally indicate that there is a negative or positive response between individuals. Distance between them could be maintained by selection through time as the individuals of a population grow. Competition for food or space, or the behavior of another species could also tend to space the individuals. Likewise, a food source or a behavioral habit related to defense or reproduction could result in a clumped pattern. Random dispersion assumes that each individual's position in space is a function of chance alone and is independent of the location of the other members of the population.

DISPERSION OF ANIMALS IN BOTTOM PHOTOGRAPHS

Submarine photography provides an opportunity to investigate spatial patterns on large and small scales. The fluctuations in populations can be observed from photograph to photograph as the ship drifts during a lowering. The distance between photographs can be estimated, and the measurement offers one way to estimate patch size. Since drift is generally parallel to bottom contours, the same populations persist throughout the lowering.

With this method, nonrandom clumping should be seen, if it exists, on several scales. With a ship drift of 0.5 to 2 knots, several kilometers are usually covered during a lowering. Patches or clumps on the order of several hundreds to thousands of meters across would be observed, if they exist. There have been few cases in which patches of this order of magnitude have been seen, except when the ship has drifted offshore rather than along given contours. When this has occurred, it has seemed obvious that the ship has crossed from one "depth-sediment-current"-regulated zone into another, rather than having transected patches of clumps of the same zone. It is admitted that this zonation in essence is a form of patchiness, but zonation is a function of a continuum of change in depth-related environmental variables (species, temperature, current direction, sediment type, etc.) and as such is narrowly and precisely aligned for long distances with the isobaths. Patchiness, as we envision it, is characterized by discontinuities in species or communities which do not correlate with depth. The problems then are (a) to assess variation in densities of this nature; (b) to ascertain statistically whether they are random, evenly dispersed, or clumped; and (c) to determine clump or patch size when species are contagiously distributed.

A smaller scale of dispersion can be identified within individual photographs. But this is rarely the case. A sea cucumber believed to be *Peniagone willemoesia* has been photographed in dense clumps in submarine canyons along the east coast of the United States (Ross, 1968, p. 169; Stanley and Kelling, 1968). As many as 150 to 300 animals, each several centimeters long, mass themselves into areas 1 or 2 m in diameter. These groupings appear to occur predominantly in submarine canyons, which may be indicative of a response to food sources that have been concentrated in the canyon (Rowe, 1971). Undoubtedly, large particulate organic matter can be an attractant that will cause this kind of dispersion. Photographs of the Hatteras abyssal plain show intense animal activity in the form of bioturbation or tracking and churning of the sediments only around isolated clumps of detrital *Sargassum* (Schoener and Rowe, 1970). Where food resources are the factors limiting animal populations, they can have profound effects on animal distribution.

Clumping is not always so intense that it can be seen visually in a photograph or measured by complete absence from parts of a series of photographs. Contagion can be much less pronounced and still be extremely nonrandom. In these cases, the patchiness must be assessed statistically.

Both clumping and vertical zonation by the sea cucumber *Scotoplanes* has been observed by Barham et al. (1967) using a submersible off California.

Poisson Distribution

If objects that are arranged randomly on a two-dimensional surface are sampled with appropriate sized quadrats, the probability that any one object will be encountered is small, but objects are encountered sometimes as the number of quadrats becomes sufficiently large. Under these conditions, the number of individuals expected per unit follows a Poisson distribution, where the probability that a quadrat will contain X objects is

$$P_x = \frac{X^x e^{-\bar{X}}}{X!}$$

where $\bar{X}$ is the mean per unit. A characteristic of the Poisson distribution is that the variance is equal to mean, or $V/\bar{X} = 1$. This ratio, called the "coefficient of dispersion," can be used to assess aggregation. Unfortunately, it cannot be used to estimate patch size unless the sample area can be easily controlled and varied.

If $V/\bar{X}$ is less than 1, a regular distribution is suggested, and if $V/\bar{X}$ is greater than 1, a contagious distribution is indicated. There are two methods of testing the significance of the difference of the ratio from unity. It can be compared with its standard error using a t test, where $s = 2/(n-1)$ and n is the number of samples.

An alternate is to use the "index of dispersion" (ID),

$$\text{ID} = \frac{V}{\bar{X}} (n - 1)$$

Significance is tested by entering the ID in a χ^2 table, with $n-1$ degrees of freedom. If the distribution is regular, it is significant when the probability is unusually high, and vice versa. If it is contagious or clumped, the probability will be unusually low.

Patchiness in *Ophiomusium, Hyalinoecia,* and *Parapagurus*

The most abundant species (maximum of about 10/m²), which also has one of the widest continuous depth (1400–3200 m) distributions was the ophiuroid *Ophiomusium lymani*. The mean, the variance, and the index of dispersion (ID) have been calculated for this species where it occurred in the photographic study off

TABLE 11-1.   PATCHINESS IN *OPHIOMUSIUM LYMANI*

Station Number	$\overline{X}$	N	V	ID	P_x2	Conclusion	Depth (m)
4667	24.5	21	64.2	52	<.001	Contagious	2085–2125
261	7.1	18	8.1	17.6	>.5	Random	2000
2495	1.6	5	0.3	0.8	>.9	Regular	2800
2740	1.4	13	0.8	2.8	>.9	Regular	2900
2760	1.0	11	2.4	24	.01	Contagious	1900
4398	0.7	21	1	32	<.05	Contagious	2000
4656	1.8	13	3.5	24.7	<.02	Contagious	2000
4659	3.1	14	4.6	21	<.1	Random	2000
4664	8.1	7	14.5	10.8	<.1	Random	1960
5107	3.8	15	4.6	16.8	<.3	Random	1580
6235	1.7	21	1.4	17.2	<.7	Random	2900
6237	1.2	16	0.6	8.0	>.9	Regular	3040
6748	0.6	11	0.5	7.1	<.7	Random	2900
7766	0.5	17	0.4	11.8	>.8	Random	
5791	13.2	16	30.9	34.5	<.01	Contagious	1750
7557	46.9	9	599	101.6	<.001	Contagious	1700
5799	4.8	37	12.1	92.5	<.001	Contagious	2545
2778	2.9	21	2.9	20.6	<.5	Random	2700
2780	0.19	21	0.16	17.0	<.7	Random	3200
5805	5.0	20	8.3	32.3	<.05	Contagious	2760
7548	0.7	3	1.3	4	<.1	Random	2775–2580
7616	45.3	21	547.4	229	<.001	Contagious	1655–2048
7553	56.6	24	304.2	124	<.001	Contagious	1926–1850
7812	0.9	23	1.4	35.2	<.05	Contagious	3060
5802	1.2	17	2.1	27.2	<.05	Contagious	2710–2670
4400	18.8	9	93.4	40	<.001	Contagious	2000
1142	12.9	8	43.6	23.8	<.01	Contagious	2000
1124	20.3	3	1.3	0.14	>.9	Regular	1500
5807	0.19	16	0.16	13.5	<.7	Random	2530–2635
4671	34.1	20	121.9	68.4	<.001	Contagious	1960–2140
7524	0.9	8	0.7	5.5	<.7	Random	3200
2487	0.2	10	0.18	8.1	<.7	Random	2480
2491	1.0	5	0.5	2.0	<.8	Random	2800
2492	0.3	12	0.2	8.8	<.7	Random	3080
2493	0.7	3	0.3	1.0	<.7	Random	3134
2494	0.5	2	0.5	1.0	<.3	Random	3006
4445	0.3	12	0.2	7.7	<.8	Random	3110
5129	0.3	4	0.3	3	<.3	Random	2480
5795	0.3	4	0.3	3	<.5	Random	1520–1575
6227	0.6	10	0.5	7.2	<.7	Random	2990
6233	0.3	3	1.0	2	<.5	Random	3000–3045
6229	0.8	12	0.3	4.4	>.9	Regular	3000–3020
1146	0.5	2	0.5	1	.3	Random	3200

North Carolina (Table 11-1). The significance level is 10%. Few definite general-izations can be made from the information. Dispersion appeared to be contagious at 15, random at 23, and regular at four of the 43 stations. All but one (8) of the high mean densities (>10/photo) appeared to be contagious; at moderate densities (1–10/photo) five were contagious, seven random, and three regular. At low den-sities (<1/photo), one was contagious, 17 random, and one regular. The exception at high densities appeared to be regularly dispersed. Although far from conclusive, this pattern suggests that the more common the individuals, the greater the probability of contagion, which is a normal feature among most animal assemblages. When the means of the different lowerings were compared, significant differences could be seen relative to depth. The abundance increased steeply from 1400 m to just over 1900 m and decreased less abruptly as depth increased to 3200 m, where the animal disappeared. Perhaps this continuum of change caused the contagion to ap-pear as an artifact of the analysis where contagion did not actually exist. This criti-cism may be justified, but no reliable trends in mean density could be picked out in the individual lowerings.

Another of the more abundant species, the polychaete *Hyalinoecia artifex*,

TABLE 11-2. CONTAGION IN *HYALINOECIA ARTIFEX*

Station	$\bar{X}$	N	V	ID	P_x2	Conclusion	Depth (m)
7561	1.0	6	0.4	2.0	>.8	Random	500
7565	2.5	2	0.5	0.2	<.7	Random	400
7574	0.2	5	0.2	4	<.4	Random	500
7586	0.3	14	0.7	31.2	<.01	Contagious	600
7588	11.5	6	7.9	4.1	<.7	Random	500
7593	5.3	17	15.3	46.4	<.001	Contagious	400
7618	10.8	23	16.7	33	.05	Contagious	500
7622	0.2	6	0.2	5	<.5	Random	295–325
7625	19.0	24	35.3	43.7	<.01	Contagious	420
7627	4.3	4	72.3	51	<.001	Contagious	665–920
965	0.8	4	0.3	0.9	>.8	Random	300
4673	1.0	2	2	2	<.2	Random	1000–1065
4930	0.1	9	0.1	8	<.7	Random	600
4937	6.4	35	19.4	102	<.001	Contagious	435–475
4434	1.8	16	0.7	6	>.9	Regular	385
4667	0.1	21	0.1	18	.3	Random	2085–2125
4659	0.3	14	0.2	9.1	<.8	Random	2000
5795	1.3	4	0.3	0.6	<.9	Random	1520–1575
6240	0.1	13	0.1	12	<.5	Random	985
6243	0.5	4	0.3	2.1	<.5	Random	480
6245	1.5	13	1.4	10.8	<.7	Random	500
6256a	4.3	3	15.6	7.4	<.05	Contagious	500
6256b	0.6	18	0.9	25.5	<.1	Contagious	500
6260	0.1	19	0.1	18	<.5	Random	1000–1225
6865	1.3	8	0.3	1.4	>.9	Regular	480–510
5811	0.6	5	0.3	2.0	<.8	Random	1500–1540
1142	0.3	8	0.2	6.3	.5	Random	2000
4671	0.9	20	2.8	61.7	<.001	Contagious	1960–2140

displayed similarly unpredictable patterns. At the 28 locations where it was present, it appeared to be randomly dispersed at 17, contagious at nine, and regular at two (Table 11-2). The latter two locations could be a product of chance, but they were at moderately low densities. Nine contagious patterns out of 26 again suggests that the species has a tendency to clump; and again comparing densities of the random and contagious patterns, there appears to be a tendency to increase contagion as density increases. *H. artifex* occurs in extremely dense, often clumped populations in a narrow zone from about 350 to 600 m. Its abundance then decreases markedly and the animal disappears near 1000 to 1200 m. It recurs at six stations between 1500 and 2000 m. Of those deep populations, all but one appear to be randomly dispersed—the one exception seems to be contagious. We are aware of no relation between the density and clumping of *H. artifex* and that of *O. lymani*, where the animals occur together.

Another abundant species with a wide bathymetric range was *Parapagurus pilosimanus*, a hermit crab. Of the 27 locations where it occurred, it appeared to be regular at only two and contagious at only three (Table 11-3). There is little reason to

TABLE 11-3.   CONTAGION IN *PARAPAGURUS PILOSIMANUS*

Station	$\bar{X}$	N	V	ID	$P_{\chi}2$	Conclusion	Depth (m)
7524	0.4	8	0.3	4.9	<.7	Random	3200
7559	1.9	13	0.7	4.8	>.9	Regular	615
7570	0.5	2	0.5	1	.3	Random	270
7574	0.2	5	0.2	4	<.5	Random	500
7577	0.5	2	0.5	1	.3	Random	600
7580	2.1	12	0.8	4.4	>.9	Regular	700
7585	1.0	14	0.6	7.8	>.8	Random	690
7586	2.0	14	1.2	7.8	>.8	Random	600
7596	0.2	6	0.2	5	<.5	Random	315
7625	0.08	24	0.08	23	<.5	Random	420
7627a	0.3	4	0.3	3	<.5	Random	665–920
339	1.0	2	2.0	2	<.2	Random	760–800
341	1.8	6	3.4	9	<.2	Random	600
4930	0.6	9	0.5	7.2	<.7	Random	600
2493	0.3	3	0.3	2	<.5	Random	3134
6235	0.1	21	0.2	40	<.01	Contagious	2900
6243	0.3	4	0.3	3	<.5	Random	480
6245	0.2	13	0.1	10.8	<.7	Random	500
6270a	3.0	12	6.9	25.3	<.01	Contagious	520
6270b	0.1	15	0.1	14	<.5	Random	750
6841	1.4	5	1.3	3.6	<.5	Random	620
6847	1.6	5	0.8	2.0	<.7	Random	750–1040
5791	0.06	16	0.06	15	<.5	Random	1750
7616	0.1	21	0.1	18	<.7	Random	1655–2048
7553	0.04	24	0.04	23	<.5	Random	1926–1850
4671	0.05	20	0.05	19	<.5	Random	1960–2140
5493	0.3	6	0.6	10	<.1	Contagious	3600–3620

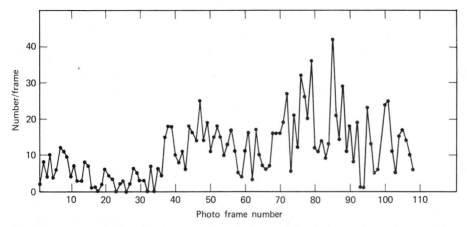

Figure 11-1. The quill worm *Hyalinoecia*, as found in the archibenthal zone along a horizontal line of depth off Peru.

suspect that these results were due to anything other than chance, and our conclusions regarding this species substantially parallel the others we have made, although densities are not nearly as high in this species.

The quill worm *Hyalinoecia* was similarly distributed along the single isobath of 1800 m off Peru. More than 100 photographs were taken over a horizontal distance of 3.24 km (Figure 11-1). The worm appeared in a band or ribbon no broader in depth than 20 m (1980 and 2000m). It was neither collected nor photographed above or photographed below that zone. Within that range, the abundance of the worm per frame varied from 0 to 42 specimens. Obviously, the worm was more abundant at some places than others and tended to clump over the horizontal distance (ID = 640 or "highly clumped"). In many respects, its distribution is quite comparable to that recorded for the closely related *Hyalinoecia artifex* off the Carolinas.

Each camera lowering off the Carolinas has been plotted (Figure 11-2), and analyses of the photographs have allowed the abundance of each species to be contoured between stations. This method allows the different populations to be compared visually with one another and with contours of ecological data.

SPECIES DIVERSITY

As early as 1962, (Clarke, 1962b; Barnard, et al., 1962) it was recognized that the deep-sea contains many more species of crustaceans and molluscs than was previously imagined. Menzies described more than 100 new species from trawls taken by the *Vema* in the Atlantic from depths greater than 2000 m. These taxonomic observations remained relatively unexploited. Meanwhile Sanders and his associates compared total numbers of individuals with numbers of species in samples from New England to Bermuda (Sanders, et al., 1965; Hessler and Sanders, 1967; Sanders, 1968; Sanders and Hessler, 1969) and concluded that deep-sea communities showed a greater species diversity than temperate shallow-water communities.

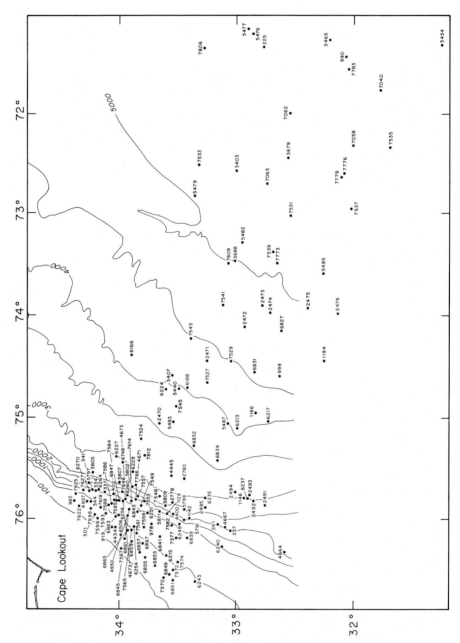

Figure 11-2. Camera lowerings off the Carolinas plotted with depth contours. Most are R/V *Eastward* stations.

Definition

Diversity, a community ecologic concept, refers to the heterogeneity (or lack of it) in a community or assemblage of different organisms. Most simply, it is the number of different species found under any given set of circumstances of time and space.

Evolution of the Concept

Fisher et al. (1943) drew attention to Gleason's (1922) observation that in most natural communities there is a linear relation between numbers of species and the logarithm of the number of individuals. In other words, as sample size is enlarged, there is only an arithmetic increase in species where individuals increase geometrically (spp./$\log^e$ individ.). Williams (1964) went on to summarize numerous natural assemblages, in addition to the insect faunas on which he originally worked, and included some of the *Challenger* deep-sea benthic stations. Although he said little about the causes of this hierarchy among species, the semilogarithmic relationship has been incorporated into almost all approaches to defining diversity, using the equation log $Y = a + X$, where the slope or rate of change in species versus the $\log_e$ of individuals is an "index" of diversity.

Diversity Indices

The major difficulty in quantifying diversity is that ecologists generally deal with samples rather than whole assemblages. For example whereas MacArthur (1965) could know the total number of species pairs of birds within a small area of forest, Margalef (1958) dealing with phytoplankton, or Williams with insects, had to be content with samples that were only presumed to represent the whole. As it happens, much of the discussion of diversity has dealt with finding "indices" that measure various aspects of the heterogeneity of an assemblage, regardless of sample size. Success has been meager, and those indices or graphical approaches which are significant to benthic studies have been considered in detail by Sanders (1968).

Dominance Versus Species Diversity

There are basically two types of diversity. One is simply the total number of species in a sample, and it is generally expressed as the fraction of species per individual. The second is concerned less with the total number of species than with the way the different species are distributed proportionally in the sample. The former, which we can call "species diversity," places as much importance on rare species as on the most abundant, whereas "dominance diversity" generally measures those common species which are numerically the most abundant.

Sanders (1968), in discussing the differences between these and listing references to approaches to their measure, demonstrated that there is, a consistent correlation between dominance and species diversity with small sample sizes (up to 400 individuals). As sample size increases, however, the correlation diminishes, and Sanders concluded, as had Whittaker earlier (1965), that the relationship is a weak one and sample-size dependent.

Rarefaction

The rarefaction method of Sanders (*op. cit.*) involves graphing numbers of species versus numbers of specimens from large benthic samples. Curves are derived by artificially rarefying actual, large abundant samples just as though the samples were smaller. The percentage of each species in the original sample contributes that percentage of individuals to the subsample, and this holds for each species at different sample sizes until a species becomes less than one. The remaining fraction, divided by the percentage the small sample is of the original, gives the additional number of species. The resulting graph, an interpolation from a real sample of thousands of individuals to a mathematically contrived subsample of none, has the shape of a hyperbola which approaches an asymptote as sample size increases. This artificial manipulation of species composition eliminates variations that would occur because of a normal patchiness in the fauna.

The Results of Sanders and Hessler

At small sample sizes, all the graphs from the Sanders–Hessler data are virtually inseparable, but as samples increase in size, the heights of the lines (i.e., the ratio of species to individuals) are grouped at various levels above the abscissa, depending on where the samples came from. Those from the deep- sea and the tropics are high above the abscissa, representing high diversity, whereas those from estuaries and the Atlantic boreal sublittoral shelf are low. The deep-sea samples were taken from depths greater than the outer margin of the shelf. There the seasonal temperature changes are greatly diminished or are no longer apparent, and this stability in time relative to other regions led Sanders (1968) to his "stability–time" hypothesis, a combination of several hypotheses suggested earlier (Klopfer, 1959; Fischer, 1960).

An alternative that Sanders and Hessler did not adequately reckon with is the probable effect of productivity or available food on diversity. MacArthur (1965) reasoned that the larger the food supply, the greater the number of species that can be accommodated in an environment. Hessler and Sanders (1967) pointed out that although their deep-sea diversity measurements were all high, they steadily dropped from the upper continental slope down to the deepest station at 4700 m. Animal densities can be predicted to drop from 10 to 100 times in this depth range, which means food supplies drop by at least the same amount (Rowe, 1972). This decrease in available food could be the cause for the decrease in diversity that was observed by Hessler and Sanders. In other words, productivity may relate to diversity, even though quite apparently biomass and diversity are not necessarily related (Chapter 10).

Information Theory

Margalef (1968) employed information theory (Shannon and Weaver, 1963) to measure diversity from samples from an assemblage of phytoplankton. Independent of sample size, the relationship is

$$D^1 = -p_i \log 2\, p_i \qquad p_i = \frac{n_i}{N}$$

where n is the number of species and p is the proportion of each species (1 to n) in the total sample, and N the total number of individuals in the sample.

This information function or "negentropy" (Odum et al., 1960) has achieved perhaps the widest usage of all indices because of the straightforward approach involved and the purported independence of sample size. It achieves independence because it is influenced more by the relative numbers of the most common species, although it is affected only slightly by the rarer ones.

Cautions in using the function have been pointed out by Sanders (1968) and Lie (1968). The data must be adequately represented in the sample, and for an ecological study we assume that sampling must always deal with the same types of faunas taken with the same types of sampling devices. It would lack meaning, for example, to compare the information in a Foraminifera assemblage with that in the polychaetes. An esturarine sample taken with a grab can be compared with those from the deep sea taken by an anchor dredge only if the dredge or grab did not selectively favor a particular type of animal. Restrictions of this type must be left to the investigator's insight. Unfortunately, statistical techniques to put confidence limits on "information" have not been used (Pianka, 1966) and to our knowledge are not available.

Sanders et al. (1965) have been able to separate the deep-sea fauna as a whole from shallow-water fauna on the basis of diversity, but as far as can be ascertained, they have not disclosed differences within the great depths. Sanders (1968), for example, refers to only seven samples from the deep sea. Of these, all but one (at 2500 m on the upper rise) are from the upper continental slope rather than from the abyss as we define it. We initially inferred from their works that diversity increases with increased depth, but this has not really been tested yet (Sanders, personal communication).

Except for the work of Buzas and Gibson (1969) with benthic Foraminifera, which showed a continual increase in information function with increased depth on the continental rise and abyssal plain off the eastern United States, trends in diversity at great depths have not been described. Theoretically, diversity gradients should be useful in delimiting ecological zones. Although Sanders et al. used polychaetes and bivalves (more than 70% of their fauna numerically), the big epibenthonts in the photographs can be assumed to make up the dominant proportion of the biomass or organic carbon. If diversity is at all related to organic production, the information in the dominant proportion by weight may be as important as that in the numerically dominant but much smaller species. We do not mean to imply that this information is superior, but rather that it deserves consideration because it may have meaning.

INFORMATION OFF NORTH CAROLINA

The values of bits of information (Information or H) versus depth for all stations were plotted (Figure 11-3) and no visible pattern emerged. Reasoning that small samples could give erroneous values, we eliminated all stations with less than 10 individuals, but still no trends could be seen. Sanders compared the Information Function with rarefaction by plotting Information (H) against number of individuals in his rarefied samples. The value of H increased sharply up to an approximate sample size of 50 to 100 individuals, where it reached a maximum and remained constant. For our photographic data, if all samples of less than 50 individuals are disregarded, there is a loose but linear increase in H with depth. Likewise, if all with less than 100 are omitted, the fit is enhanced further. We can conclude that when population densities of large invertebrates are great enough to allow photographing 100 or more individuals, the information function increases with depth.

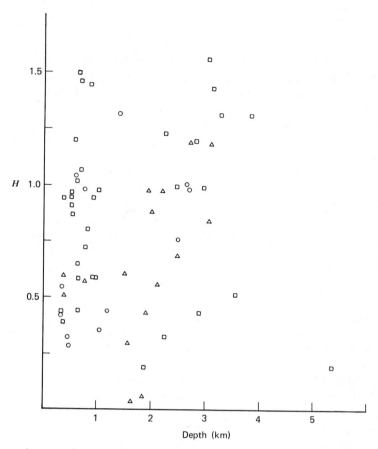

Figure 11-3. Information function H from camera stations off North Carolina versus depth: Squares, stations with less than 50 individuals; circles, stations with 50 to 100 individuals; triangles, stations with more than 100 individuals.

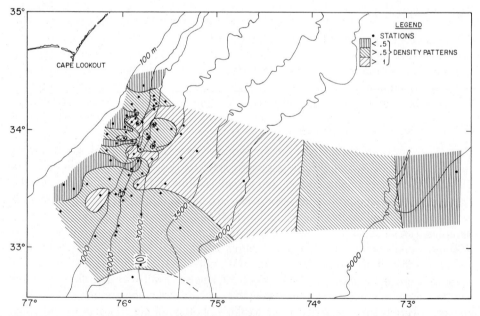

Figure 11-4. Values of H plotted as indicated for stations where calculation was practical.

274

Because emphasis here is on possible forms of zonation, contours were also drawn at the arbitrarily chosen 0.5 and 1.0 intervals (Figure 11-4) of H. There is considerable spatial variability, and it does not seem reasonable to make precise conclusions from the location of each contour. The lower values (0.5), however, tend to be distributed from shallow water into the depths, whereas the high values predominate in deep water. Intermediate values extend far seaward along the south but slope to the north at the slope base, where they are replaced by the region with the greatest information (1.0).

RELATIONSHIPS OF SAMPLE SIZES, CLUMPING, AND INFORMATION FUNCTION

With small samples, H can be overestimated between clumps of the common species if the rarest species happen to be included; on the other hand, diversity may be underestimated with subsamples, as in the other cases when rare species are not captured. In our use of the information function on photographic data, each station is subsampled by each photograph. With this approach, patchiness in relation to diversity can be assessed. Sander's rarefaction method, using only large samples, disregards distributions during the interpolations, so that patchiness present in actual small samples would be evened out by the artificial interpolation. Small samples, then, dealing for the most part with clumps, would always underestimate diversity. The proximity of this estimate, of course, would be related to the degree of patchiness and size of the subsamples. Likewise, H could be overestimated by taking subsamples between clumps that include rare species. It can be concluded that samples must be big enough to include "clumps" of rare species as well as common ones.

We must admit that although the indices reviewed appear to overcome the difficulty of varying sample sizes, they do nothing toward deciphering the relation of diversity to natural aggregations of species. Clumps, for example, are changing the physical environment for the rare species. Does this change in "habitat" in and out of clumps therefore force us to consider each a separate within-habitat diversity? If so, to combine them would be to consider interhabitat diversity.

Once sufficient data are available, we expect that the comparative diversity of genera and species can be used to discern the relative stabilities of wide geographic areas such as basins. For example, in our data, there were 12 genera of isopods in the Arctic versus 111 in the Antarctic. Dominance diversity, on the other hand, can indicate what has been happening in the more recent history of an environment, as in the diversity in and adjacent to a deep-sea trench or submarine canyon.

The use of a diversity index as an indication of stability has many inherent problems. First, it is unlikely that the deep sea has been a constant environment, and therefore we see little reason to believe that high deep-sea diversity is an index of environmental constancy, as Sanders has claimed. Lacking from any stability–diversity model are two factors which we call recruitment and death, or extinction.

In the shallow marine environment, life cycles are much shorter than in a rain forest, and recruitment and death must play a highly significant role in determining diversity at any given time or place. This is best shown in the temporary aquatic environment or in temporary tidal pools, where a rich and diverse population develops annually or seasonally and is entirely dependent in its structure on recruitment from nearby populations. No one would attribute a long term environmental stability to such communities, no matter how diversified they appeared to be at the time of sampling. During Pleistocene sea-level changes, the intertidal communities must have shifted in topologic situation from one sea level to another over short periods of time, each sequential stage being a function of recruitment from the next. This cannot be denied. Whether a similar process is occurring in deep-sea, level-bottom communities remains to be determined, but the possibility should be taken into account before any measure of high diversity is attributed to long-term stability.

At the outset we made reference to Thorson's view regarding "life form" instead of species as the crucial determinant of the composition of a marine community. To date, no one has approached diversity from the standpoint of "life forms," and it will be interesting to see whether the two approaches have any degree of similarity. The more the species of the deep sea are studied, the less reliable becomes the definition of a deep-sea species. Some "species" may be hybrids, as is commonly the case in deep-sea echinoderms according to Hyman (1955), and others may be mutants as a consequence of the high level of radiation in the deep-sea sediments (Menzies, 1965). It is axiomatic that the value of a diversity index of communities as an ecologic tool depends to a great extent on taxonomic knowledge, and unfortunately knowledge of the taxonomy for most animal groups in the deep sea is still in its infancy.

CHAPTER 12

Adaptive Morphology and Organic Complexity

About the beginning of the nineteenth century many distinguished men of science seem independently to have developed the idea that the structure of animals and their occurrence in various localities are determined by external conditions. Lamarck in his *Philosophie Zoologique* (1809) writes as follows: "The external conditions always and strongly exert their influence on all living beings. This influence is, however, difficult to ascertain because its effects only appear, and may be recognized, after a very long time." [Murray and Hjort, 1912, p. 660].

There is no a priori question that the adaptive morphology of a taxon has had a significant role in the selective success of an animal group in the deep sea. Feeding type, reproductive mode, eyes, pigmentation, size, and the like are all examples of adaptive morphology and show a relationship to abyssal living. An animal living in a particular habitat usually exhibits functional adaptation that may or may not be directly related to or associated with successful life in that environment.

The general studies that have been made suggest an absence of external and adaptive morphology among the deep-sea types, and especially among the Crustacea (Dahl, 1954). Only recently have examples of abyssal adaptive morphology come to light, and these involve internal and not external structures.

MOLLUSCA

The abyssal bivalve species *Abra profundorum* (Smith) (Figure 12-1) has been studied by Allen and Sanders (1966), who found that the abyssal species has the smallest gills, the largest palps, and the greatest hind-gut storage area. Those authors related this structure to abyssal habits of the species.

The ratio of gill area (G) to palp (P) times 100 was used as an index of measure.

Animal	G/P × 100	Remarks
Abra ovata (intertidal)	52.7	Largest gills, smallest palp
A. tenuis (intertidal)	36.9	
A. alba (11–1200 m)	16.7	
A. nitida (50–4500 m)	15.1	
A. profundorum (2861–5001 m)	10.4	Smallest gills, largest palp

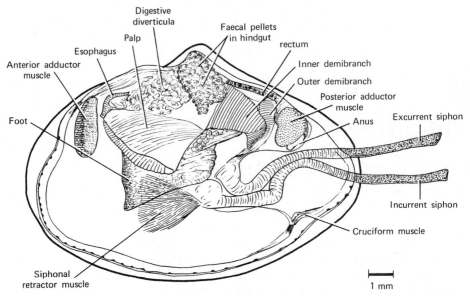

Figure 12-1. The gross anatomy of *Abra profundorum* as seen from the left side with the left shell valve removed. Note that this is drawn from a preserved specimen with the siphons partially retracted. (After Allen and Sanders, 1966.)

Thus as the palp area increased, the gill area decreased with increasing depth of habitat. The genus *Abra* belongs to the Scrobicularidae, a family whose reduction in gill size is correlated with life in mud substrates.

The increase in hind-gut storage area in *Abra* was associated with the needs to store organic matter (fecal matter) in the digestive tract when living in an environment low in organic supply and to extract nutrients from this storage supply through the aid of bacterial commensals.

Unfortunately Allen and Sanders did not specify the depth from which illustrated examples were taken in their studies, and this makes interpretation difficult. The comparisons of *Abra profundorum* (Smith) with *Scrobicularia plana* are not particularly revealing, because since they belong to different molluscan genera, they can hardly be considered to be closely related species. Generally the coiling of the gut increases with animal size in these bivalves, and there seem to be at least two strong internal morphologic changes associated with depth of habitat in *Abra*.

TABLE 12-1. COMPARISON OF RADULAR SPACE, LISTED AT 100% IN TWO SPECIES OF CHITONS AND THE PERCENTAGE OF THAT AREA OCCUPIED BY RADULAR MUSCLES, VESICLES, AND SHEATH IN THE INTERTIDAL AND DEEP-SEA CHITONS[*]

	Percentage			
Animal	Radular Space	Radular Muscle	Radular Vesicles	Radular Sheath
Ischnochiton (intertidal)	100	50	18	9
Lepidopleurus (*Velero IV*, Station 7234)	100	19	18	10

* Unpublished data, Layton and Menzies.

Polyplacophora

Menzies and Layton (unpublished data) have studied the radular morphology of a variety of shallow-water limpets, chitons, as well as their deep-sea counterparts. The radular musculature, the muscles that operate the radula, show a decrease in size with increasing depth in species of shallow water chiton *Ischnochiton* from California (intertidal), and deep-sea chiton *Lepidopleurus* from the wall of the Cedros Trench in Mexico (Table 12-1).

Monoplacophora

The "shallow-water" Cambro-Devonian genus *Pilina* is the presumed precursor of the modern genus *Neopilina*. The modern *Neopilina* have the same eight pairs of segmental foot muscles as the fossil *Pilina*, but the deep-sea *Neopilina* exhibit a great reduction in muscular development. The following tabulation compares the muscle scar area of the shell of *Pilina* and *Neopilina* from 6200 m:

Animal	Shell Area	Area Muscle Scars	Shell Covered by Muscles (%)
Pilina	100	1.66/13	12.5
Neopilina			
N. vemae	100	1.22/38	3.2
N. veleronis	100	.75/46	1.6

The weak development of the segmental muscles in *Neopilina* (*Neopilina*) *veleronis* Menzies and Layton is revealed in Figure 12-2.

ECHINODERMATA

Information about the echinoderms is extracted mainly from the monograph by Hyman (1955). Like other phyla, the echnioderms show scarcely any recognizable special external adaptation to abyssal life.

Crinoidea

The comatulids, which have ciri but no stalk characterize the shelf environment mainly in warm water in the Indo-Pacific. They have a lower bathymetric limit of 1500 m. The Millericrinida are stalked crinoids that lack ciri and contain most of the deep-sea species. It might be claimed that having a stalk would be detrimental in the active shallow waters. Here ciri could become important for attachment. It is difficult to see, however, why stalkless forms could not get along just as well and perhaps better than the stalked forms in deep water. Curiously, stalked millericrinids are in shallow water, but these have a heavy basal disc for attachment and none gets any deeper than 1700 m. (this depth is archibenthal in tropical seas).

The Bathycrinidae contain the deep-sea species, and these stalked forms have basal roots that entwine about foraminiferan skeletons, forming a netlike attachment that appears to be quite secure (Figure 1-8). The Bathycrinidae include *Rhizocrinus*,

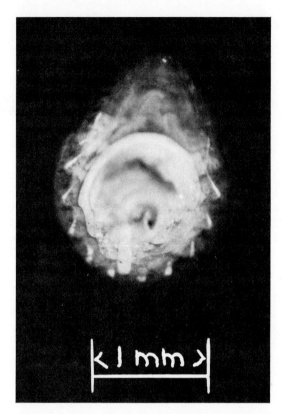

Figure 12-2. Photomicrograph of dorsal body surface of *Neopilina* (N.) *veleronis* with shell removed, showing weak segmental muscle development (Menzies and Layton, 1963).

Bathycrinus, Monachocrinus, and *Democrinus;* these genera are poorly defined, however, and species are shifted from one genus to another. The suborder Cyrtocrinida has short stalked forms. Members of one family, the Hyocrinidae, have a disc instead of roots, and at least one species (*Hyocrinus bethellianus*) penetrates as deep as 4936 m in the Antarctic. The other species are known from depths less than 2860 m.

Holothuroidea

The holothurians would seem to be remarkably adapted to deep-sea life. They are tubular and often burrowers which usually feed on the substratum, eating the oozes. Naturally the algae feeders such as *Synaptula* are unknown from the abysses, but more astonishingly, the sand burrowers such as *Synapta* and *Leptosynapta* are also absent from the deep sea. The order Dendrochirota, which in shallow waters has species that feed on plankton, contributes the Psolidae to the shelf depths and the genus *Sphaerothuria* (Figure 5-100) to the abyss. The order Aspidochirota has several genera that either are entirely deep-sea or have entered the abysses, such as:

Pseudostichopus (Figures 5-26*h*, 7-15*i*)	*Synallactes*
Bathyplotes	*Mesothuria*
Pelopatides	

The order Elasipoda is predominantly a deep-sea order. Most genera have species that walk over the mud and ingest it. They do not appear to burrow and hence eat only the surface sediments, spewing out behind their characteristic fecal masses. Many species in these genera are found in dense assemblages, where they likely constitute most of the biomass such as *Scotoplanes* (Figure 5-18*d*, 7-15*g*), *Psychropotes* (Figure 5-26*a*, 7-15*i*), *Peniagone* (Figure 5-27*a*–*c*, 7-15*b*), and *Elpidia*. Others are seldom seen and when visible are usually alone, such as *Benthodytes* (Figure 4-34), *Euphronides* (Figure 4-41), *Kolga*, or *Scotonassa*. None of the deep-sea species of holothurians, except for *Sphaerothuria* and *Molpadia*, burrow into the sediments, in contrast to popular belief illustrated by Wolff (1962) that they plow through the sediments. To judge from underwater photographs, they disturb the surface of the sediment very little, and the most one can see is a delicate trail and the markings left by the "feet." Actually, the species that lack the "feet" of the elasipodids make a more obvious track and have been as successful as the elasipodids in penetrating the deep sea. Deep-sea holothurians may be characterized as mud walkers or mud skimmers, and as far as we know few are infaunal. There appears to be no single external characteristic that distinguishes the deep-sea holothurians from their shallow-water counterparts, hence there is no obvious superficial deep-sea adaptation.

Asteroidea

The sea stars have deep-sea species that come from each of the four living orders. There is no single external morphological feature unique to the deep-sea forms. On the deep-sea floor most asteroids are quite widely dispersed from one another and are often seen only on the surface of the mud and not burrowing. One of the most abundant sea stars at the Milne-Edwards Deep was never seen in photographs; it was so commonly taken in trawls, however, that it seems likely that the animal burrows out of sight of the camera. One of the most successful deep-sea penetrants belongs to the Cribellosa. These lack both an intestine and an anus. But the Notomyota, which have the abyssal Benthopectinidae, have both anus and intestine and are common in the deep sea. The strange cribriform organ that characterizes the Porcellanasteridae might have some significance to deep-sea life, but in shallow-water species it seems to function mainly as a filter to keep large particles out.

Echinoidea

Two major groups of echinoids enter the deep sea. These are the Lepidocentroidea, which have a lantern chewing apparatus, and the Spatangoidea, which lack a lantern. The former tend to browse on the surface of the sediments, whereas the latter burrow into the sediments. Lepidocentroidea entering the deep sea belong to the family Echinothuridae, including:

Phormosoma	*Aerosoma*
Hydrosoma	*Hemiphormosoma*
Calverisoma	*Kamptosoma*

The sand dollars are absent from the deep sea, even though their ability to burrow might make them prime candidates for deep-sea life.

Cidaroids are mainly tropical shelf animals, but two Antarctic species are found at depths greater than 2725 m; and one of these, *Aporocidaris milleri,* ranges from Alaska to the Antarctic as a deep-sea endemic between 300 and 4000 m.

Most urchins are scavengers or carnivores. Cidaroids eat bryozoans, but *Brissopsis* and *Meoma* ingest sand. The food of sand dollars is not known. The Echinothuridae and the Aspidodiadematidae eat bottom ooze. Genera among the Spatangoidea that are either abyssal or wholly abyssal include

> *Urechinus*
> *Pilematechinus*
> *Calyme*
> *Pourtalesia*
> *Echinosigra*
> *Helgocystis*
> *cpatangocystis*
> *Ceratopyga*

No deep-sea species of echinoid is cosmopolitan. Many hybridize, moreover, making species difficult to characterize.

Ophiuroidea

No single morphological feature sets a deep-sea ophiuroid apart from a shallow-water one. Both the Ophiurae and the Eryalae have deep-sea species. Some deep-sea ophiurids burrow in the sediments like *Asteronyx,* whereas most browse over the bottom. Some of the commonest deep-sea ophiuroids are *Ophiomusium lymani* (700–4000 m) and *Amphiophiura bullata* (4800–5000 m). Some species occur in great number in one region, whereas others are quite solitary and widely dispersed. They appear to combine ingestion of bottom matter with a carnivorous habit.

From this survey of the echinoderms, one feature stands out rather boldly — namely, that the sand burrowers, such as *Synapta* and *Leptosynapta* in the Holothuroidea, the clypiasteroids of the Echinoidea (sand dollars), and the astropectinids of the Asteroidea, are generally confined to sandy shelf depths and do not enter the deep sea. Those that do enter the deep sea are either mud walkers or mud crawlers and are seldom infaunal. Apparently echinoids have entered the deep sea by first achieving the ability to live on or in mud, but not necessarily the ability to burrow.

The only deep-sea echinoderms to which a filter feeding habit can be assigned are the stalked Bathycrinidae. These also are the only ones showing a rootlike attachment to fasten in the mud.

CRUSTACEA (ISOPODA)

Zharkova (1966) reported on the relation between increase in size and depth of habitat as well as on cell-size increases in selected abyssal isopods. With *Haplomesus* and *Hydroniscus* she found that size increase was accompanied by an increase in cell diameter of the intestinal cells. With *Storthyngura* and *Eurycope,* cell size and cell diameter and depth were correlated for animal sizes of up to 12 mm. Larger animals in those genera showed an increase in number of cells as well as in cell size. Her data suggest a pressure cell-size and cell-number relationship which should be examined further, especially in light of the report by George and Menzies (1967) that large size is not a rule for abyssal species.

Among the Crustacea and probably also among other deep-sea organisms there occur a considerable number of morphological changes that probably have no relation to hydrostatic pressure and depth *per se* and are nonadaptive as far as we know today. They have in the past been claimed to have some relevance to depth.

Color

A loss of pigment is a common feature of deep-sea invertebrates. Archibenthal species are often variously pigmented from reds and blacks, but the abyssal forms usually are white or dull gray in color. The reason for lack of pigment may be related to one, all, or none of the following variables: the peculiarities of abyssal nutrition, hydrostatic pressure, and low temperature. At the moment, pigmentation is best left as a nonadaptive characteristic.

Eyes

The loss of eyes falls in the same category as loss of pigment (see also pp. 294–300).

Spines

A minimum development of body spines has been considered by one investigator (Wolff 1962) as a special character of trench fauna (isopods). This has since been shown to be incorrect (Menzies and George, 1967). Because spines and the absence of spines in the isopods are widely distributed traits, no special adaptive significance can be attached to them.

Long Appendages

Long appendages have been attributed to abyssal Crustacea as an adaptation to prevent sinking into soft oozes, but this peculiarity (if real) is mainly associated with bathypelagic genera such as *Munnopsis* and not with benthic species.

FORAMINIFERA

The abyssal foraminifera are characterized by a high proportion of arenaceous species. This relationship results mainly from increased solubility of carbonates at low temperatures and is not necessarily depth related. The foraminiferan communities cited by Thorson (1957b) for the high latitudes in shallow water are also quite typical of those deep-sea species having arenaceous and fibrous tests.

FEEDING AND TROPHIC RELATIONSHIPS

It has been shown by Sokolova (1959, 1966), Vinogradov (1961), Filatova and Zenkevitch (1960), Neiman (1966), and various other workers that the vast majority

of the abyssal invertebrates from below 3000 m are either detritophages or sestonophages, including a few carnivores but no herbivores. Sokolova (*op. cit.*)
utilized the weight of one trophic group to the total biomass to give the measure of
the biocoenosis. In this manner Sokolova and others (cited) identified the biocoenosis of the abyssal depths as *eutrophic* or *oligotrophic*. In the eutrophic biocoenosis the organic content of sediments is high, ranging from 0.25 to 1.5%
These conditions prevail near continents or under current systems of high productivity where detritus is abundant and detritophagous animals are dominant.

The oligotrophic biocoenosis is characterized by sediments with organic contents below 0.25%, containing little detritus. In these conditions, which usually
occur far at sea and at great depths, sestonophagous types seem to predominate.

Sokolova's (1959) classic study on the trophic distribution of communities along
the floor of the Kurile–Kamchatka Trench (Figure 12-3) revealed a repetition of
feeding type independent of depth. The animal types utilized by Sokolova are
generalizations that require verification.

There appears to be some continued attraction to the concept of ladder migrations of zooplankton as provided first by Vinogradov (1961) as a means of feeding the
abyssal benthos (Figure 12-4). Zooplankton biomass at abyssal depths is measured in
milligrams per cubic meter (Table 12-2), whereas the biomass of the benthos is
measured in grams per square meter. With this kind of trophic pyramid it seems more
likely that the migrants in the ladder near the lowest rung depend on the abyssal
benthonts for food, and not vice versa. This has been shown by Barnard (1962) for

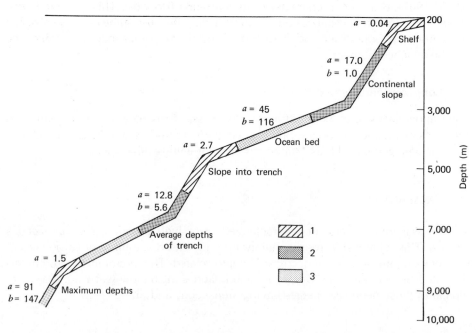

Figure 12-3. Correlation between benthos groups and bottom topography of the ocean: 1, Sestonophage
zone; 2, zone of a considerable development of all three feeding groups; 3, zone of development of detritus feeders, either only roughly sorting the soil or swallowing it whole — a, ratio by weight of detritus
feeders to carnivores, b, ratio of detritophages to carnivores. (Sokolova, 1959; see Zenkevitch, 1963).

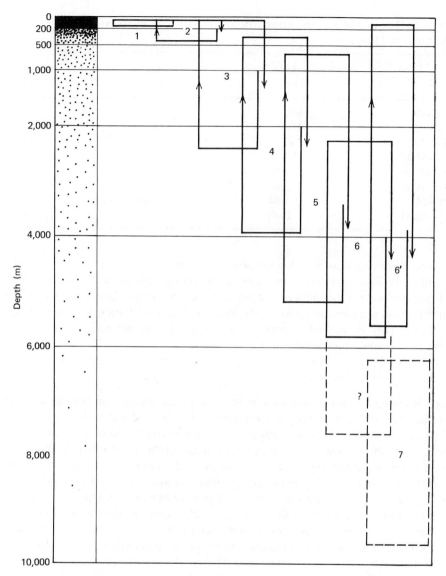

Figure 12-4. Scheme of vertical migrations of the deep-sea plankton: 1, Migrations of the surface species; 2, migrations extending over the surface and transition zone; 3, migrations extending over the surface, transition, and upper layers' deep-sea zones; 4, migrations extending over transition and part of the deep-sea zone; 5, migrations within the whole deep-sea zone; 6, regular migrations of some species extending through the whole water column; 7, range of distribution of ultra-abyssal animals. To the left: variations of the plankton abundance with increasing depth (frequency of points in each layer is proportional to the biomass of the plankton). (Vinogradov, 1961.)

certain amphipods. Menzies (1962c), who reviewed the sources of food (organic matter) for abyssal invertebrates, concluded that primary production in surface waters is probably less important than "fall-out" from the continental shoals and shelves through the action of turbidity currents. As has been seen, the highest concentrations of abyssal benthos occurs along the margins of continents and the least away from continental influence.

TABLE 12-2.   VERTICAL DISTRIBUTION OF PLANKTON BIOMASS (mg/m^3),
ZENKEVITCH (1963)

Depth (m)	Kurile–Kamchatka Trench	Mariana Trench	Bougainville Trench
0–50	508	24.0	127
50–100	376	14.9	107
100–200	288	10.9	32.8
200–500	59.3	2.1	9.4
1000–2000	21.8	1.0	2.4
4000–6000	2.64	—	0.09
4000–8000	1.84	0.012	—
6000–8000	0.48	—	0.01

The demonstration of a higher abyssal benthic biomass under the equatorial current of the Pacific at depths in excess of 2000 m (Filatova and Zenkevitch, 1960) is strong evidence of the influence of plankton rain on benthic biomass (Figure 10-1). In this case all samples are far from any influence from land; thus only the impact of the productivity of upwelled water in the tropical currents can account for the increase in benthic production on the sea floor.

Exceptions to the Rule

There is strong evidence that certain bryozoans (catalogued by Sokolova as ses-tonophages) feed directly on the sediments of the abyssal sea floor. For example, *Kinetoskias* is a bushy bryozoan on a hyaline stalk which is imbedded in the sediments. This animal shows zoaria touching and disturbing the sea bed in what we interpret as a feeding response (Figure 5-26e). A similar but more vigorous behavior is observed on the part of the bryozoan *Farciminaria* (or *Levensella*) in the Antarctic. These are only two examples of a classic filter-feeding type of organism which in the abyss most probably feeds on the sediments rather than by filtering seawater, as do the shallow-water bryozoans. Except for the relatively few sestonophages in the deep sea, the great bulk of the abyssal fauna is made up of animals that live in and on mud, eating it as well. Thus the sediment and its organic constituents must play a strong role in the development and structure of abyssal communities.

Importance of Bottom Sediments

As was shown by Thorson (1957), the bottom sediments are highly significant in the development of communities. Sanders (1956, 1958) found little correlation between mean particle size and mean species distribution. Purdy (1964) reworked Sander's data from Long Island Sound and Buzzards Bay by plotting the percentage of deposit feeders against the percentage of clay content. He found a good rela-tionship—namely, that the percentage of deposit feeders increased as the clay content increased (Figure 12-5a) and the percentage of suspension feeders decreased with an increase in clay content (Figure 12-5b). Purdy (1964) then related the increase in deposit feeders to the particle size and to the amount of adsorbed organic matter.

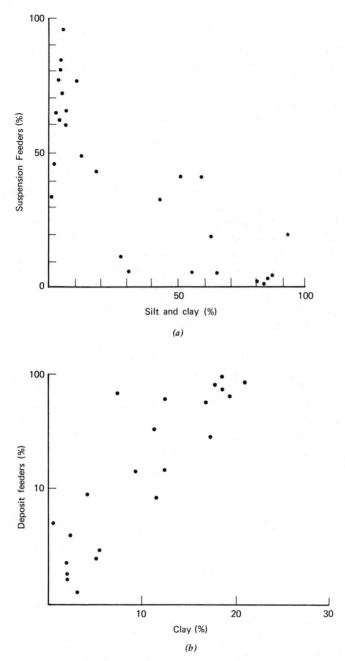

Figure 12-5. (a) Correlation between the proportion of deposit feeders in the bottom fauna (infauna) and the clay content of the substrate (data from Sanders, 1956, 1958) (from Purdy, 1964; $r_s = +0.85$); (b) relation between the proportion of suspension feeders (infauna) comprising the bottom fauna and the silt and clay content of the substrate (data from Sanders, 1956, 1958) (from Purdy, 1964).

The findings of Sanders (*op. cit.*) as interpreted by Purdy (*op. cit.*) do not fit the picture of oligotrophic abyssal distribution demonstrated by Sokolova, Filatova, Zenkevitch, and others because the increase in sestonophages in the red clays of the abyss is contradictory to the following statement by Purdy:

> As the texture of sediment becomes finer the proportion of suspension feeders decreases and the abundance of deposit feeders increases until finally the deposit feeders are more numerous than the suspension feeders.

Red clay is after all one of the finest sediments known. Doubtless what Purdy states is generally true for shallow water, and it is only likely that red clays are outside the limits of his observation where other forces come into play.

Menzies (1962c) showed that abyssal isopods could not be assumed to have entirely one type of feeding or another. Even so, 93% of the 35 species were deposit feeders with or without added carnivorous activity. The deposit feeders were divided almost equally between those with selectivity for uniform particle size and those consuming particles of differing size. Wherever hard substrates (inexplicably, manganese nodules are the exception) are available, there appears a rather flourishing amount of epifaunal organisms that are largely sestonophages.

BREEDING IN DEEP–SEA BENTHOS

Deep-sea animals must breed for survival and the perpetuation of the species through reproductively accommodating to the special environmental conditions in the abyss. The absence of sunlight and consequent loss of primary autotrophic organic production has resulted in a general deficiency in the nutrients in the benthic deep-sea environment. Highly restricted food supply and lack of seasonal oscillations of physical characteristics in the deep-sea impose certain demands on reproduction not encountered in surface or shoal-water environments.

The most obvious departure is the poor representation in the abyss of species producing pelagic larvae. The dominant nature of nonpelagic development in the deep sea was originally pointed out by Wyville Thomson (1874). A comparison of the life history and breeding patterns of deep-sea invertebrates with those of the Arctic and the Antarctic was made by Thorson (1950). The three most common modes of reproduction among shallow marine species are as follows:

1. Vegetative mode or asexual reproduction by fission, laceration, or budding. This is common among protozoa, flatworms, polychaetes, and nemertines.
2. Direct development or viviparity in which fertilization is internal and larvae are incubated in the female brood pouch or marsupium to a fairly well-developed juvenile stage. This is common among the Peracarid Crustacea.
3. Pelagic larvae. This mode of reproduction is the usual and most frequent pattern of breeding among shallow marine animal communities. The species often are dioeceous. During the spawning period the females in breeding phase discharge numerous eggs which are externally fertilized by the sperm discharged by the males into the ambient water. The fertilized eggs develop into larvae which lead a free-swimming life for a certain period and then metamorphose and grow into adults. The three common types of pelagic larvae known are: (a) lecithotrophic larvae from large, yolky eggs, which develop independently of planktonic food sources (this type constitutes approximately 10% among shallow communities);

(b) planktotrophic larvae with short pelagic life, constituting about 5% in shallow communities; and (c) planktotrophic larvae with long pelagic life, constituting 85% of the fauna of shallow water communities. Long pelagic larval development can be supported in shallow depths in the uppermost water layer, where phytoplankton form an abundant source of food for the swimming larvae.

Knowledge of the reproduction of abyssal species is fragmentary. Polychaetes represent one of the dominant groups in the benthic deep-sea environment. Several species of tubiculous worms dwell in abyssal depths, but very little is known of their reproduction and development. Hartman (1967) reported brood protection in two Antarctic deep-water species of Onuphidae, *Nothria notialis* and *Paronuphis antarcticus*. These two worms exhibit a highly evolved pattern of incubation of young.

Nothria notialis constructs long, cylindrical tubes that are internally chitinized and externally coated with silt. The tube of the female shows six to fourteen lateral capsules, in alternate series. In each capsule the worm deposits two ova, which develop simultaneously into larvae. The capsules at the proximal end of the tube contain the oldest and the most developed larvae, having a maximum of 24 parapodial segments and a well-developed cephalic region. Successive capsules harbor younger larvae and the distal-most have undifferentiated yolky masses. Each capsule is separated from the parent tube by a weblike membrane, which evidently prevents the entry of larvae into the parent tube but may permit the inflow of nutrients and water transport. This mode of incubation keeps the larvae from all external adversities including food scarcity and predation. The oldest larvae, which emerge from the proximal capsules, resemble the adults except in body dimension and number of segments. They all have well-developed pharyngeal armature and alimentary tract. A similarly advanced pattern of brood care occurs in *Paronuphis antarcticus*.

Another tube-dwelling onuphid polychaete *Hyalinoecia* sp. is known from the archibenthal zone of the Atlantic and Pacific. Hartman (1967) compared the abundance of this quillworm in deep water with *Onuphis teres*, which is a giant worm attaining a length of 200 cm. There is no published information on the mode of reproduction of these aggregating giant polychaetes. Generally polychaetes are labile in their mode of reproduction and development, and they exhibit, besides the asexual mode, several sexual reproductive types. Viviparity and brood protection are common. They have pelagic as well as nonpelagic larvae.

There seem to be many exceptions to the generalization that the deep-sea environment is devoid of larvae. There have been reports of free-swimming larvae from the ocean surface of the brachiopod *Pelagodiscus* in which the adults live as deep as 2,490 m in the Indian Ocean (*Valdivia* Expedition Station 238; Helmcke, 1940) and at greater depths in the Antarctic (3397 m, Gauss Station).

Ashworth (1915) reported the occurrence of free-swimming larvae "probably referable to the species *Pelagodiscus atlanticus*" in the plankton off South Brazil in the Atlantic; near Misaki, Japan, in the Pacific; and a few miles off Cape Comorin, Indian Ocean. The adults are known from all major oceans with a wide bathymetric distribution from 400 m to the abyssal plain at 5400 m (Odhner, 1966).

Some authors have discussed the possibility of vertical currents aiding the migration of pelagic larvae between the surface and the bottom (Mortensen, 1921). It seems likely that certain migratory species, such as the fish *Reinhardtius hippoglossoides* and the Antarctic krill *Euphausia superba*, are capable of spawning in selected areas in the deep water where vertical currents exist. In the North Atlantic we have photographed and collected the barnacle *Scalpellum* at depths as great as 5300 m. Assuming that these deep-sea species produce planktotrophic larvae having a prolonged pelagic life, the larvae would be required to swim upward to 5 mi across the water column to reach the photic zone; they would also have to take a descending journey to the deep ocean floor for permanent settlement, "a hopeless task," as Thorson had stated. Nevertheless, the larvae succeed in this "hopeless" undertaking.

The breeding behavior of echinoderms ranges from free-swimming larvae to viviparity. No one kind is dominant among deep-sea species. Both cold-water and deep-sea asteroids tend to brood their young and to have large yolky eggs. Many species of ophiuroids are hermaphroditic, and 25% show protandry. Of the 53 species that are known to brood their young, half come from polar regions. True viviparity is known only from one Antarctic species. Among the Holothuroidea, the habit of brooding the young dominates in cold-water species as well as deep-sea species, where the young are carried in external pockets of the skin.

Curiously, the epitome of brooding in which the young develop in the coelom is restricted to shallow-water tropical species. The most specialized brooding of young is carried out by the shallow-water Antarctic species *Taenigyrus contortus* in which the ovaries become modified as brood areas. A number of echinoids, cidaroids, and spatangoids brood their young, and all members of the shallow-water Antarctic genera *Abatus, Amphineustes,* and *Tripylus* brood their young.

The reproduction of deep-sea fishes has been studied. Rass (1941) reported that the Arctic species *Gadus (Eleginus) navaga* produced a few large, demersal eggs instead of small and numerous pelagic eggs. Aside from a few notable exceptions such as the Antarctic deep-water fish *Pleuragramma antarcticum,* pelagic eggs are rarely known for polar and deep-sea fishes. Demersal eggs are laid by almost all species of the families Cottidae, Zoarcidae, Liparidae, Cyclopteridae, Anarrhicharchidae, and Agonidae. Some Antarctic species (e.g., *Notothenia simla*) show parental care in which the parent coils around the eggs.

Marshall (1953) discussed the ecological significance of large eggs and larvae in polar and deep-sea fishes. There are definite advantages for bottom-dwelling fishes to produce demersal eggs large in size and few in number, which eventually hatch into larvae of fairly advanced stage. This affords greater chances of survival and the least intraspecific competition for food. In the abyss, where food scarcity is often the rule, brood protection, viviparity, and abbreviated larval life may be of assistance in the propagation of species.

The most comprehensive account on the reproduction of major deep-sea fish groups was the review of Mead et al. (1964), which contained examples from deep-sea species living as adults below 2000 m. The class Chondrichthyes is represented in the deep sea below 2000 m by about 15 species of sharks, skates, rays, and chimaeras. But these abyssal species do not differ reproductively in any way from their epipelagic or coastal relatives. In all members of this class, fertilization is internal and development is oviparous, ovoviviparous, or viviparous. One interesting aspect is the question of how dioecious fish species in the dark abyss locate their mates to insure fertilization of gametes. Mead et al. (1964) suggested that in the two most speciose and abundant groups of abyssal benthic fishes, the Macrouridae and Brotulidae, sound production and reception play an important role in sexual attraction, thereby aiding in the formation of aggregations. Bioluminescence is another important mechanism, and some elasmobranchs, stomiatoids, and the clupeoid fishes—searsiids as well as some alepocephyalids—display luminescence that is often sexually dimorphic.

Fish species of the family Brotulidae are chiefly benthic deep-sea inhabitants; the viviparous brotulids are barely represented below 2000 m, and oviparous members are more common in the deep depths. There is no information on the young stages of any benthic brotulid. In these dioecious demersal fishes, mate location is accomplished by acoustical means. The family Zoarcidae includes about 150 species, they are generally found in the cold, high-latitude polar and temperate regions of both hemispheres, although they also occur at great depths throughout the world oceans. The zoarcids contain both oviparous and viviparous species; eggs of deep-sea forms are very large and few. This low fecundity of zoarcids suggests that some form of parental care may take place.

The order Iniomi includes an array of benthic abyssal species belonging to the families Bathypteroidae, Bathysauridae, and Ipnopidae, known to occur mainly below 2000 m. Many of these monoecious species are known to be hermaphroditic. In the absence of a mate, these deep-sea Iniomi are capable of self-fertilization and produce a relatively small number of eggs, which develop on or near the bottom. Several benthic eels of the order Apodes are represented in the abyssal depths, especially species of the genus *Synaphobranchus*. One factor that reduces the problem of mate location among deep-sea Apodes is their tendency to form relatively dense populations. Furthermore, the eels tend to spawn in selected areas and during particular seasons. The youngest larvae of most North Atlantic species are confined to, or are most abundant in, the Sargasso area, whereas the older leptocephali and adults are much more widely distributed. This evidence suggests that distant spawning migrations take place, and the occurrence of young larvae indicates that duration of spawning season of each species may be a few months, varying in time of year from species to species.

The olfactory organs are large in the benthic deep-sea eels, and their sense of smell appears to be rather sensitive, as evidenced from experiments with *Anguilla* (Mead et al., 1964). No luminescence or other special adaptation to species recognition has been reported. The deep-sea eels are dioecious and viviparous, like all other eels. All known eggs are epipelagic. The epipelagic larval stage is of long duration and the leptocephali reach a considerable size during metamorphosing stages before descending to deeper depths.

Adult specimens of approximately 100 known species of deep-sea angler fishes have been trawled from depths between 1000 and 3000 m. These ceratioids are unique in their extreme sexual dimorphism in which the dwarfed males are permanently attached to and parasitic on the large female. This is indeed an adaptation to the living conditions in the abyss. The male obtains as a parasite its nutriment from the blood of the female, and the single function of the male is to produce semen. This practice makes the female appear as a kind of self-fertilizing hermaphrodite (Bertelson, 1951). A female ceratioid *Ceratias holboelli* 65 cm long, with attached ripe male, was found to contain nearly 5 million eggs in the ovary (an egg being ca. 0.3 mm in size). Ovoviviparity or viviparity seems to be unlikely in ceratioids, and no information suggesting internal fertilization is reported. The ceratioids pass their larval life in epipelagic waters and the young ceratioids sink to about 3000 m during metamorphosis. The species are generally bathypelagic and solitary living.

In general it appears that the deep-sea fishes tend toward one of three available options: (a) production of many small eggs that develop in productive surface waters, (b) production of few but large eggs that develop into advanced young in the deep sea, or (c) viviparity.

Breeding Cycles in Deep-Sea Environment

In general, the abyssal benthic environment is characterized by stable uniform physical features with hardly any noticeable seasonal changes. However, the fauna inhabiting this uniform deep-sea environment tends to show definite reproductive periodism, as evidenced from the data on abyssal isopods (George and Menzies 1967, 1968a) and echinoderms (Schoener 1968).

Deep-sea benthic samples were taken in different months of the year from the same general area off Antarctica by the R/V *Eltanin*. The Scotia Sea was subjected to intensive investigation between 1961 and 1964, and more than 100 trawls were hauled from the bottom of the sea. Several hundred isopods were captured, but in the collections there persisted a definite paucity of specimens in reproductive phase. Samples taken from July to November yielded more females in breeding stages, suggesting cyclic reproductive activity. Data on six sub-Antarctic species of the typical abyssal genus *Storthyngura* clearly demonstrated the existence of seasonal breeding activity in deep-sea organisms (Figure 12-6) (George and Menzies 1967). This finding raises the question of how, in constant physical environment such as the deep sea, reproductive rhythmicity prevails as a seasonal biological phenomenon.

Schoener (1968) ascertained that some abyssal brittle stars show a period of increased reproductive activity during the summer months in the northwestern Atlantic. In winter, samples of ophiuroids were found to have well-developed gonads and eggs, whereas in May this was not the case. Although Schoener had inferred from the data on *Ophiura ljungmani* that reproduction is not completely halted during any part of the year, there seems to be an increase in gonad development and presumably in reproduction during the late winter.

Schoener also examined the population structure of these two ophiuroid species during the year and found very similar reproductive trend and size-frequency histograms at depths between 1100 and 2200 m. In July and August there is a distinct increase in the smaller size (0.5 mm). In December the gravidity is low, as it was in May. If an annual reproductive cycle is assumed, it would follow that large population increases for *Ophiomusium lymani* occur around July and August, whereas in other months of the year individuals increase in size and decrease in abundance.

Further evidence for seasonal breeding cycles in deep-sea environments came from the intensive collections along the continental slope and the Hatteras abyssal plain of the North Atlantic off the Carolinas (George and Menzies, 1968a). Among the 134 benthic trawls made from R/V *Eastward*, only 13 contained female isopods in reproductive stages. More than 90% of the collections with gravid females were obtained after the end of the surface summer season, from August to November. There appears to be a pronounced increase in the reproductive activity of isopods in the month of November. It is striking to note that 14 of 15 female specimens collected in five stations in November were found to be in the reproductive phase. In striking contrast, 152 isopods from a single abyssal trawl in March contained 74 females; but none was gravid, none had an empty brood pouch, and none developed oostegites (Figure 12-6b).

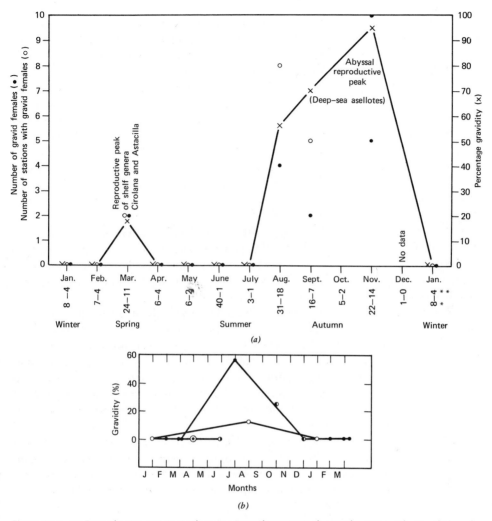

Figure 12-6. (a) Reproductive activity in deep-sea isopod species in the Northwestern Atlantic; (b) breeding activity in sub-Antarctic abyssal species of *Storthyngura* (Crustacea, Isopoda): ●, *S. birsteini*; ◑, *S. scotia*; ○, *S. robustissima*; ⊙, *S. eltaniae* and *S. sepigia*, and ◑, *S. triplispinosa*. (After George and Menzies, 1967, 1968a) * successful trawls, ** with isopods.

Earlier studies (George and Menzies 1967) on Antarctic deep-sea species indicated that the "abyssal reproductive peak" reached the maximum (57%) in July and declined to 25% in October. However, the duration of the reproductive period falls between July and October at the southern high latitude (60°S) and between August and November at 34°N.

Although there is a slight shift of one month ahead in the onset of breeding in the Antarctic abyss, it is interesting to note that the breeding period for abyssal isopods in both deep-sea regions seems to endure for four months during the same part of the year (July–November). It is quite possible that the typical deep-sea isopods, mostly asellotes, are not summer breeders. From the information, we may suppose that breeding activity in selected deep-sea animals decreases or ceases in summer months and increases in the fall and winter seasons of the year. Such a generalization may not apply to all abyssal animals. It might be possible that some abyssal groups or species propagate the whole year. Orton's view (1920) of continuous breeding as a general rule for all animals inhabiting stenothermal regions, such as the polar seas, the tropics, and the deep-sea environment, may not be entirely correct.

Another interesting observation on reproductive periodism in deep-sea animals is related to breeding trends exhibited by species belonging to typical shallow-water genera. To this category belong usually nonasellote isopod genera such as *Cirolana, Rocinella, Astacilla, Serolis,* and *Arcturus*. The abyssal reproductive peaks of species belonging to two such shallow water genera, *Cirolana* (590 m) and *Astacilla* (3100 m) are out of phase with the typical deep-sea reproductive cycle. These breed in March, unlike the species of asellote deep-sea genera. At least two kinds of breeding cycles are in operation in the deep-sea environment—a fall—winter cycle for true deep-sea species, and a spring cycle for the more typically shallow-water genera now living in the archibenthal zone and the deep sea.

The apparent relationship of increased reproductive activity seen in the same part of the year between two latitudinally widely separated deep-sea regions suggests that biological phenomena such as reproductive periodism will operate in the same period for any given group in the abyss, everywhere in the world oceans, and irrespective of latitude.

The cause of reproductive periodicity in the physically uniform deep sea is yet unanswered, and only speculations can offer possible reasons for the cyclic trend. That endogenous solar and lunar rhythmicity persists in organisms even under constant laboratory conditions is amply supported by recent evidence. It is perhaps reasonable to suppose that deep-sea animals show persistent reproductive cycles in the absence of seasonal environmental changes that might reflect their early origin from shallow-water organisms. An alternate explanation is that undetectable periodic nutrient energy enters the abyss by way of inflow.

Deep-sea endemic genera could well have been derived from Antarctic shelf hypopsychral species where the winter months of the year (northern hemisphere) are summer months of temperature (southern hemisphere). In this case they would show a northern hemisphere winter increase in production corresponding to the Antarctic summer, as seems to be the case.

VISION AND LIGHT

The fauna in the benthic abyssal province is almost uniformly without eyes and blind. There are few exceptions to this rule, and any exceptions that do exist are all related to cases of eurybathial species that come from shallow water in the Antarctic or the Arctic.

TABLE 12-3. EXAMPLES OF INTERTIDAL ISOPODS
WITH COMPLETE OR NEARLY COMPLETE EYE-LOSS

Interstitial	*Caecianiropsis* Menzies and Pettit
Commensal	*Caecijaera* Menzies
Parasitic	Bopyroidea in general, males and females
Burrowing	*Limnoria*

The shallow waters of the intertidal and shelf provinces are not without examples of species that show degeneration or loss of eyes. This occurs among the interstitial fauna, commensals, parasites, and burrowing species (Table 12-3). On land the cave fauna is parallel to the abyssal fauna in eye degeneration and eye loss.

In the sunlit parts of the ocean, where vision can play a part in survival, eyes are present (98–99% level) in most of the species. In the abyssal province, eyeless species predominate at a level of 98 to 99%, and only an occasional species bears well-developed eyes.

It is exceedingly difficult to rationalize this picture with reference to the evolution and development of abyssal species. The possibility of gene control related to nutrition and light could be examined in relation to melanin formation and rho-

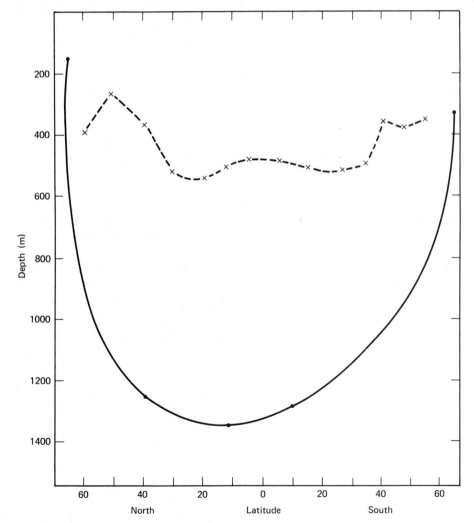

Figure 12-7. The 50% level of fauna bearing eyes (Ol$_{50}$) for marine isopod species in relation to depth and latitude and light extinction curve with circles, Ol$_{50}$; curve with crosses, light extinction coefficient. (After Menzies et al., 1968, p. 96.)

dopsin chemistry, because in every known instance, loss of pigment correlates very well with eye loss.

A determination of the proportion of the isopod fauna with and without eyes, at increasing depths at various latitudes, was made by Menzies, George, and Rowe (1968). In one example of tropical submergence, it was discovered that the depth at which half the fauna did and half did not have eyes increased as low latitudes were reached (Figure 12-7). This submergence generally followed the curve for light penetration, but the depth at which no animals had eyes was considerably below the depth of light penetration.

The amount of visible light and the degree of development of eyes in a fauna are closely related. This phenomenon applies also to other groups of animals, such as the bathypelagic crustaceans.

Eye loss is not related to a definite taxanomic level. Certain marine genera are eyeless. Eye loss appears to occur independently in various general almost as though simple disuse is a cause of eye deterioration and loss. The blind species living in total darkness would be at an advantage over an eyed species if other sensoria showed compensatory development, but there is no evidence that this has happened. The principal conclusion that might be justified is that vision appears to play no significant role in the lives of deep-sea marine isopod crustaceans as far as their survival is concerned.

Complete pigment loss is not always related to lens loss because species without eye pigment retain their lenses (Figure 12-8), whereas others show considerable lens reduction in the presence of eye pigment (yellow).

Abyssal Eye-Shine

Abyssal eye-shine, the reflection of light back from the eye, is known to occur among the abyssal shrimps and the abyssal serolid isopods. These examples demonstrate the presence of functional eyes at great depths (Figure 12-9).

The external expression of eye deterioration in the genus *Serolis* is shown in lenses of the compound eyes of species from different depths of the sub-Antarctic and the Antarctic. Here the loss of eye pigment or change in eye color precedes the loss of lenses.

Province	Eye Color
Shelf	Black
	Black
Archibenthal	Black
	Red
	Pink-blue
Abyssal	Black
	Pink-blue
	Pink
	Golden yellow
Endemic abyssal	Colorless

Serolis nearaea Beddard, is distributed around Antarctica between 239 and 3794 m depth. The species is eurybathial but stenothermic. It has well-developed

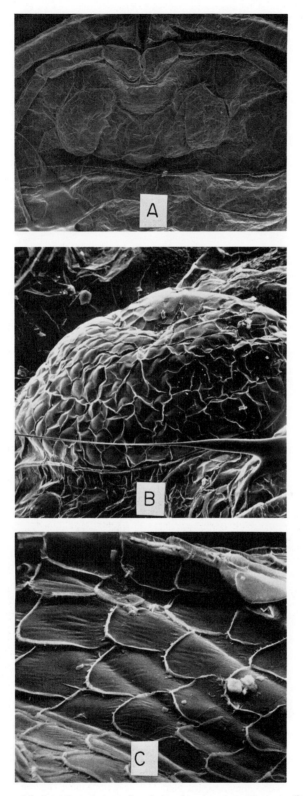

Figure 12-8. *Serolis* sp., blind: (a) Dorsal view of cephalon showing eyes, 20× magnification; (b) eye at 100× magnification; (c) scales on antenna at 1000× showing their similarity to "lenses" of eye. Photos courtesy of the Biology Department, Florida State University; Scanning Electron Microscope Laboratory.

297

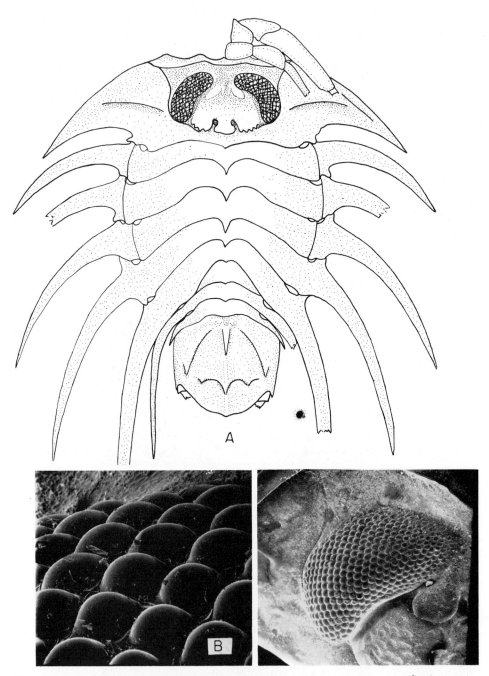

Figure 12-9. *Serolis nearaea* Beddard: (a) dorsal view of animal, (b) lenses at 180 × magnification; (c) eye dorsal view, 20 × magnification; (d)–(g) *Serolis bromleyana* Suhm: (d) dorsal view of animal, (e) posterior edge of eye, (f) enlargement of posterior part of eye showing loss of facets, 50 × magnification, (g) enlargement of dorsal part of eye showing reduction in lens height, 550 × magnification. Photos (b), (c), (e)–(g) courtesy of Biology Department, Florida State University, Scanning Electron Microscope Laboratory.

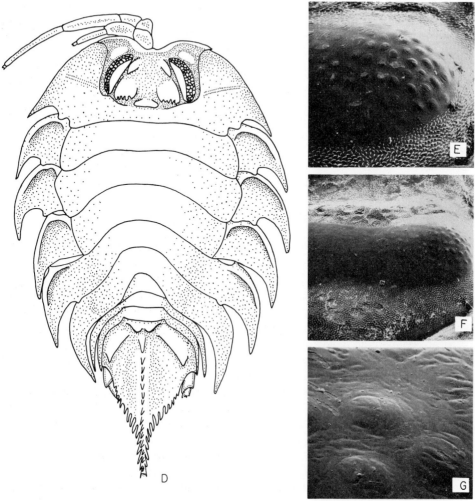

Figure 12-9. (*Continued*)

ocular facets and lenses, but the eye color is pinkish blue. Figure 7-14*d* shows the animal on the abyssal sea floor at 3020 m depth, where it lives in close contact with the surface sediments.

Serolis bromleyana Suhm, like *S. nearaea,* is a eurybathial stenothermic animal with a depth range between 744 and 4886 m. Unlike the latter, however, it shows a deterioration of the lenses of the eye even though the eyes are yellow. Specimens showing eye-shine have been photographed at a depth of 4000 m, Figure 12-9.

Serolis sp. is a stenobathial, stenothermic abyssal endemic. Under low power of the dissecting microscope the eyes did not become visible and the head was colorless. But under the scanning electron microscope it was evident that eye lobes were present, although the lenses had deteriorated to little more than the scales around the head. This specimen was from a depth of 4643 m in the Antarctic, Figure 12-8.

In addition to examples of the Serolidae showing progressive eye deterioration with increasing depth of dwelling, some excellent examples are found in the

Cirolanidae and the Arcturidae (Beddard, 1886). With only a few exceptions, both families in shallow water have multifaceted black eyes and in deep water have pink, yellow, or colorless eyes, with or without eye facets.

ORGANIC COMPLEXITY

Of the 22 known animal phyla, 50% are represented in fresh water, 100% in the marine environment, and probably 95% in the deep sea. Thus the deep-sea environment contains almost all the known phyla of the animal kingdom and, therefore, each known level of organic complexity for these phyla. This organic diversity at the highest taxon level merely serves to illustrate the exceptional environmental plasticity occurring in the members of the higher taxa. It is cited here to emphasize that the deep-sea apparently is a less special environment for colonization than is the fresh water.

This picture only suggests presence or absence and says nothing about numerical or mass dominance of one level of organization in the deep sea. Likewise, it does not illustrate that only one or two subgroups (classes or orders) of a given phylum are represented in the deep sea. Representation by one order of a phylum appears to be mainly related to the habits of animals belonging to those taxa. The deepsea representation appears to be controlled to a large extent by the biotope. For 90% of the abyssal benthic biota, mud is the biotope. Thus the general biotic composition of an abyssal benthic environment may or may not be similar to the composition of a shallow-water marine, level-bottom mud community.

There is no question that almost all marine phyla are represented in the deep sea. The proportion of biomass or species number at a given depth changes so that at one depth sponges dominate, whereas at another foraminiferans or ophiuroids or holothurians or polychaetes dominate.

In each animal class that lives in the ocean, there are always one or more families, often belonging to different orders, from which abyssal species are drawn. One frequent common denominator of these species is the habit of successful mud eating.

Endemic Abyssal Families

There exist some families and some orders which are almost exclusively abyssal; others have only one or two abyssal species. Thus we have two implications—that there is and that there is not a "family" level of organization that has significance to deep-sea life. The number of endemic abyssal families in any marine phylum is few. If there is any rule that applies, as evidenced by the Echinodermata and the Isopoda, it is that deep-sea endemism functions mainly at the species or genus level and not at higher levels of organic complexity.

Loss of eyes, decrease in coloration, protandry, and brooding of young are all common features for most deep-sea animals, but not one of these characteristics is adaptive to the deep sea alone nor unique to deep-sea species, nor a cause of deep-sea existence.

CHAPTER 13

Distribution of Marine Fauna

> Every species has three maxima of development, in depth, in geographic space, in time. In depth we find a species first represented by few individuals, which become more and more numerous until they reach a certain point, after which they again diminish, and at length altogether disappear. [Forbes, 1854, p. 173.]

Forbes was among the first of the marine biologists to recognize the bell-shaped curve of normal distribution of marine life with depth. Marine animals respond in their distribution to a variety of biotic and abiotic factors, both in depth and along a horizontal plane. The existing terms applied to any one pattern of distribution are not entirely adequate to describe interaction. For example, endemic, circumtropical, pantropical, and circumpolar describe the geographic range of a species in the broadest sense but do not associate this distribution with any other related or causal factors. In contrast, terms such as polar, subpolar, temperate, and tropical describe the temperature or marine climate requirements but not the geographic features. For this reason we prefer to discuss the distribution of an animal using three or more terms, which provide a more exacting image of the distribution of an animal than any single term can do.

GEOGRAPHY

Geographic distribution refers to the geographic range of an animal, but whether this is discontinuous is most often not taken into account. Within this framework, there are three types of species.

1. *Stenographic species* are endemic to a region and hence have a limited geographic range.
2. *Eurygraphic species* have a geographic range that extends over many degrees of latitude or longitude. Such species bear classifications including worldwide, pantropical, circumtropical, and circumpolar.
3. *Disjunct eurygraphs* are found in several oceans but are not continuous in distribution throughout the "range" (e.g., antitropical and bipolar species). Many estuarine species of the tropics belong to this category.

MARINE CLIMATE

The capability of a species to withstand one or more different marine climates is characterized in terms of temperature tolerance.

A *stenothermal* species is confined to one marine climate (e.g., hypopsychral, polar, subpolar, boreal, temperate, subtropical, tropical). A *eurythermal* species extends broadly over two or more marine climates.

DEPTH

Depth distribution can also be described in terms of the two extreme situations — stenobathy and eurybathy.

We arbitrarily restrict the term *stenobathial,* indicating a species with a narrow depth range, to those species having a range of less than 300 m or 10% of the average depth of the ocean. Most stenobathial species have a lesser range of bathymetric distribution. Species having a depth range in excess of 300 m are termed *eurybathial.*

SALINITY

The range in salinity throughout most of the oceans is so narrow that the marine species can only be considered as stenohaline in distribution; indeed, not one euryhaline oceanic species is known. Euryhaline species for the most part are confined to estuaries, rivers, or the Low Arctic. Catadromous fishes are the exception — for example, the European eel, which spawns in the Sargasso Sea but develops in the European rivers. Many estuarine species have stenohaline larvae and euryhaline adults. Salinity tolerance, osmoregulation, and the distribution of estuarine forms are not considered. It is sufficient to say that most estuarine species are stenographic in distribution and eurythermal and, often, endemic to only one estuary. For this there are many examples of fishes and a few examples of marine invertebrates. For instance, the species *Synidotea laticauda* is endemic to San Francisco Bay, and a similar but distinct species *Synidotea marplatensis* is endemic to the Rio del Plata in Argentina. Some estuarine species have a disjunct eurygraphic distribution which appears to have been brought about through transportation around the world on ship bottoms. Several species of barnacles belong to this category.

Analysis of the various kinds of extremes in distribution with reference to marine climate, geography, and depth allows classification of distribution patterns into eight categories. For each of these it is possible to assemble examples. We have chosen examples mostly from species of isopods that meet the following criteria:

1. Species that were captured several times, preferably known from 10 or more finds.
2. Species with a stabilized nomenclature. Varieties of species are considered to represent one species.
3. Species whose bathymetric range is based on records of adult specimens, not on shells, fragments, larvae, or juveniles.
4. Species that are truly benthic, either as a component of the epifauna or infauna, leading a holobenthic life thus spending the entire life history on the sea floor.

5. Species whose thermal, bathymetric, and geographic ranges are fairly well known and recorded in published reports.

DISTRIBUTIONAL TYPES

Type	Depth-Related	Temperature-Related	Geographic Range
1.	Eurybathial	Stenothermal	Stenograph
2.	Stenobathial	Stenothermal	Stenograph
3.	Eurybathial	Eurythermal	Stenograph
4.	Stenobathial	Eurythermal	Stenograph
5.	Stenobathial	Stenothermal	Eurygraph
6.	Stenobathial	Eurythermal	Eurygraph
7.	Eurybathial	Stenothermal	Eurygraph
8.	Eurybathial	Eurythermal	Eurygraph

Examples to illustrate these eight basic types of the two-dimensional distribution (Figure 13-1) are chosen from known marine species that adequately meet the five criteria just listed.

Eurybathial, Stenothermal Stenographs

Animals with the eurybathial, stenothermal, stenographic type of distribution are found at the poles and in the accessory seas such as the Mediterranean Sea and the Red Sea, where there is little change in temperature from the sea surface to the deep ocean floor. The range of temperature in vertical profile is 5°C or less, and hence

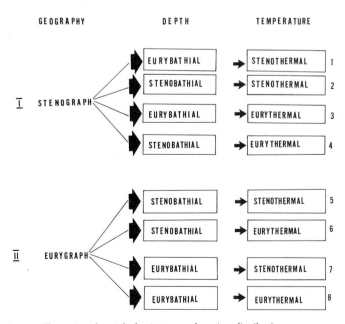

Figure 13-1. Diagram illustrating the eight basic types of marine distribution.

animals within such distribution may be called stenothermal. When the animals are also restricted geographically to the accessory sea or to a trench, they are also stenographic in distribution. Not all animals in the polar regions or the Mediterranean and Red Sea or trenches are stenographic, but many examples can be cited. Similarly, not all but many of the species have wide depth ranges in excess of 2000 m and are hence eurybathial in distribution. The polar eurybathial, stenothermal stenographs belong generally to a hypopsychral temperature regime of minus temperatures, whereas those from the Mediterranean are temperate in temperature requirements (ca. 13°C) and those of the Red Sea are tropical (ca. 22°C).

OF THE ARCTIC

As a consequence of the nearly isothermal conditions of the High Arctic, many species from various phyla are known to have a eurybathial, stenothermal, and stenographic distribution in the North Polar basin (Table 13-1).

OF THE ANTARCTIC

Several Antarctic species that occur at very shallow depths on the shelf also penetrate to deep water, showing a bathymetric range of about 1000 m. Among the isopod Crustacea there are several Antarctic species typifying this type of eurybathial, stenothermal, stenographic distribution from shelf to archibenthal depths. The most striking examples are the circumpolar Antarctic endemic species *Glyptonotus antarcticus* (2–1000 m) and *Antarcturus adareanus* (11–600 m). The ten-legged pycnogonid *Decolopoda australis* exhibits a similar pattern of distribution from near-shore depths to about 1000 m (Hedgpeth, 1969b). The large epifaunal species illustrating this type of distribution from the inner shelf to archibenthal depths

TABLE 13-1. ARCTIC EURYBATHIAL STENOTHERMAL STENOGRAPHS

Group	Species	Depth Range (m)
Polychaeta	*Antinoella sarsi*	22–3800
Polychaeta	*Apomatus globifer*	20–1900
Isopoda	*Eurycope hanseni*	460–2750
Isopoda	*Ilyarachna hirticeps*	12–2600
Isopoda	*Haploniscus bicuspis*	610–2600
Isopoda	*Macrostylis subinermis*	1000–2000
Isopoda	*Gnathia stygia*	20–2550
Porifera	*Arrhis phyllonyx*	23–1180
Porifera	*Thenea muricata*	300–2363
Pycnogonida	*Colossendeis proboscidea*	41–425
Mollusca	*Cirroteuthis mulleri*	550–2340
Mollusca	*Pecten frigidus*	1000–2800
Asteroidea	*Tylaster willei*	590–2400
Asteroidea	*Solaster squamatus*	90–1600
Holothuroidea	*Elpidia glacialis*	100–2800
Ophiuroidea	*Ophiopleura borealis*	51–1542
Ophiuroidea	*Ophioscolex glacialis*	300–2500
Pisces	*Raja hyperborea*	200–2800
Pisces	*Rhodichthys regina*	1100–2340

TABLE 13-2.   ANTARCTIC EURYBATHIAL STENOTHERMAL STENOGRAPHS

Group	Species	Depth Range (m)
Isopoda	*Glyptonotus antarcticus*	2–1000
Isopoda	*Antarcturus adareanus*	11–600
Isopoda	*Serolis neaera*	280–4000
Isopoda	*Serolis meridionalis*	1266–2725
Pycnogonida	*Decolopoda australis*	shelf–1000
Pycnogonida	*Pycnogonum rhinoceros*	154–1115
Pycnogonida	*Pycnogonum gaini*	shelf–2495
Holothuroidea	*Echinocucumis hispida*	50–1400
Crinoidea	*Phomachocrinus kerguélensis*	10–1080
Crinoidea	*Bathycrinus australis*	2514–4636

are the Antarctic species of the holothurian *Echinocucumis hispida* (50–1400 m) and the crinoid *Phomachocrinus kerguélensis* (10–1080 m).

Many Antarctic species extend from the shelf to depths in excess of 1000 m along the continental slope, but these are all geographically restricted to the Antarctic latitudes and within the polar–subpolar climate (stenothermal). A selected list of species (Table 13-2) illustrates this trend.

OF THE TRENCHES

A cold psychral climate ranging only 2 degrees in temperature covers the deep trenches of the world, and the fauna is likely to be stenothermal as a consequence. These trenches often have species that are restricted as far as is known to only one trench and are hence stenographic in distribution. Furthermore, although many of the species show depth ranges of 2000 to 3000 m and are hence eurybathial, few are known outside of one trench or another. The examples of ultra-abyssal or hadal eurybathial, stenothermal stenographs are numerous, but because the collections are few, the stenography of the species might be subject to error. The listing in Table 13-3 could be greatly increased.

OF THE MEDITERRANEAN

A wide vertical distributional range as seen in the cold stenothermal regions is also encountered for some warm-water species in the Mediterranean, where the 13.5°C regime extends down to about 4000 m. Several bathypelagic species that occur at relatively shallow depths elsewhere in the North Atlantic penetrate into the Mediterranean abyss. Among the endemic species both *Desmosoma chelatum* (600–2769 m) and *Ianirella bonnieri* (1237–2769 m), known only from the Mediterranean deep regions (George and Menzies, 1968d), exemplify the eurybathial, warm stenothermal stenographic type of distribution.

Stenobathial, Stenothermal Stenographs

The marine species that have a narrow bathymetric range in a restricted geographic region within a limited thermal range, belong to the stenobathial, stenothermal,

TABLE 13-3. TRENCH EURYBATHIAL STENOTHERMAL STENOGRAPHS

Animals	Area	Depth Range (m)	Remarks
Isopoda			
Storthyngura benti	South Pacific Kermadec Trench	5340–7000	Abyssal to trench depths
Storthyngura chelata	North Pacific Kurile–Kamchatka Trench	5345–6860	Abyssal to trench depths
Eurycope eltaniae	Southeast Pacific	4526–5943	Abyssal to trench depths
Desmosoma similipes		3909–6154	Abyssal to trench depths
Mesosignum multidens	Peru–Chile Trench	3470–6330	Abyssal to trench depths
Ilyarachna peruvica	Southeastern Pacific Peru–Chile Trench	1927–4925	Archibenthal to abyssal depths
Desmosoma coalescum	Southeastern Pacific Peru–Chile Trench	2335–4925	Archibenthal to abyssal depths
Haploniscus bruuni	Southeastern Pacific Peru–Chile Trench	3940–6300	Abyssal to trench depths
H. concavus	Southeastern Pacific Peru–Chile Trench	4520–6300	Abyssal to trench depths
Macrostylis longifera	Southeastern Pacific Peru–Chile Trench	4520–6300	Abyssal to trench depths
Nannoniscus perunis	Southeastern Pacific Peru–Chile Trench	4520–6300	Abyssal to trench depths
Haploniscus belyaevi	Northwestern Pacific	3102–5495	Abyssal depths
Ischnomesus andriashevi	Northwestern Pacific	4000–6560	Abyssal to trench depths
Bathycopea ivanovi	Northwestern Pacific	2867–4070	Abyssal depths
Crinoidea			
Bathycrinus australis	Antarctic Near Crozet Island	2514–4636	Abyssal depths

stenographic type of distribution. There are two major kinds: the stenobathic-shallow-water kind and the stenobathic-deep-sea kind.

STENOBATHIC SHALLOW-WATER SPECIES

Stenobathic shallow-water species are either pelagic or shallow-water organisms that evidently are tied to the sunlit layers of the sea. The surface dwellers lead a planktonic mode of life and are widely distributed by the surface currents. Among benthos, several intertidal or eulittoral species that are endemic in certain geographic regions belong to this type of stenobathic distribution. The regional study of the intertidal fauna off Peru and the Carolinas indicates that a very high percentage of the intertidal isopod species are confined within the tide limits. These species have not successfully colonized even the shallow depths beyond the low-tide mark. Pressure may or may not be the factor controlling their stenobathic distribution within the intertidal zone.

The intertidal offers a strikingly diversified ecological situation with greater food availability and shelter. Recent studies (Menzies and Wilson, 1961; Schlieper, 1968) have also revealed the ability of several intertidal organisms to tolerate pressure as great as 100 to 400 atm, equivalent to the pressure prevailing at 1000 to 4000 m. It is not known whether such endurance to experimental high pressure by intertidal organisms can be interpreted as their ability to survive, feed, and reproduce normally as in their natural habitat. Since these organisms are not found in high-pressure environments in nature, perhaps their type of distribution is stenobathic. In the Table 13-4 this kind of marine distribution is illustrated by selected examples from shallow stenothermal environments from different geographic regions and different climates. The shallow circumpolar species from the Arctic and Antarctic may also be included in this distributional pattern.

STENOBATHIC DEEP-SEA SPECIES

Species inhabiting certain great depths of the ocean, always living under high pressure, illustrate the stenobathic deep-sea kind of distribution. Zobell (1953) demonstrated that the bacteria from the Philippine Trench reproduce best in an appropriate culture medium at 2 or 3°C and 1000-atm pressure (i.e., pressure equal to that prevailing at 10,000 m, where the species normally live). There is no published record of a single species being found alive when trawled from depths exceeding 1500 m. A high percentage of abyssal species, particularly the infaunal members, seems to show both a restricted geographic distribution and a narrow bathymetric range of less than 1000 m within the deep-sea stenothermal environment. Many abyssal endemics are known from one ocean basin and one sample, but those which have been captured more than once are cited (Table 13-5) to represent this type of distribution.

TABLE 13-4.   STENOBATHIAL STENOTHERMAL STENOGRAPHIC TYPE OF
DISTRIBUTION, SHALLOW WATER

Species (Isopoda)	Geographic Region	Depth (m)	Climate (°C)	Species endemic to
Munna arnholdi Gurjanova	Bering Sea	0–4 (intertidal)	−1.6 Polar	Bering Sea
Antias charcoti Richardson	Antarctic	0–20	Polar	Antarctic
Antias hofsteni Nordenstam	South Georgia Isl.	0–4	Subpolar	South Georgia Isl.
Iathrippa multidens Menzies	Southern Chile	0–4	Antiboreal 4–7	Southern Chile
Ianira capensis Barnard	South African coast	0–4	Temperate 14–21	Southern Africa
Bagatus algicola Miller	Hawaii	0–4	Subtropical 23–25	Hawaii
Bagatus longimanus Pillai	Indian coast	0–4	Tropical 27	India
Serolis paradoxa	Circumpolar Antarctic	2–50	Polar– subpolar	Antarctic
Bathynomous doederlini	Sea of Japan	446–675	Temperate	Japan
Parabathynomous natalensis	Indian Ocean	756	Temperate	Indian Ocean
Bathynomous giganteus	Gulf of Mexico	400–600	Temperate	Gulf of Mexico

TABLE 13-5. STENOBATHIAL STENOTHERMAL STENOGRAPHIC TYPE OF
DISTRIBUTION, DEEP SEA

Animals	Area and Position	Depth (m)	Temperature (°C)	References
Isopoda				
Storthyngura scotia George and Menzies	Scotia Sea 55–64°S 58–67°W	2450–2816	−1.8	George and Menzies (1968b)
Stylomesus hexaspinosus Birstein	Northwestern Pacific	3874	+1.8	Birstein (1963)
Macrostylis affinis	Northwestern Pacific	5441–5695	+1.8	Birstein (1963)
Macrostylis vemae	N. Atlantic	5410–5684	2.1	Menzies (1962)
Porifera				
Thenea delicata	Antarctic deep sea	2926–3400	−1.8	Koltun (1969)
Cladothenea andriashevi	Antarctic deep sea	3000	−1.8	Koltun (1969)
Cladorhiza mani	Antarctic deep sea	3700	−1.8	Koltun (1969)

Eurybathial, Eurythermal Stenographs

The type of distribution known as eurybathial, eurythermal, stenographic is un-
common and is seen only in a few shallow-water organisms extending into moderate
deep regions. These eurybathial species occupy a thermal regime whose range
exceeds 10°C, but they are endemic to a geographic region. Evidently populations of
these species are metabolically adapted to withstand significant seasonally induced
temperature changes. Abyssal endemics do not exhibit this type of distribution
because of their restricted occurrence within the stenothermal deep-sea depths.

Most intertidal species are endemic to the intertidal, and the percentage that
enters deeper waters in the shelf is exceptionally low. Those rare eurybathic shallow-
water species which represent this type of eurybathial, eurythermal, stenographic
distribution are presented in Table 13-6.

TABLE 13-6. EURYBATHIAL EURYTHERMAL STENOGRAPHIC TYPE OF
DISTRIBUTION

Species (Isopoda)	Area	Depth (m)	Temperature (°C)	Remarks
Desmosoma laterale G. O. Sars	Davis Strait Skagerak	50–1096	2–15; subpolar to temperate	Shelf to archibenthal depths
Ilyarachna acarina Menzies and Barnard	Off California	73–1120	3.5–12.8; subpolar to temperate	Shelf to archibenthal depths
Jaeropsis bidens Menzies	Off Chile	0–400	9–16	Intertidal to shelf depths

Stenobathial, Eurythermal Stenographs

The stenobathial, eurythermal, stenographic distributional pattern is exclusively seen in middle or lower latitude shallow marine environments, where significant changes occur in the ambient temperature. The life history of these organisms is so synchronized with the regular seasonal temperature oscillations that their spawning or peak reproductive activity usually coincides with a particular thermal level. Intertidal endemics and inner shelf species with a restricted bathymetric range in a given geographic region show this type of distribution. Deep-sea species are totally excluded from this category because the classification involves a wide thermal range. The species in Table 13-7 typify this type of distribution.

Stenobathial, Stenothermal Eurygraphs

A species with a narrow depth range, a wide geographic distribution, and a narrow temperature range belongs to the stenobathial, stenothermal, eurygraphic category. Because most deep-sea species are endemic to either one ocean or one basin, only a few have a broad geographic range. One example apparently is the ophiuroid *Amphiophiura bullata* from the lower abyssal red clays of all oceans. The polychaete worm *Trophonia wyvillei* extends from the North Pacific at 60°N into the Indian Ocean at 56°S, but it has been found only within the narrow depth range of 3330 to 3566 m. Species of this type appear to be genuine abyssal endemics.

Stenobathial, Eurythermal Eurygraphs

A large number of shelf and intertidal benthic species, as well as many planktonic forms, are found in several oceans as amphiboreal species or tropical–subtropical cosmopolites, or "bipolar" species. For instance, the ophiuroid *Amphipholis squamata* is cosmopolitan, having a depth range of only 250 m throughout its range, and *Ophiactis savignyi* is circumtropical in the littoral zone. The sea star *Ceramaster patagonicus*, a subpolar–temperate shallow-water species, is known from the Bering Sea, southern Alaska, the Gulf of California, and Patagonia. The species *Ctenodiscus crispatus* is known from the North Pacific, southern California, the Gulf of Panama, and the Atlantic coast of North America as a shallow-water subpolar–temperate species.

TABLE 13-7. STENOBATHIAL EURYTHERMAL STENOGRAPHIC TYPE OF DISTRIBUTION

Species (Isopoda)	Area	Depth (m)	Temperature (°C)
Caecijaera horvathi	California	Eulittoral	12–21
Uromunna nana	Hawaii, Chilean coast	Intertidal	4–19
Janiralata occidentalis	California, Washington	0–70	5–21
Uromunna mediterranea	Mediterranean	2–30	12–25
Uromunna ubiquita	Southern California to Washington	0–37	5–24

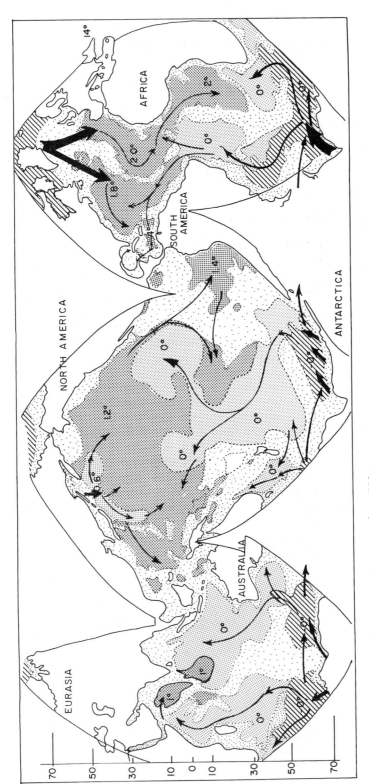

Figure 13-2. Temperature distribution at depths greater than 4000 m.

Eurybathial, Stenothermal Eurygraphs

The uniform temperature (2°C) and constant salinity of the abyssal regions of the world oceans (Arctic, Antarctic, Indian, Atlantic, and Pacific–Figure 13-2) led the zoogeographer Ekman (1953), and Murray and Thomson of the *Challenger* expedition, to believe that all deep-sea species are worldwide in distribution. The abyssal fauna, particularly the infauna, is predominantly endemic at the species level. Nevertheless, a great number of both epifaunal and infaunal species do show a wide geographic and bathymetric distribution within the stenothermal environment of the deep sea.

We consider these eurybathial species inhabiting the abyss of two or more oceans to be representative of the eurybathial, stenothermal, eurygraphic type of distribution. The invertebrate species that are commonly encountered in different oceans are given in Table 13-8. Several species occur in both the Arctic and Antarctic oceans but are not known from the other oceans in the middle and lower latitudes. In early literature, primarily owing to misidentification, vast numbers of species were cited as examples of bipolar distribution. Not one isopod species illustrates biopolar distribution. Recently Koltun (1969) listed the following six species of sponges exhibiting bipolar distribution: *Suberites montiniger*, *Sphaerotylus schoenus*, *S. borealis*, *Artemisina apollinis*, *Hymedesmia simillima*, and *H. longurius*. The eurybathial ascidian species *Corynascidia suhmi* is known from the deep-sea regions of both the Arctic and Antarctic oceans.

Eurybathial, Eurythermal Eurygraphs

Although a high percentage of species are endemic to the Arctic or Antarctic oceans, some low polar species have successfully "invaded" adjacent geographic regions of boreal (or antiboreal) and temperate marine climates, or vice versa. There are several species in common between the coastal waters of South America and the Antarctic continental shelf and slope depth.

Pawson, (1969), while discussing the distribution of holothurians, referred to the Magellanic region as an important portal of entry for the Antarctic species. Kott (1969) discussed the distribution of the ascidian *Styela nordenskjoldi*, which extends from the Antarctic shelf to the North Pacific off California. Similarly, several Low Arctic and sub-Arctic species have been found to occur in the Pacific and Atlantic areas of temperate climate. These species show a vertical range from intertidal to archibenthal depths or from shelf depths to the abyss. It seems most likely that such species are relict from former warm periods instead of invaders from the poles.

POLAR PHENOMENA

Polar Submergence and Polar Emergence

In addition to the special characters of the Arctic and Antarctic faunas, two salient faunal features appear to operate equally at high latitudes near the poles. The first distributional pattern refers to the tendency of shallow-water eye-bearing genera to

TABLE 13-8. EURYBATHIAL STENOTHERMAL EURYGRAPHIC TYPE OF DISTRIBUTION

Group	Species	Geographic Range	Bathymetic Range (m)	Temperature (°C)
Isopoda	*Serolis bromleyana*	34°S–62°S	800–4000	<2
Isopoda	*S. trilobitoides*	40°S–69°S 56°W–161°E	80–920	<2
Isopoda	*Storthyngura triplispinosa*	12°N–45°S 06°E–88°W Atlantic and Pacific oceans	2856–5843	<2
Isopoda	*Haploniscus bicuspis*	Arctic and off South-western Africa; east of Argentina	698–5024	0
Isopoda	*Haplomesus quadrispinosus*	Arctic South of Japan	520–4150	0–1.8
Isopoda	*Munneurycope murrayi*	Atlantic Ocean, Pacific Ocean, Indian Ocean, Bering Sea (Arctic)	530–7800	(Bathypelagic)
Isopoda	*Bathyopsurus nybelini*	North of West Indies, Puerto Rico Trench, Kermadec Trench, Tasman Sea	4400–7900	(Bathypelagic)
Isopoda	*Mesosignum usheri*	Caribbean to S.E. Pacific	1533–4065	1 to 4
Pycnogonida	*Pallenopsis calcanea*	35°S–55°N 51°W–139°E	650–2000	(Bathypelagic)
Cirripedia	*Scalpellum abyssicola*	Northern Atlantic, Indian Ocean, Pacific	2000–5540	(Bathypelagic)
Asteroidea	*Fryella tuberculata*	Northern Atlantic, Indian Ocean, Pacific	3365–5300	
Holothuroidea	*Benthodytes typica*	Southern Pacific, Indian Ocean, North Atlantic	1955–4600	
Ophiuroidea	*Amphiophiura convexa*	Northern Atlantic, Indian Ocean, Northern Pacific	2920–5270	
Tunicata	*Culeolus murrayi*	Northern Pacific, Antarctic, Indian Ocean 60°S–32°N	3390–4640	
Tunicata	*Corynascidia suhmi*	62°N–64°S Arctic, Antarctic	2434–4636	Bipolar species
Coelenterata	*Scleroptilium grandiflorum*	Northern Atlantic, Indian Ocean, Northern Pacific	500–4000	
Gastropoda	*Basilissa simplex*	Southern Atlantic, Northern Pacific	3475–5450	
Polychaeta	*Nephtys elamellata*	Northern Atlantic, Southern Pacific	4255–7000	
Pycnogonida	*Nymphon procerum*	Northern Atlantic, Northern and Southern Pacific	2470–4600	

penetrate into the abyss (polar submergence of shallow genera). The second pattern involves a simultaneous tendency of the eyeless abyssal genera to emerge from the abyss into shallow water (polar emergence of abyssal genera). These two kinds of generic behavior are distinguishable from the standpoint of morphology.

Polar Submergence of Shallow Genera

In the Arctic, five eye-bearing isopod genera, *Munna, Calathura, Pleuroprion, Mesidothea,* and *Gnathia,* show submergence. Forbes (1846) recognized this distribution in the Arctic as an example of boreal outliers, that is, assemblages of northern species occupying the deeper areas of about 80 to 100 fathoms. The majority of the known species of these genera bear eyes and inhabit only shallow depths. The minimum and maximum depths of the same five genera in the Arctic are as follows:

Arctic (68°N)	Minimum Depth (m)	Species	Maximum Depth (m)	Species
Munna	12	*M. fabricii*	869	*M. acanthifera*
Calathura	162	*C. brachiata*	1200	*C. brachiata*
Pleuroprion	12	*P. frigidum*	1200	*P. frigidum*
Mesidothea	12	*M. sabini*	1200	*M. sabini*
Gnathia	12	*G. stygia*	2600	*G. stygia*

Among the five genera, three are endemic to the northern high latitudes but the other two genera, *Munna* and *Gnathia,* are cosmopolitan. In the lower latitudes, however, these two genera, consisting of more than 50 species each, tend to show a restricted shallow depth distribution in intertidal and shelf faunal zones. These two genera have species well represented in the abyssal fauna in the Antarctic and Arctic seas. Species of *Munna* and *Gnathia* are known to occur from 3 m down to 3800 m in the Antarctic. Besides these two, there is a number of Antarctic shallow genera exhibiting this polar trend of submergence into the abyss, as shown in the following tabulation:

Antarctic Shallow Genera (Examples of Polar Submergence)	Minimum Depth (m)	Eye Bearing	Maximum Depth (m)	Eye Bearing
Cilicoea	5	×	600	×
Cymodoce	5	×	210	×
Antias	3	×	210	×
Aega	5	×	800	×
Glyptonotus	3	×	900	×
Austrosignum	5	×	2600	×
Microarcturus	12	×	3800	×
Munna	3	×	3800	×
Gnathia	3	×	3800	×
Cirolana	3	×	4005	×
Antarcturus	3	×	4005	×, ○

Antarctic Shallow Genera (Examples of Polar Submergence)	Minimum Depth (m)	Eye Bearing	Maximum Depth (m)	Eye Bearing
Serolis	5	×	5446	×, ○
Astacilla	3	×	1600	×
Rocinella	3	×	1900	×
Neojaera	3	×	1900	×
Janthopsis	50	×	3475	×
Edotea	100	×	800	×
Cassidinella	100	×	600	×
Stenetrium	100	×	3475	×, ○
Jaeropsis	51	×	81	×
Iathrippa	5	×	800	×
Paramunna	5	×	800	×

All these genera are known to live in intertidal or littoral waters in tropical and temperate marine environments, where all possess well-developed eyes. Among these 22, six genera—namely, *Serolis, Cirolana, Rocinella, Aega, Astacilla,* and *Gnathia*—are represented in the deep sea of low latitudes by blind species that are usually not allied to the eye-bearing shallow species in the same latitude. The presence of such abyssal species suggests an origin from the poles. For example, *Serolis carinata,* occurring in the shelf off the Carolinas, and *Serolis vemae,* occurring in the abyssal plain in the North Central Atlantic, are so different in morphological character that no argument supports the concept of a shallow-water species of the genus evolving into a deep-sea species from lower latitudes.

Kott (1969) listed the following seven species of Antarctic ascidians demonstrating tropical submergence in deeper waters north of the subtropical convergence: *Molgula pyriformis, Didemnum tenue, Ascidia meridionalis, Styela nordenskjoldi, Molguloides immunda, Didemnum studeri,* and *Bathypera splendens.*

Polar Emergence of Abyssal Genera

Several investigators, on the basis both of benthic and pelagic fauna, remarked that certain taxa inhabiting archibenthal or purely abyssal zones in equatorial and lower latitudes tend to emerge to shallower depths in the poles. Haeckel (1887) described the same phenomenon using pelagic Radiolaria. Ekman (1953) referred to this very common phenomenon as *equatorial submergence* because the submergence of a species increases toward the lower latitude. Most of the examples illustrating equatorial submergence are cases of populations of a single species, rather than various species of one genus. For this reason a special term is used.

Emergence by abyssal genera in both North and South polar regions may be illustrated by isopods (Figure 13-3). Abyssal isopod genera such as *Macrostylis, Haploniscus, Storthyngura, Ilyarachna, Eurycope, Desmosoma,* and *Munnopsis* occur at deep zones near lower latitudes, whereas they enter shallower zones in the Arctic and Antarctic (Figure 13-3).

All these asellote genera are widely distributed in the world oceans, and each

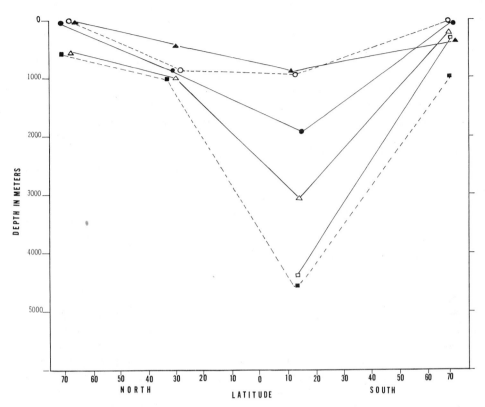

Figure 13-3. Upper depth limits of isopod genera illustrating the phenomenon of polar emergence of abyssal genera: △, *Haploniscus;* ▲, *Munnopsis;* ○, *Desmosoma;* ●, *Ilyarachna* and *Eurycope;* □, *Storthyngura;* ■, *Macrostylis.*

genus consists of more than 30 species. All are blind and have no indication of eyes. Both *Macrostylis* and *Haploniscus* represent typical components of the benthic infauna, whereas the other genera are moderate swimmers, possessing paddlelike appendages. It is noteworthy that species of *Macrostylis* are distributed near the equator between 5800 and 6200 m. This genus emerges across lesser abyssal zones at the poles to reach the shallow water. The upper depth limits of genera exhibiting the polar emergence tend to fall at different depth levels in the lower latitudes. These upper limits at low latitude do not correlate with any isotherm but are always located far below the vertical extinction of sunlight.

The most positive correlation in the distribution of these genera is the coincidence between the upper limits of the genera showing the polar emergence and the start of the abyssal faunal province. In the Antarctic, *Macrostylis* reaches a minimum depth of 900 m at the uppermost limit of the Antarctic abyssal faunal province. Several species of *Haploniscus* are found within the abyssal faunal province off Peru, but none is yet known from the archibenthal or shelf zones There the upper bathymetric limit is identical also with the start of the abyss. Similarly, 90% of the Atlantic genera showing polar emergence make their appearance at the commencement of the abyssal faunal province off the Carolinas at 35°N.

ZOOGEOGRAPHY OF ABYSSAL SPECIES

The horizontal range of marine animals over the earth describes their zoogeographic distribution. Throughout the history of deep-sea studies, a wide variety of conflicting statements have been made about the zoogeography of abyssal animals. It happens that each statement is more or less correct, depending on the depth of distribution of the animals. For example, in the 1800s Murray, Thomson, and Agassiz held to the view that abyssal species were worldwide in distribution. As late as 1957 Bruun supported this idea with examples of widely dispersed fishes. In contrast, Ekman (1953) contended that deep-sea species were not worldwide in distribution.

The reason for these diammetrically opposing views appears to reside in a genuine difference in zoogeographic patterns that is a function of the depth involved. Thus Murray, Thomson, and Bruun were all dealing with species from the continental margins in archibenthal and upper abyssal zones, whereas Ekman was specifically referring to species of the fauna living at depths below 4000 m. The species that comprise the fauna of various deep-sea zones are different. Also the reproductive patterns and the capacity for larval dispersal diminishes with increasing depth.

Vinogradova (1956a, 1956b, 1959) was the first to demonstrate numerically the distinction between these two basic kinds of distribution in the deep sea (Table 13-9). She found that 40% of the abyssal species that lived at depths less than 2000 m were common to both sides of the Atlantic but that none of the species from 4000 m and below were in common. All cases of eurygraphic abyssal distribution belong to the upper parts of the sea, and many cases of stenographic distribution are found in the greatest depths. Tracing the zoogeographic trends from the surface of the ocean downward we find the following regions and characteristics.

1. *Intertidal Faunal Province:* Species are often stenographic in the sense that they

TABLE 13-9.   THE NUMBER OF DEEP-WATER SPECIES COMMON TO DIFFERENT OCEANS IN RELATION TO DEPTH AT WHICH THEY OCCUR[*]

Depth at which Species Occur	Species Common to Antarctic and All Other Oceans (% of Antarctic Species)			Species Common to Eastern and Western Atlantic (% of Species in Total Area)	Species Common to Western, Eastern, and Northern Pacific (% of Species in Total Area)
	Atlan-tic	In-dian	Pa-cific		
Deep-sea species rising to depths less than 2000 m	70	60	27	40	49
Species from depths below 2000 m	15	40	4.3	12.9	12.6
Species from depths below 3000 m	6	10	2.5	7.2	15.6
Species from depths below 4000 m	0	0	0	0	1.2

* After Vinogradova (1959, p. 206).

occupy one of the marine climate divisions of the earth. These climates are distributed latitudinally north to south. Pantropical species that range across the divisions of latitude are eurygraphic in the sense that they are distributed around the circumference of the earth; they are stenothermal in so far as they are confined to the tropics. Today their distribution is made disjunct by water currents, large expanses of ocean, and land barriers. It seems likely that the proportion of pantropical benthic species is smaller than the proportion of pantropical pelagic species.

2. *Shelf Faunal Province:* Species in the shelf faunal province tend to be more eurygraphic than the species of the intertidal faunal province and are less subject to the marine climate division at the water surface. Nevertheless, it is obvious that the shelf fauna of the tropics is decidedly different from that of the temperate, boreal, and polar marine climate regimes. These discrepancies have been abundantly illustrated by Thorson (1956) in his treatise on level-bottom communities.

3. *Archibenthal Zone of Transition:* The proportion of eurygraphic species reaches its maximum among the larger species and larvae-producing species of the AZT. It is to the species of this zone that Murray, Thomson, and Bruun referred. We call this eurygraphic distribution parallel continental margin distribution.

4. *Abyssal Faunal Province:* As the abyssal faunal province is approached, the proportion of eurygraphic species diminishes. Finally it approximates zero in the deep basins of the oceans and in the deep trenches. It is on the basis of these species that Vinogradova (*op. cit.*) divided the world abyss into zoogeographic compartments.

5. *Distribution Along Continental Margins:* The distribution of species in the upper continental margins beyond the shelf province is characterized by a high proportion of species that are found on both sides of an ocean and often in all oceans. It is this kind of distribution that led Agassiz (1888) and others to believe in a worldwide distribution of abyssal fauna.

The best known example of the similarity between members of such parallel continental margin populations concerns a comparison of the archibenthal fauna of the Northeastern Atlantic with that of the Northwestern Atlantic. We call these the parallel continental margin species. At least 26 species, which are either identical or closely related, occur on both margins.

Parallel Continental Margin Species in the Atlantic

Northeastern Atlantic (Ireland to Bay of Biscay)	*Northwestern Atlantic* (Carolinas)
Sponges	
1. *Hyalonema boreale*	*Hyalonema boreale*
2. *Thenea muricata*	*Thenea muricata*
3. *Cladorhiza abyssicola*	*Cladorhiza* sp.
4. *Tentorium semisuberites*	*Tentorium semisuberites*
Coelenterata	
1. *Anthomastus agaricus*	*Anthomastus grandiflorus*
2. *Pennatula aculeata*	*Pennatula aculeata*
3. *Umbellula* (3 spp.)	*Umbellula* sp.

Parallel Continental Margin Species in the Atlantic (*Continued*)

Northeastern Atlantic (Ireland to Bay of Biscay)	Northwestern Atlantic (Carolinas)
4. *Bolocera longicornis*	*Bolocera tuediae*
5. *Epizoanthus paguriphilus*	*Epizoanthus paguriphilus*
6. *Lophohelia prolifera*	*Lophohelia prolifera*
7. *Flabellum* (2 spp.)	*Flabellum goodei*
Echinodermata	
1. *Ophiomusium lymani*	*Ophiomusium lymani*
2. *Amphiura opandia*	
3. *Ophiomusium planum*	*Amphiophiura bullata*
4. *Phormosoma placenta*	*Phormosoma placenta*
5. *Psychropotes* (2 spp.)	
6. *Rhizocrinus* (2 spp.)	*Rhizocrinus* sp.
7. *Mesothuria intestinalis*	*Mesothuria intestinalis*
8. *Scotonassa translucida*	*Scotonassa* sp.
Worms	
1. *Notomastis latericius*	*Notomastis* sp.
2. *Hyalinoecia tubicola*	*Hyalinoecia tubicola*
3. *Hyalinoecia tubifex*	*Hyalinoecia artifex*
Crustacea	
1. *Polycheles grimaldii*	*Polycheles* sp.
2. *Parapagurus pilosimanus*	*Parapagurus pilosimanus*
3. *Geryon tridens*	*Geryon quinquedens*
4. *Bathyplax* sp.	*Bathyplax typhlosa*
5. *Euryonicus* (4 spp.)	*Euryonicus* sp.
6. *Munida* (4 spp.)	*Munida valida*

The Possibility of Worldwide or Eurygraphic Continental Margin Distribution

Within the provinces or zones that we delimit, it is abundantly clear that many species tend to have a much narrower distribution—for example, as a ribbon or bands of varying width that might be a few meters to many meters wide, each might be continuous around the continents of the world oceans. Such species are eurygraphic eurybathials that are either stenothermal or eurythermal.

At a particular geographic place, these species seem to have a most restricted range in depth at a given time. It is most likely that the depth restriction depends on the interaction among factors such as reproductive mode, food supply, temperature, and predators. Distributional records are incomplete for most species that are found in more than one ocean. As we have seen, they are better for those occurring along the continental margins of the North Atlantic. The record is today based on too few samples to justify a continuous distribution as a general rule along the continental margin. We present this idea only as a working hypothesis that can be subjected to testing.

For a species to maintain itself as a distinctive morphologic unit from one ocean to the next, there remains the necessity for gene flow. We visualize this as being continuous in the ribbon or bandlike distribution, facilitated by larval stages that are dispersed by deep geostrophic currents. With this type of distribution it is likely and even highly probable that a series of clines would develop in time at the extreme ends of distribution of a given species and that subspecies would ultimately develop. This seems to be the case in the quill worms, *Hyalinoecia* and in the hermit crabs *Parapagurus;* it is likely that the same thing occurs in the ophiuroid *Ophiomusium lymani* and its synonyms, as well as in many other genera. The six species or species complexes are recognized as examples of worldwide continental margin distribution. Many other species could be added to the list.

WORLDWIDE CONTINENTAL MARGIN SPECIES

A few species and many genera having a multitude of rather closely related species show worldwide continental margin distribution. Species are:

1. *Scotoplanes globosa* Thiel, which is found along the continent of Antarctica, the South Atlantic, the Indian Ocean, and the North and South Pacific.
2. *Parapagurus pilosimanus* complex, which is found along the continental margin of the South Atlantic, the North Atlantic, and the North and South Pacific (being absent from the Poles).
3. *Ophiomusium lymani,* which is found in the North Atlantic, the North and South Pacific, and the Antarctic (absent from the Arctic?).
4. *Umbellula* and its species complex, which occur in all oceans, including the Arctic and Antarctic.
5. *Ypsilothuria* (=*Sphaerothuria*), which is found in the North Atlantic, the Southeast Pacific, the Gulf of Mexico, the Antarctic, and elsewhere.
6. *Hyalinoecia tubicola* complex, which is found in the North Atlantic, the North Pacific, and the Antarctic.

This includes only large epifaunal species and not the small species of the infauna.

The coincidence between the fauna of the archibenthal zone of transition from geographic locations far removed from each other leads us to suspect that many, if not most, species in the AZT either are the same from place to place or are very close relatives to one another. For example: Horikoshi (1970) during the JEDS expeditions provided a composite photograph (Figure 13-4) of an archibenthal assemblage off Japan. The genera are remarkably similar to those reported by Le Danois from the Northeastern Atlantic.

Horikoshi's photograph includes *Hyalinoecia, Parapagurus, Epizoanthus pagurophilus, Hyalonema, Polycheles,* and perhaps other genera and species similar to or identical with those from the continental margins of the eastern and western North Atlantic!

Distribution in Basins and Trenches: Vinogradova's View

Vinogradova related the restricted distribution of abyssal species to the macrorelief of the sea floor below 4000 m and elaborated a scheme that is generally consistent with the seabed topography (Figure 13-5).

Figure 13-4. An archibenthal assemblage of more than 22 species from the North Pacific (33°47'N ×
140°33'E), between 1280 and 1380 m depth: 1, Pycnogonid *Colossendeis colossea*; 2, *Parapagurus* sp.; 3,
Epizoanthus; 4, *Dibranchus*; 5, *Lithodidae*; 6, *Polycheles*; 7, *Pandalidae*; 8, *Ethusa*; 9, *Ophiolepidae*; 10,
Goniasteridae; 11, *Cribrella*; 12, *Brisingidae*, 13, *Kuphoblemnon*; 14, *Neptunea*. (Courtesy of Dr.
Horikoshi, JEDS expedition S, Station F-22.)

A. Pacific–North Indian area (deep-water region)
    A-1. Pacific subarea
        A-1a. North Pacific province
        A-1b. West Pacific province
        A-1c. East Pacific province
    A-2. North–Indian subarea
B. Atlantic area
    B-1. Arctic subarea

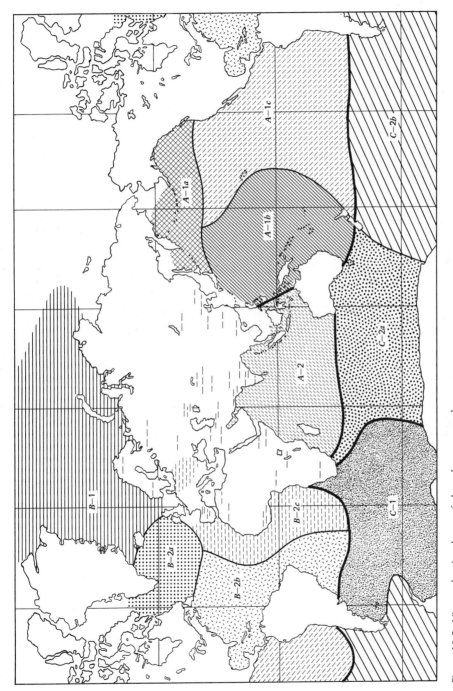

Figure 13-5. Vinogradova's scheme of abyssal zoogeography.

B-2. Atlantic subarea
    B-2a. North Atlantic province
    B-2b. West Atlantic province
    B-2c. East Atlantic province
C. Antarctic area
    C-1. Antarctic–Atlantic subarea
    C-2. Antarctic–Indian subarea
        C-2a. Indian province
        C-2b. Pacific province

In our opinion Vinogradova's areas, subareas, and provinces are quite useful first-order approximations, but they should be subjected to critical examination as new facts of topography, hydrography, and taxonomy become known, and the entire scheme should also relate to the past history of the fauna. We do not accept the Arctic as a subarea of the Atlantic and believe it to be an area in its own right (see Chapter 6). The use of province as a subdivision of a subarea and area also seems to be inappropriate.

A HYPOTHETICAL SCHEME

The scheme which we propose here is an alternate to the one drawn up by Vinogradova and it uses, where possible, temperatures as well as topography as determinants for province definitions. Like Vinogradova's scheme, ours may be subjected to testing whenever sufficient data exist and hence is also a useful working hypothesis. In many ways our hypothesis is quite similar to that proposed by Ekman, differing in the consideration of the Indian Ocean as more related to the Antarctic than to the Pacific. The outline follows:

A. Pacific Deep-Water Region
    A-1. Northwest Pacific province
    A-2. Central Pacific province
        A-2a. Northern Mid-America trench area
        A-2b. Southern Mid-America trench area
        A-2c. Peruvian area
        A-2d. Easter Island area
        A-2e. Tuamoto–Marquesas area
        A-2f. Northern New Zealand area
        A-2g. New Guinea–Borneo–Philippine area
        A-2h. China Sea area
B. Arctic Deep-Water Region
    B-1. Norwegian province
    B-2. Greenland–Fram province
    B-3. Eurasian province
    B-4. Siberian province
    B-5. Canadian province
C. Atlantic Deep-Water Region
    C-1. Northwestern Atlantic province
    C-2. North–South Eastern Atlantic province

A. Pacific Deep-Water Region

The first region includes depths of the Pacific (Figure 13-6) in excess of 4000 m, south to 10°S, containing bottom water all above +1°C. The deep water has originated in the main from the North Pacific. At the west this deep-water region is strongly separated from the Indian deep-water region by higher elevations of submerged land stretching from New Guinea in an arc toward China. This region is divisible into approximately 10 virtually isolated areas or provinces either of distinctive topography or water temperature.

A-1. NORTHWEST PACIFIC PROVINCE

There is a uniform area of water colder than 0°C, located between Kamchatka and the southern part of Japan. No topographic boundary separates the Northwest

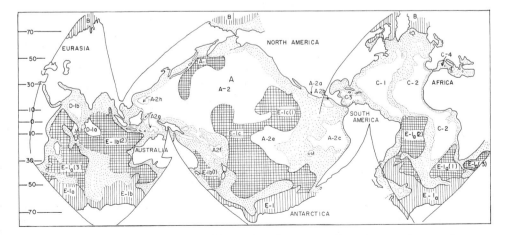

Figure 13-6. A new hypothetical scheme of lower abyssal zoogeographic regions, provinces, and areas.

Pacific province from the next and largest province, but its low temperature makes it distinctive.

A-2. CENTRAL PACIFIC PROVINCE

The largest province in the Pacific deep-water region is characterized entirely by temperatures in excess of +1°C. The Central Pacific province extends to 30°S on the eastern border and to 30°S on the western border; a large western bulge reaches out to 150°E. Its north central portion extends across the Pacific Ocean, stopping at roughly 120°E. In the central southern portion this province is bordered by 0°C bottom water from the Antarctic. The areas of the Central Pacific province are given lower-case Roman letters starting clockwise from the American coast.

A-2a.   Northern mid-America trench area off Mexico; bottom water temperature 1.4°C.

A-2b.   Southern mid-America trench area off Costa Rica; bottom water temperature 1.6°C.

A-2c.   Peruvian area between Peru and Chile and extending to the Antarctic–Pacific rise; bottom water temperature 1.8°C.

A-2d.   Easter Island area in the Pacific Antarctic rise; bottom water temperature 1.6°C. This is the smallest of the areas.

A-2e.   Tuamotu–Marquesas area bordered on the west by the Pacific–Antarctic rise and on the east by the Antarctic–Pacific province; bottom water temperature 1.0°C.

A-2f.   Northern New Zealand area, located northeast of New Zealand and bordered by elevated land everywhere; bottom water temperature 1.7°C.

A-2g.   New Guinea–Borneo–Philippines area–a distinctive area of three deep basins all having water above 2.7°C: the basin off the southern Philippines is reputed to have bottom water of +9.9°C.

A-2h.   China Sea area, located between China and Luzon (Philippines); with bottom temperatures of 1.8°C.

There is in addition a series of four small areas southwest of New Guinea with bottom water temperatures of 1.6°C; these may or may not be related to the northern New Zealand area. They are not numbered.

B. Arctic Deep-Water Region

Deep basins of the Arctic Sea all have minus-temperature water and are divided into several provinces according to bathymetry (Chapter 6).

The two faunal divisions proposed by Menzies (1963), based on the distribution of *Mesidothea megalura megalura* and *Mesidothea megalura polaris*, reflect the limited data available on which divisions can be made. Utilizing Vinogradova's reasoning, the Arctic should be divided into six provinces corresponding to the basins of the Arctic. From the Atlantic to the Pacific these are: Norwegian basin, Greenland basin, Fram basin, Eurasian basin, Siberian basin, and Canadian basin.

B-1.   Norwegian province (3000 meters). The Norwegian province is a basin located under the Norwegian Sea and bounded on the Atlantic end by the 560-m deep Faero–Iceland–Greenland Ridge and toward the Arctic by the deep Nansen Sill (Figure 6-1).

B-2.   Greenland–Fram province. A basin of 3000-m depth exists that may or may not connect with the Fram basin toward the east. The two basins are separated from the Norway basin by the Nansen Sill and from the Eurasia basin by the mid-ocean ridge.

B-3.   Eurasian province. There is a central High Arctic province coincident with the Eurasia basin; it extends to 4500-m depth and is separated from the Fram basin by the mid-ocean ridge and from the Siberian basin by the Lomonosov Ridge.

B-4.   Siberian province. A central High Arctic province of 3500-m depth occupies the Siberia basin and is separated from the Eurasia basin by the Lomonosov Ridge and from the Canada basin by the Alpha ridge.

B-5.   Canadian province. The Canadian province, the largest Arctic province, is 3500 m deep. It is separated from the Pacific by the Chukchi shelf and Bering Strait (50-m depth) and from the Siberia basin by the Alpha Ridge.

C. Atlantic Deep-Water Region

The third major region includes all the Atlantic, the accessory seas of the Caribbean and the Gulf of Mexico, and the Mediterranean, having bottom temperatures above +1°C. The Atlantic deep-water region is bounded in the southwestern Atlantic by the Antarctic bottom water less than 0°C and on the southeastern Atlantic by the Walvis Ridge. It may be divided into four provinces.

C-1.   Northeastern Atlantic province: Situated off North America from south of Greenland to the equator, the Northeastern Atlantic province has bottom temperatures between 1.2 to 1.8°C and is separated from the next province by the mid-Atlantic Ridge.

C-2.   North–South Eastern Atlantic province. This province extends from off Spain to the Walvis Ridge in the South Atlantic and is separated from C-1 by the Mid-Atlantic Ridge. Bottom temperatures range from 1.6 to 2.2°C.

C-3.   Caribbean–Gulf Province. This province includes the two Caribbean basins and the Gulf of Mexico. It is separated from province C-1 by the Antilles. Bottom temperatures are in excess of 3°C. The three isolated basins could be considered as areas of the province.

C-4.   Mediterranean Province. The basin of the Mediterranean Sea is separated from the Atlantic by the Strait of Gibraltar; bottom water temperatures are above 13°C.

D. Indian Deep-Water Region

The Indian deep-water region involves the basins of the Red Sea and the Andaman Sea, which are all regions of positive temperature.

D-1.   Arabian province. Located between western India and Africa and having bottom temperatures between 1.0 and 1.35°C, but separated from the 0°C water of Antarctic origin, the Arabian province may be divided into three regions.

D-1a.   Southern Indian area, located off the southern tip of India as a large, isolated body of positive-temperature water.

D-1b.   Arabian area located below the Arabian Sea and separated from the areas "a" and "c" by elevated submerged land; bottom temperature is 1.35°C.

D-1c.   Afro-Indian area. The Afro-Indian area is an isolated body of plus-tempera-
ture water between India and Africa and separated from the Andaman area by
an elevation of land.

E.   Antarctic Deep-Water Region

The Antarctic region includes all deep water around Antarctica and water of 0°C or
less which sweeps into the other oceans.

E-1.   Antarctic Circumpolar province. The Atlantic circumpolar province is characterized by water below 0°C and is generally circumpolar south of 50° latitude. It
has at least three areas.

E-1a.   Atlanto–Indian–Antarctic area. The Atlantic area extends as minus-degree-
temperature water up to 45°S latitude on the eastern side and up to 48°S latitude
on the western side. In the western Indian Ocean it extends up to 50°S latitude
and splits into two parts: an eastern South Atlantic subarea extends up to the
Walvis Ridge and the other branch, toward the east, continues into the
southeastern Indian subregion.

E-1b.   Austro-Indian Antarctic area. This area contains minus-degree-centigrade
water up to 45°S latitude in the eastern part of the Indian Ocean. It gives rise to
the southwestern Indian subarea as far as 10°N latitude and also to the eastern
Australian subarea up to 30°S latitude.

E-1c.   Southeastern Pacific Antarctic Area. Extending westward toward South
America reaching up to 45°S latitude, the Southeastern Pacific area gives rise to
the vast South Central Pacific subarea, which extends as far as 10°N latitude in
an eastern branch and as far as 20°N latitude in a western branch.

The abyssal isopod genera known from depths exceeding 4000 m in eight test
regions from the foregoing hypothetical scheme were tabulated to determine the
degree of homogeneity that existed among these geographically isolated deep-sea
regions. It appears that 18% of the abyssal genera tend to be truly cosmopolitan in
having species in the selected test regions representing all the five major oceans. Furthermore, 16% of the genera are known from four oceans, 18% are known from
three oceans, 25% are known from two oceans, and 33% are known from only one
ocean. In other words, 67% of the abyssal genera are cosmopolitan in having species
in two or more world oceans in the lower abyssal zone at depths exceeding 4000 m.

The data on isopod genera suggest that the Southeastern Pacific, which has positive temperature, shows the least relationship (38%) to the Antarctic Pacific, which
has a negative temperature. This comparison is made here with one abyssal zoogeographic region of the Pacific having a positive temperature +1.8°C. from
4000 to 6200 m and with one region of the Antarctic. A comparison between
three negative-temperature abyssal regions—namely, the Antarctic Pacific (E-1),
Southwestern Atlantic (E-1a), and the Southeastern Atlantic (E1a-1)—shows a significantly high percentage of similarity (77%) of genera among these three regions.
The three selected test regions are widely separated from one another, but they
are all subject to the influence of Antarctic bottom water. From this preliminary
examination of generic relationship between negative and positive temperature
abyssal regions, we infer that our partition of abyssal faunal geography has certain merits over that of Vinogradova.

At the species level, the deep-sea regions from different oceans seem to be 100% isolated from one another at depths exceeding 4000 m. This feature of high species endemism was revealed in Vinogradova's demonstration analysis that no species are in common between the Pacific, Atlantic, and Indian oceans of the abyssal depths from below 4000 m (Table 13-9). The high infaunal species endemism in different ocean basins was earlier pointed out by Menzies (1965) and, therefore, species endemism as an index for delineating boundaries between abyssal faunal geographic regions may not be a valid criterion. Genera and species groups, however, do illustrate degrees of affinities between ocean basins and zoogeographic provinces, as demonstrated by the distribution of groups, and subgroups in the abyssal genus *Storthyngura* (George and Menzies, 1968a, 1968b).

COMMUNITIES WITHIN ZONES

Thus far we have tried to avoid dealing with the ecology of communities in the abyss. We have discussed communities, patchiness, dominance, and dispersion, but we have not elaborated on our own views. Obviously we accept the existence of communities of organisms in the sea as a reality, and we recognize that each zone has its own set of communities. Thorson (1956) advocated the concept of parallel level-bottom communities throughout the world oceans. His idea was that a community of one sediment type in one geographic place was dominated (in mass or in numbers) by one or more species. He believed that this community was repeated in other geographic locations by similar types of organisms belonging to ecologically parallel forms in similar sediments. For example, the boreal level-bottom community of Norway with *Macoma baltica* was replaced by a *Venus-Spisula* community in the Persian Gulf. Thorson's (1957b) studies included only large species of 1.75-mm diameter or larger, and because of this peculiarity of methodology, Thorson missed the more numerous smaller animals.

Others have tended not to accept Thorson's view, disagreeing with his idea of dominance by one or two large species. There seems to be no question that Thorson's idea is correct when applied to the fauna of the continental shelf. The concept does not appear to hold in archibenthal or in the abyssal faunal province. We have seen for instance, that certain large species of the AZT and the upper abyssal zone are distributed in quite narrow ribbons or bands of contagious distribution. The level bottom embraced by an entire zone is only a small fraction of the area of a given zone, and any one large animal in the zone characterizes only its own band or circle of distribution, and not the entire zone as an example of community dominance. Essentially, then, the fauna comprising a zone is composed of few to many substantially distinct communities, and a zone does not represent one community even though it has mainly one type of sediment. In abyssal depths, a mosaic distribution of communities such as that advocated by Hessler and Sanders (1967) seems to be most likely. However, large epibenthos may be distributed in parallel communities with the minute infaunal species showing more random associations.

CHAPTER 14

Origin and Evolution of the Abyssal Fauna

> The author proceeds from the concept that the opinion about the very great antiquity of the ocean is most widely recognized at the present time. This opinion is shared by many geophysicists, geochemists, geologists and biologists. . . . [Zenkevitch, 1966, p. 207, English summary.]

The quotation by Zenkevitch is presented to demonstrate our essential agreement with his basic premise regarding the age of the ocean and the marine fauna. It is in subsequent details that we disagree and especially with his view that:

> the temperature and chemical regime of the ocean [is] similar in its main characteristics to the present one. . . for at any rate several million years. . . . [and] the low temperature of the near-polar regions and of the deep ocean waters may be considered a geophysical constant.

It is to the questions of temperature and sediment (chemistry) that we devote some special effort because of the importance attached to their "stability" through time by those who advocate a deep-sea fauna of considerable antiquity. We do not reach a solution to all problems, but we believe that we can safely state that the deep-sea temperatures and sediments (and hence sedimentary chemistry) have not been constant throughout geologic time. Once this idea is accepted, there remains no reason to believe that the deep-sea fauna is in the main one of great antiquity.

ORIGIN OF MARINE LIFE AND FOSSIL RECORD

Many theories have been proposed to explain the origins of the earth, the water, and life itself. The abyssal fauna apparently came about long after life had formed on earth and much longer after the land and water had evolved. Although our inquiry commences with the advent of the fossil record, it is of some utility to identify points in the time scale of the evolution and development of the earth in order to maintain perspective.

The earth presumably originated, through a process not yet thoroughly understood, some 3.5 to 5 billion years ago (Cloud, 1968). During this process the land developed and the age of the atmosphere and hydrosphere is often placed around

1.8 to 3.0 billion years, although some persons believe that erosion by running water happened 4 billion years ago (Donn et al., 1965), which implies the presence of an atmosphere and a hydrosphere at this early time. Proponents of this theory emphasize that the temperatures were near freezing.

Life probably originated in water. Many scientists adhere to the belief that life originated in the seas, but whether this water was then salty and hence marine is not certain.

It is known that life originated sometime between 2.0 and 0.8 billion years ago and that life has since evolved into the life forms known today and into those that are now extinct. Most of the phyla seen in ancient rocks can be identified with phyla living in the oceans today. This is the dominant evidence that life originated in the seas. The other bit of evidence comes from the saline nature of protoplasm. These bits of evidence do not constitute proof that life originated in a marine environment because there is a 2.5 to 0.4 billion year hiatus between the origin of the hydrosphere and the formation of the earliest identifiable fossils. At first the water must have been fresh and not salty, and the ocean volume must have increased. It is not known when the oceans became deep, and hence any argument for a Pre-Cambrian origin of a deep-sea fauna is not supported by evidence.

Cloud (1968) placed the advent of metazoan evolution near the development of Pre-Cambrian glaciation between 0.6 and 0.7 billion years ago. His reconstruction of events appears in Figure 14-1. A modern geologic time scale based on geochemical age-dating is presented in Figure 14-2, as elaborated by Kulp (1961).

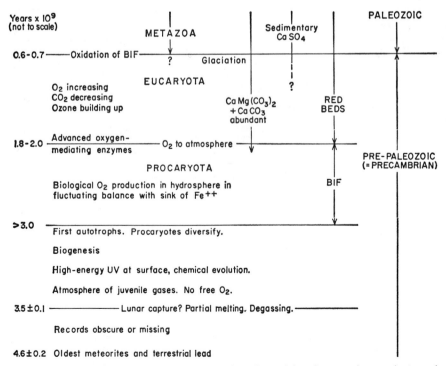

Figure 14-1. Postulated main features of interacting biospheric, lithospheric, and atmospheric evolution on the primitive earth. (After Cloud, 1968.)

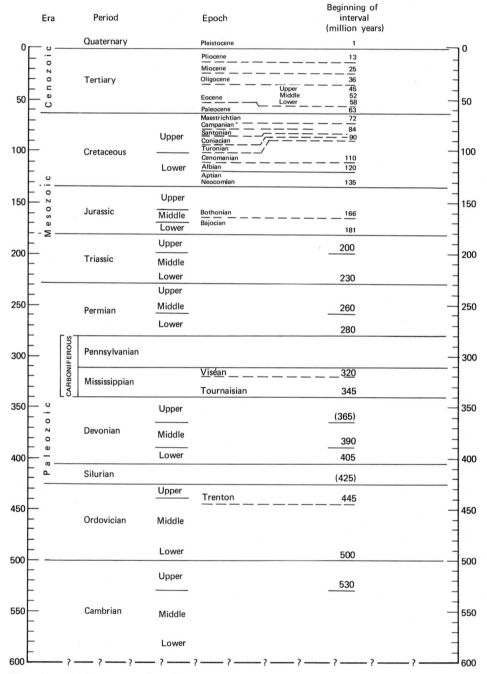

Figure 14-2. Geologic time scale (Kulp, 1961).

Origin of the Hydrosphere

If water originated on earth after the formation of the atmosphere, as is commonly believed, then the earliest seas must have been at one time quite shallow and fresh, and their volume must have been less than it is today. By the Paleozoic time it is believed by Kuenen (1950) that the ocean volume was 1300×10^6 km³, or only 70×10^6 km³ less than the present volume, and that the salt concentration was very close to what it is today (i.e., ca. 32–34 parts per thousand in salinity). This concept is that of permanence, and not all theorists are in agreement.

For example, Walther (1926) believed in an ocean volume increase by 50% or more from the start of the Mesozoic era, only around 160 million years ago (Figure 14-3). The big problem with this concept is in accounting for some 600×10^6 km³ of water. Usually it is easier to accept the concept of permanency than it is to worry about the missing water. No knowledge is available regarding the depth of the ocean at these times, although the existence of regionally restricted 10,000-m-deep geosynclines has been confirmed, and the idea of the existence of Paleozoic deep-sea deposits now seems to be quite acceptable to geologists (Dietz and Holden, 1966). The fossils in these deep-sea deposits consist in the main of radiolarian and other pelagic fossil remains; no genuine deep-sea benthic animal remains have been reported yet.

CONTINENTS AND CONTINENTAL DRIFT

Whether continental drift has taken place is important to an understanding of the existing abyssal fauna and its origin. Sea floor spreading to produce continental drift, as advocated by Du Toit (1937), Dietz (1962), Runcorn (1962), and others, is now generally accepted as a fact of geologic history. Earlier many of the data on distribution of evaporites, sedimentary structures, paleobotany, paleontology, and paleoclimatology were in varying degrees in support of the Wegnerian concept of a protocontinent of Gondwana Land or Pangaea. Now, however, even the most avid antagonists of continental drift concede, if reluctantly, that drift separated the Old World from the New World with the creation of the Atlantic Ocean (Figures 14-4, 14-5). Allard and Hurst (1969) (Figure 14-6) have shown the great stratagraphic similarity of the continental margins of Africa and Brazil, and this is most convincing as to the prior connection made between Africa and South America. The thinness of the sedimentary cover in the Atlantic Ocean, as the mid-Atlantic Ridge is approached, is a similarly strong argument in support of the past connection between these continents.

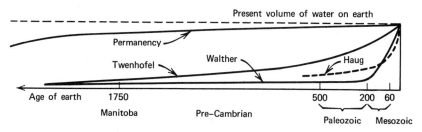

Figure 14-3. Volumetric increase in hydrosphere according to various workers. (After Kuenen, 1950.)

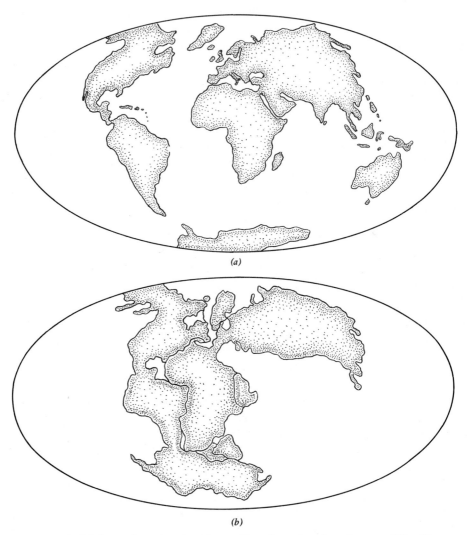

Figure 14-4. (a) Existing configuration of continents, (b) continental configuration some 150 million years ago. (After Wilson, 1963.)

The absence (today) of widespread oceanic Paleozoic fossil deposits and the discovery of the continuous Mesozoic seismic horizons in the Atlantic and the Pacific may be looked upon as an argument that a Paleozoic deep sea did not exist. The Mesozoic layers are identified as A, β, and B in the Atlantic and as A', β', and B' in the Pacific. These have since been identified as Cretaceous sediments.

Figure 14-7 is a copy of a seismic profile from the Atlantic as given by Windisch et al. (1968). Presumably the horizons are identical in all oceans, although the absolute thickness differs from place to place. The horizons may suggest indeed that the ocean basins are no older than the Mesozoic, as has been indicated by Beloussov (1960), Saito, Burckle and Ewing (1966), Saito, Ewing, and Burckle (1966).

The concept of sea-floor spreading advocated by Dietz (1961) results in the progressive destruction of any preexisting Paleozoic rocks by their incorporation into

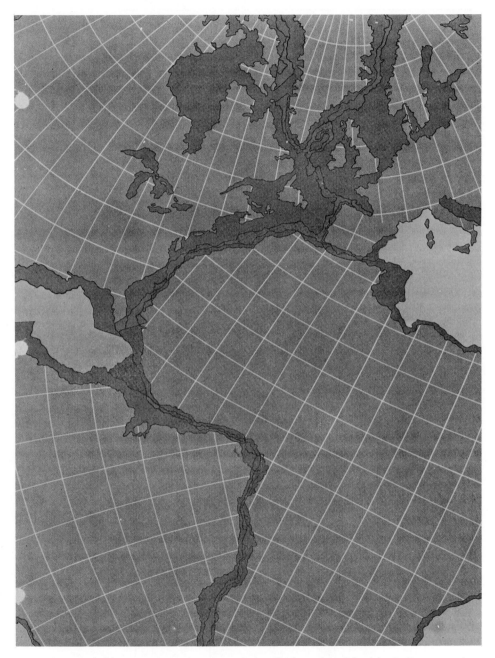

Figure 14-5. Probable arrangement of continents before the formation of the Atlantic Ocean was determined by Bullard (1969) with the aid of a computer.

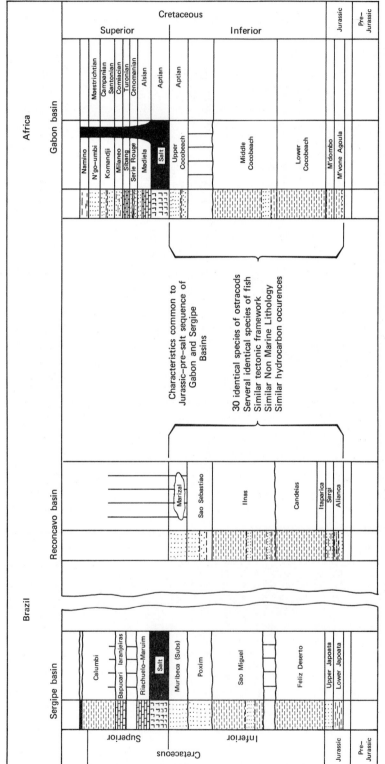

Figure 14-6. Schematic stratagraphic and lithologic correlation between the Sergipe–Reconcavo Basin of eastern Brazil and the Gabon Basin in Africa. (After Allard and Hurst, 1969.)

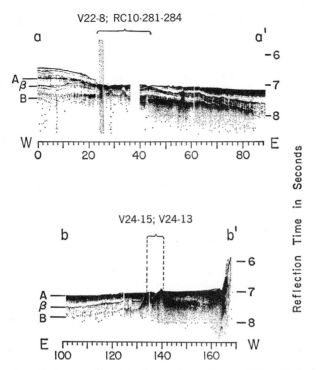

Figure 14-7. Atlantic subbottom profile of horizon outcrop β area. (After Windisch et al., 1968.)

the continents as they drifted, obliterating any Paleozoic evidence on the ocean basin floor. On the basis of sedimentary rates and a spreading of 1 cm/year, Ewing et al. (1967) placed the time of separation of the continents to form the Atlantic at 250 million years ago—somewhere between the Permian and the Triassic. Sea-floor spreading rates have been calculated by Vine (1966) and Pitman and Hertzler (1966) at 4 cm/year for the Pacific, 1 to 2 cm/year for the Atlantic, and 2 to 3 cm/year for the South Atlantic and Indian oceans.

Rifting to produce the Gulf of Aden was suggested by Laughton (1966; see Ewing and Ewing, 1967) as a periodic event associated with major movements during the Eocene and the Miocene. Atlantic spreading cycles are attributed to the Cretaceous and Miocene orogenies by Ewing and Ewing (1967).

Active and quiescent periods of spreading and of sedimentation make the placement of the precise times of activity difficult and hence leave the age of any ocean in doubt, as well. A Triassic origin for the Atlantic seems to be the best bet at the moment, but no matter where we place this event in geologic time, it appears clear that the Atlantic was once much smaller and shallower than it is today.

Similarly, the Pacific may well have been shallower than it is today. Fisher et al. (1970) suggested that the basement rocks of the North Pacific increase in geologic age as the observer proceeds westward from North America (Figure 14-8), and that the oldest basement rocks are Jurassic or older between New Guinea and Japan.

To us, the existence of worldwide fossil species during the Tethys seas implies that these seas were continental and were also connected to one another as shallow

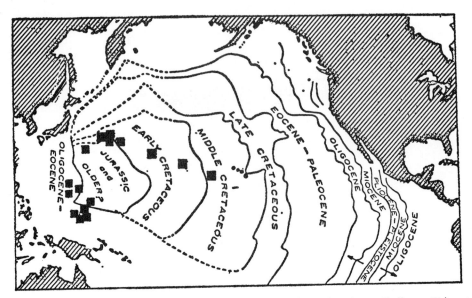

Figure 14-8. Basement ages in the North Pacific and drill sites of leg 6 of the *Glomar Challenger* (Fisher et al., 1970).

seas by the conjoined continents. Continental drift implies that the Atlantic is younger than the Pacific basin. Had the Pacific basin and its deep-sea been populated before the Atlantic (i.e., in Paleozoic times), then the Pacific should show more endemic families and genera in the deep sea than the Atlantic. The Atlantic should be, on an average, populated by geologically younger elements. Because there is no evidence that this is the case, it is reasonable to assume that the existing deep-sea fauna of the world is no older than that of the Atlantic fauna.

A Mesozoic origin for the abyssal fauna has been postulated by several workers, and most recently by Madsen (1961). Even Zenkevitch was not in fundamental disagreement with this idea because he considered Mesozoic relicts as evidence of great antiquity. Because there is at least some evidence that the Atlantic is largely a Mesozoic and Post-Mesozoic event and that even the Pacific fauna of the abysses could be of Mesozoic origin, it is useful to see how conditions on the sea floor with reference to sediment and temperature have changed between the Mesozoic and the Recent. If temperature and sediments have shown no changes, then the argument that the abyss is a refugium of ancient types has much theoretical merit. If, on the other hand, change can be demonstrated, there is little reason to believe that the deep sea offers a prime unchanging home for the preservation and conservation of deep-sea types, as advocated by many workers.

What we wish to emphasize in this brief survey of oceanic (Atlantic) marine sediments is that the *texture, kind,* and *thickness* of the sediment cover *has changed repeatedly throughout geologic time* and cannot be taken as a geophysical constant. *Change* is the constant feature of oceanic sedimentation regardless of whether it was achieved by way of periods of quiescence, orogenic period, sea-floor spreading, or any other means.

EVOLUTION OF MARINE SEDIMENTS

A dominant feature of marine sediments today is their distribution over the surface of the sea floor in quite regular patterns corresponding to distance from land, surface biogenic productivity, water depth, bottom water temperature, and bottom currents (erosional or depositional). The following chart of their distribution is after Kuenen (1950).

Sediment Type	Average Depth (m)	Percentage of Ocean Area	Area (millions of km²)
1. Shelf	100	8	30
2. Hemipelagic	2300	18	63
3. Pelagic			
Pteropod	2000	1	2
Globigerina	3600	35	126
Radiolarian	5300	2	7
Diatom	3900	9	31
Red clay	5400	28	102

The main geographic features of sediment distribution appear in Figure 14-9.

Changes with Time

It is perhaps axiomatic by now that marine sediments have changed in texture, in chemical composition, and in thickness with the passage of geological time. It is the unusual core or seismic profile that shows no change or layering of sediments on the sea bed with time. Pleistocene cores from the Atlantic basin exhibit alternating layers of red clays with sand layers with great regularity and with a frequency that reflects the source of the terrigenous sands below and above which are the abyssal red clays. The Atlantic core of Figure 14-12 reveals not less than 12 sand layers interspersed with mud layers in the past 2 million years. Such is the typical picture of Pleistocene abyssal sediments from the North Atlantic, where the thickness of the Pleistocene mud and sand layers is between 2 and 6 m depending on the source of the sediments and distance from land.

Ericson, Ewing, and Wollin (1963) emphasize that there is no change in the lithology at the Plio-Pleistocene contact but that during the Pleistocene the zones of mild climate were brown, whereas those containing cool-water Foraminifera were shades of gray. The browns are interpreted as representing times of continental desiccation and the grays as characteristic of fluvial continental times reflecting the increase in detrital (organic and inorganic) recruitment from the land.

The oceanic sediments of the Cenozoic are 300 to 400 m thick. It has taken them some 72 million years to accumulate at a rate of 4 to 5 meters per 1 million years. Another 500 m of sediment was deposited in the Late Cretaceous during a period of 78 million years, yielding a faster average rate of sedimentation than during the Cenozoic. The sedimentary cover over the basement rocks thins in the Atlantic as the mid-Ocean Ridge approaches and thickens near the continents. Upper Cre-

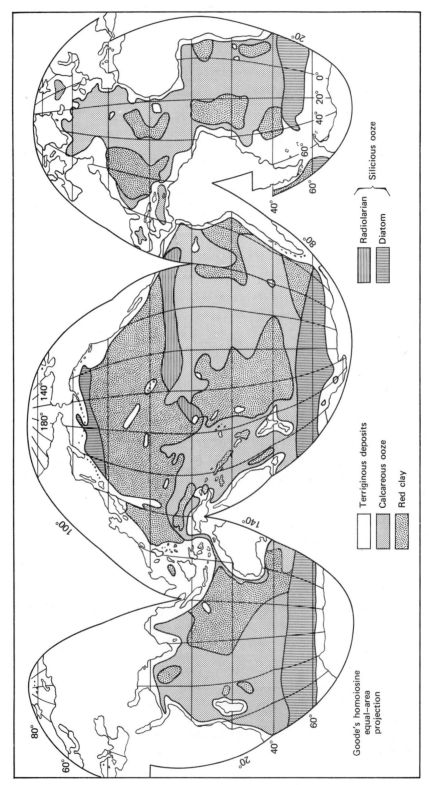

Figure 14-9. Distribution of the various types of pelagic sediments. (After Kuenen, 1950.)

Terriginous deposits

Calcareous ooze

Red clay

Radiolarian

Diatom

Silicious ooze

Goode's homoiosine equal–area projection

338

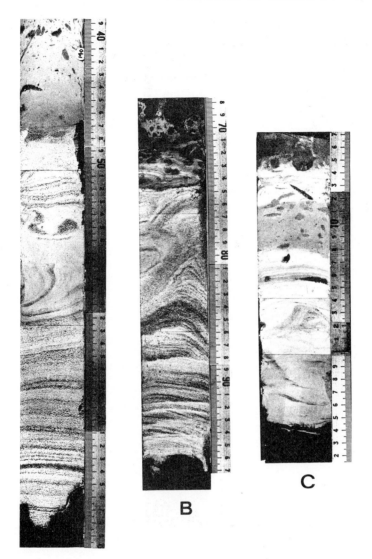

Figure 14-10. Three calcarenite beds from core 7 on Caicos outer ridge. Various structures and worm burrows, delineated by clay mineral concentrations, are apparent. (After Schneider and Heezen, 1966, Plate 3.)

taceous sediments are highly calcareous and often coarse grained. Miocene, Eocene, and Oligocene oceanic sediments are similarly chalky or cream-colored, and they have carbonate contents as high as 70%, consisting often of thin layers of calcilutite (a calcareous clay).

Changes in Pleistocene sediments from tropical Atlantic abyssal depths in grain size, carbonate content, and texture have been described by Schneider and Heezen (1966). Figure 14-10 presents a cut of one core in which the stratification of the sediments is plainly visible. Seismic stratification of the sediments appears in Figure 14-7 and the changes in grain size and carbonate content are given in Figure 14-11. The repeated layering indicates changes in sediment types while the thickness demon-

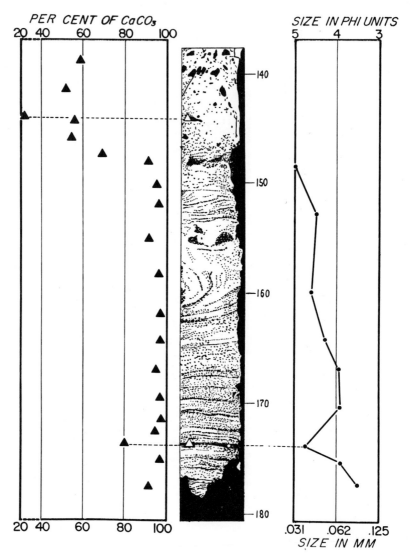

Figure 14-11. Calcium carbonate and grain size analyses of calcarenite bed A in core 7. Note lower mean grain size and lower carbonate content of lamina at 174 cm and the lower carbonate content of the upper gray lutite facies on top of the calcarenite. (After Schneider and Heezen, 1966.)

strates changes in rate of sedimentation. No matter what reasoning is called upon to account for the different sedimentary layers, it is abundantly clear that the layers are different and hence that at any given time the sedimentary home of the abyssal benthos was different at that location.

In Figure 14-12 we have reconstructed the sedimentary history of the Cenozoic–Recent seabed of the Atlantic Ocean from the core descriptions provided by Ericson et al. (1963). This reconstruction is a composite of different depths and locations and is not a single sequence. In each instance except the first, the Pleistocene cover is omitted. The length of any sediment in Figure 14-12 is artificial and bears no relation

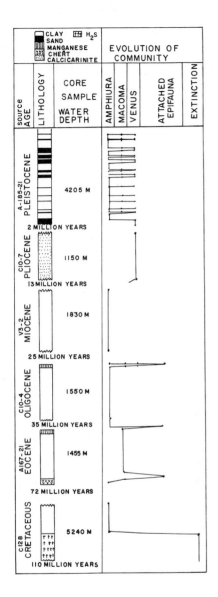

Figure 14-12. Idealized evolution of abyssal level-bottom communities associated with sediment types in cores of appropriate geological age. (Data from Ericson et. al., 1961, and other sources.)

to the thickness of a sequence. The deep-sea core descriptions on which this reconstruction is based follow.

Cretaceous (Habib, 1968)

Core V21-143 is shown in Figure 14-13. The white chalk layer between 24 and 45 cm is the Cretaceous portion below the Pleistocene red clay and the black manganese cap. Below 45 cm the sediments are black. Habib (1968) has described the black Cretaceous layers (Figure 14-14) as being rich in sulfides.

Miocene Era: core no. X3-2; water depth, 1830 m	Core Length (cm)
Pleistocene dark-brown lutite and sand consisting of Pleistocene Foraminifera and pteropod shells	26
Basal contact without manganese oxide	21
Uniform light calcilutite	412
Oligocene Era: core no. C10, 4; water depth, 1550 m	
Unsorted Pleistocene biogenous sand	2
Manganese oxide cap	1
Light-tan foraminiferal calcilutite and abundant discoasters	354
Eocene Era: core no. A167-21; water depth, 1455 m	
Manganese oxide cap	0.5
Uniform light-cream calcilutite	365.0
Pliocene Era: core no. C10-7; water depth, 1150 m	
Calcilutite	25
Abrupt change in lithology	—
Coarsely granular calcarinite	262
Faunal community change and sediment change	

Every biologist who has studied some aspect of the marine biota has learned that the fauna of a mud flat is different from the one living on gravel, and that this fauna differs from that living on sand. Unfortunately, as Thorson (1957b) has pointed out, the usual techniques of soil analysis are too crude to explain the distribution of communities with reference to soil analyses. Indeed, Thorson was finally led to state.

> Bottom animals are so sensitive that they tell us much more about the substratum than the substratum, treated according to recent coarse methods, may tell us about the animals.

But he points out that the *Venus* community is always associated with sand (Davis, 1925; Jones, 1952), whereas the *Amphiura* community is always associated with muddy bottoms. To demonstrate this, it is useful to list selected Thorson-type communities as they are affiliated with sediments. Briefly, the following outline illustrates the faunal changes to be expected with change in sediment type.

Sediment	Community
MUD	*Amphiura*
MIXED	*Macoma*
SAND	*Venus*

Amphiura Communities: (MUD), Powell (1937)
Species
1. *Amphiura rosea*
2. *Echinocardium australe*
3. *Nucula hartwigiana*
4. *Dosinia lambata*
5. *Onuphis aucklandensis*
6. *Glycera americana*
7. *Lembriconereis brevicirrata*
8. *Pectinaria australis*
Soft mud, or mud with some sand. Thorson (1957b)

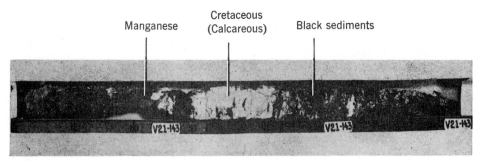

Figure 14-13. Core V21-143: Cretaceous portion is the white layer between 24 and 5 cm, below the black manganese cap. From Pacific, horizon β, Lower Cretaceous. (From Ewing et al., 1966.)

Macoma Communities: (MIXED), Thorson (1933, 1934)

Species
1. *Macoma calcarea*
2. *Mya truncata*
3. *Cardium ciliatum*
4. *Cardium* (=*Serripes*) *groenlandicum*
5. *Ophiocten sericeum*
6. *Pectinaria granulata*
7. *Astarte borealis*
8. *A. elliptica*
9. *A. montagui*

Mixed bottoms. Increasing amounts of silt may lead to a dominance of actinians (*Halcampoides*), or to a transition to a *Portlandia* community. Increasing amounts of sand lead to a dominance of *Cardium groenlandicum* and *C. ciliatum*, or to a transition to a *Venus fluctuosa* community (Zenkevitch, 1930).

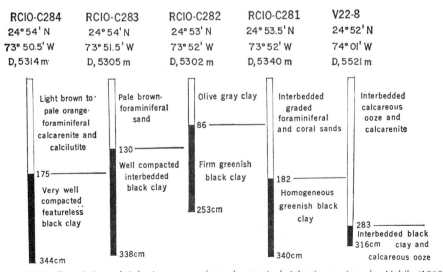

Figure 14-14. Description of β horizon cores from the tropical Atlantic as given by Habib (1968).

Venus Communities: (SAND), Thorson (1934)
Species
1. Venus fluctuosa
2. Thracia truncata
3. Pandora glacialis
4. Euchone analis
5. Scoloplos armiger
6. Astarte borealis
7. A. elliptica
8. A. montagui
Pure sandy bottoms. Increasing amount of silt lead to Macoma community.

The writers do not question that salinity, temperature, predators, food supply, and other factors may be at times more important than sediment type in the establishment of communities. We do state that, holding other environmental factors equal, the sediment type in its broad features plays a great role in determining the community that will be associated with it.

With animal communities and gross sediment type a correlated reality, in broad features at least, it should be possible to connect deep-sea sediment types with expected faunal change, as has been done in Figure 14-12. Quite obviously, the use of Venus and Amphiura communities as examples is only theoretical, because these genera do not in fact live in the deep sea—a Spinula community paired with some other genus might be more appropriate. But Venus and Amphiura are examples of the kind of community changes that might be expected when abyssal sediment types change.

For sediments of Pleistocene age, the rule of abyssal depths is the alternation of eupelagic sediments (clays) with sands derived from shallow water. And in the core that is used as an example, not less than 12 community changes in fauna are expected to have occurred during the last 13,000 years. For most of the ocean basins between the Eocene, late Pliocene, and Pleistocene, turbidites rich in sand are much rarer; thus we conceive of a theoretical dominance of an Amphiura-type community during much of the Cenozoic. During the lower Cretaceous a Venus-type community is expected from the multilayered coarse sediments in horizon A.

The foregoing description of probable abyssal faunal change is clearly somewhat fanciful because it is based on theoretical interpretation, which could, however, become a satisfactory working hypothesis for future studies.

Impact of Changes in Abyssal Sediments on A Deep-Sea Fauna

Each geologic period or epoch is separated from the other by changes in fauna in time and space. To our knowledge, no one has sought to correlate major geologic periods and their major sediment changes with changes in faunal composition in the deep sea. The evidence for sedimentary and environmental changes in deep-ocean sediments is accumulating so rapidly, however, the possible impact of these on a fauna can be deduced.

The abyssal sediments of the Paleozoic (if indeed that was a deep ocean at that time) were believed by Kuenen to consist of inorganic precipitated lime; were this so, no reasonable biologist could imagine a flourishing deep-sea fauna that was in any way comparable to the one existing today.

The hard cherts of the Cretaceous and the Jurassic of the Pacific basins tell of an undeniable change in deep sea sediments during those eras (Fisher et al., 1970), and these sediments, when first deposited, must have altered the composition of the abyssal fauna.

At least during part of the Cretaceous there is evidence for stagnation of the Atlantic deep water as suggested by the sediments of horizon β. It will be recalled that Murray had suggested that the deep water of the ocean of the Cretaceous was probably stagnant and hence unfit for benthic life. This being the case, we could not expect an earlier deep-sea fauna to have survived this catastrophic event.

Manganese caps or coverings in the Atlantic appear to separate the Oligocene from the Miocene, and the Eocene from the Oligocene and the Lower Cretaceous from the Cenozoic. These caps have not been demonstrated with great enough regularity to prove their existence from all ocean basin sediments. However, the regularity with which they occur is of a degree sufficient to constitute presumptive evidence of a significant change in sedimentation between these geologic periods. During such times both the *Venus* and the *Amphiura* communities would be absent and would be replaced by the characteristic epifaunal organisms in the manganese-nodule biotope (e.g., the craniate brachiopod *Pelagodiscus* and the scyphozoan *Stephanoscyphus*).

Most changes in deep-sea sediments relate to gradual changes in texture, grain size, and color. Such changes are more subtle than the catastrophic ones cited earlier. We believe that even changes of this small magnitude would affect community composition as the sediments changed.

MARINE CLIMATES

Kinds

The existing surface marine climates (Figure 14-15A) are arranged in broad bands around the earth from north to south. These are polar or hypopsychral (less than 0°C), subpolar (up to 5°C), boreal (5−10°C), temperate (10−25°C), subtropical (15−30°C), and tropical (25−30°C) (Hedgpeth 1957). The subpolar and boreal faunas are poorly distinguished in the north but are quite distinctive in the southern hemisphere. The surface marine climates are almost entirely surface phenomena, and their geographic extent is influenced strongly by surface currents (Figure 14-15A).

At the surface the banding of the climates mentioned previously shows the tropical zone occupying close to 50% of the surface of the earth (Figure 14-15A).

Variations in Recent Marine Climates with Depth

By 100-m depth (Figure 14-15B)—half the average depth of the continental shelf—the tropical–subtropical zone is restricted to a small area in the Indian Ocean, the Red Sea, a few smaller areas in the North American tropical Atlantic, and to North Australia and Borneo, extending fingerlike with the equatorial current. The temperate climate now occupies most of the area of the oceans, and the extent of the boreal similarly has been increased.

By 400-m depth (Figure 14-15C) there is no tropical–subtropical marine climate

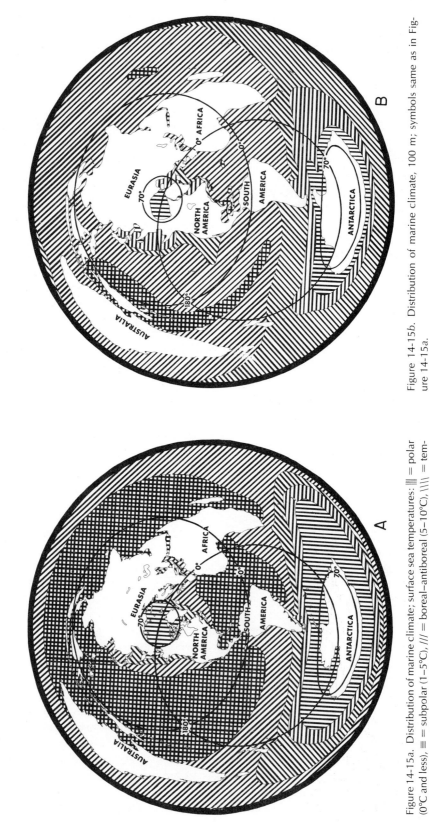

Figure 14-15a. Distribution of marine climate; surface sea temperatures: ▤ = polar (0°C and less), ☰ = subpolar (1–5°C), /// = boreal–antiboreal (5–10°C), \\\ = temperate (10–20°C), ▦ = tropical and subtropical (20–30°C).

Figure 14-15b. Distribution of marine climate, 100 m; symbols same as in Figure 14-15a.

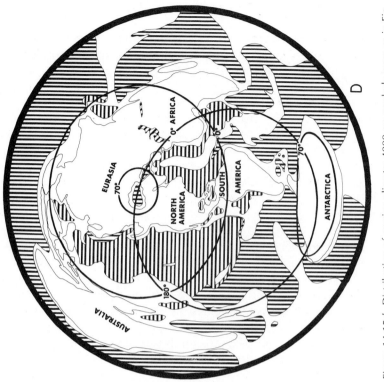

Figure 14-15d. Distribution of marine climate; 4000 m; symbols same as in Figure 14-15a. Note the absence of boreal, temperate, subtropical, and tropical temperatures. Maximum temperature, in the Red Sea, is above 15°C. Otherwise, temperatures are less than 0°C or 1 to 2°C subpolar.

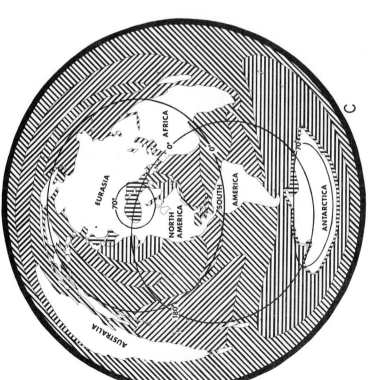

Figure 14-15c. Distribution of marine climate, 400 m; symbols same as in Figure 14-15a. Note the absence of tropical and the restriction of temperate temperatures to Caribbean and tropical North Africa, Central America and Northern Australia, Southern Japan, India, and tropical Indian Ocean.

347

on earth except in the Red Sea. The temperate is reduced in areal extent to the region of the Indo-Pacific, whereas the subpolar and boreal now occupy most of the area of the ocean. Below 400 m the importance of subpolar and polar marine climates increases.

By 4000 meters (Figure 14-15D) only the polar and subpolar marine climates exist in the oceans, and these are separated by submerged land areas. The Mediterranean maintains its temperate character. Polar water from Antarctica dominates most of the area of the earth as a subpolar marine climate.

Today it is only in the Mediterranean, the Red Sea, the Arctic, and the Antarctic that any single marine climate extends directly to the deep-sea floor from the surface. Thus it is only in these specialized environments that we can seek to find the eurybathial stenothermic organisms that would be capable of penetrating the abyss without added stepwise temperature adaptation.

The major conclusion to be reached with reference to marine climate with depth is that the tropical marine climate is a surface or near-surface phenomenon that has no influence below 400-m depth and very little below 100 m. The climates of importance to faunal zonation in the deep sea are the *polar,* the *subpolar,* and to a lesser extent the *boreal.* The submerged temperate climate is important only in the Mediterranean. The subtropical climate is important only in the Red Sea.

If the origin of the deep-sea fauna has an orderly relationship to marine climates, the origin should be sought in the polar, subpolar, and boreal marine climates of the past ages. Here we assume that it is easier physiologically and genetically for an animal to evolve into a polar one from a subpolar (or vice versa), than for a tropical species to evolve into a polar one without becoming first subtropical, then temperate, boreal, subpolar, and finally polar.

Variation in Marine Climate Through Time

Paleoclimatology has been a subject of interest to scientists for many years, and several symposia (Nairn, 1961, 1964) have been devoted to the subject. The importance of marine climate to the distribution of plant and animal life today needs no emphasis, being irrefutably evident in the surface climate zones. Here we are concerned with average climate conditions over periods of years and not with seasonal fluctuations. In the following marine climate reconstruction, isotopically determined temperatures have been given preference over other less objective sources. These are determined from the ratio of ^{16}O to ^{18}O from the carbonates in the fossil skeleton. Sources of data for various regions of the earth are given in Table 14-1 and summarized data appear in Figure 14-16.

				Temperature (°C) or Climate				
Time[1]	Million Years	Arctic	Northeastern Asia	North America Northern Europe	Tropical Surface	Tropical Deep-Seat	New Zealand–Australia	Gondwana or Antarctic
Paleozoic								
Pre-Cambrian	600	(0–1.8)[4]	—	Glacial[4]	Cold (?)	2	Glacial	(0 to −1.8)
Eocambrian	580	(0–1.8)[4]	—	Glacial[4]	30			Glacial[6]
Cambrian	530	Warm[4]	—	Warmer[4]	30	15	Warm	Warm (?)
Ordovician	500	Warm[4]	—	Warmer[4]	30	15	Warm	Warm (?)
Silurian	425	Warm[4]	—	Warmer[4]	30	15	Warm	Warm (?)
Devonian	405	Warm[10]	—	Warmer[10]	30	less than 15	Warm[7]	Cool (?)
		Warm[4]	—	Warm[4]	30		Cold[6]	Cold
Mississippian	345	Warm[4]	—	Warmer[4]	30	2+	Cold[6]	Colder
Pennsylvanian	310	Warm[4]	—	Warmer[4]	30	2+	Cold[6]	Glacial
Permian								
Early	280	Warm[4]	—	Warmer[4]	30	2+	7.7[8]	Cold
Late	—	Warm	—	Warmer	30	15	17–26†	Temperate[6]
Mesozoic								
Triassic								
Early	230	Warm	—	Warmer	30	15	Tropical	Temperate[6]
Late	—	Warm[4]	—	Warmer	30	15	Tropical	Temperate[6]
Jurassic								
Early	181	Warm[4]	—	Warmer	30	15	Tropical	Temperate[6]
Late	—	Warm[4]	—	17–23[4,12,13]	30	20†	Tropical	Tropical[6]
Cretaceous	135		—	Tropical	30	15	Warm–temperate†	Temperate†
Aptian	130	15[2]	—	16–23[4]	30	15	—	Temperate[6]
Santonian	84	16–17[2]	—	20–27[2]	30	17[2]	Warm–temperate†	Temperate
Cenozoic								
Paleocene	63	25[3]	Warm[11]	—	30	14[2]	10[5]	10† Boreal
Eocene	58	20[3]	Warm[11]	21[4]	30	12[2]	11[5]	8† Boreal
Oligocene	36	18[3]	Temperate[11]	18[4]	28[4]	10[2]	15–17[5]	6† Subpolar
Miocene	25	13[3]	Warm[11]	16[4]	24[4]	7[2]	17–15[5]	3† Glacial
Pliocene	13	10[3]	Temperate[11]	14[4]	30	4[2]	15–9[5]	0, Glacial
Pleistocene	2	0*	5	10	30	2[2]	8–13	0 to −1.8, Glacial
Recent	0	0	10	15	30	2	10–20	0 to −1.8, Glacial

* High Arctic only, doubtless the Low Arctic was higher in temperature.
** All temperatures cited are approximations within a given marine climate.
† Interpolated.
Sources
1. From Kulp (1961).
2. Lowenstam and Epstein (1959).
3. Craig after Durham (1961).
4. Schwartzbach from various sources (1961).
5. Gill (1961).
6. King (1961).
7. Lowenstam and Epstein (1959).
8. Lowenstam (1964), *in* Nairn.
9. Maack (1964, *in* Nairn) for South America.
10. Shirley (1964, *in* Nairn) for Northern Europe.
11. Kobayashi and Shikama (1964, *in* Nairn) for Northeastern Asia.
12. Imlay (1965).
13. Briggs (1970).

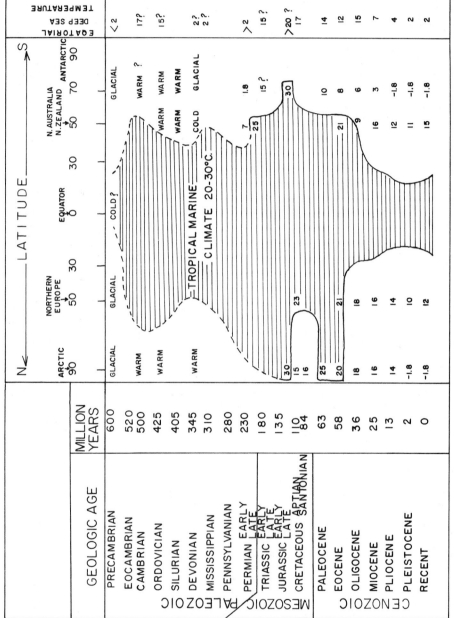

Figure 14-16. Evolution of surface marine climates.

PRE-CAMBRIAN TO EOCAMBRIAN

Both the present Arctic and Antarctic show evidence of Pre-Cambrian glaciation. New Zealand and North Europe similarly appear to have undergone glaciation. From this it is reasonable to surmise that a bipolar glacial marine climate existed and that a surface climate stratification with latitude existed comparable to that of today.

CAMBRIAN TO SILURIAN

The Cambrian Arctic was warm (Schwarzback, 1961), as was the Antarctic. No precise temperature assignment is possible, but there is evidence neither of glaciation nor of tropical environments. Accordingly, the period is considered to have been temperate. This condition continued at the poles through the Ordovician and the Silurian. During the Devonian there is some evidence of cold conditions in South Africa and in the Falklands (King, 1961), where the fauna appears to be boreal (Shirley, 1964). Conditions at the North Pole are not known but are suspected to be warm. A cold climate appears to have persisted at the South Pole in between the Devonian and the Pennsylvanian.

PENNSYLVANIAN TO EARLY PERMIAN

The Pennsylvanian period showed glaciation at the South Pole according to King (1961), but the northern hemisphere was hot—humid or arid in the view of Schwarzbach (1961). The North Pole was apparently no colder than temperate. Tropical conditions extended over much of the earth. Both New Zealand and Australia were cold. This unipolar glaciation continued into the Early Permian and any bottom water that was cold in character would have had to be generated in the southern oceans.

LATE PERMIAN

The Late Permian resulted in a replacement of glacial conditions by a temperate environment in Antarctica. Temperatures in New Zealand and Australia ranged between 24 and 26°C, or nearly tropical conditions, according to King (1961). The Antarctic can be presumed to have been only a little cooler than New Zealand, leading to a subtropical climate. The Arctic was similarly warm, and northern Europe warmer, both being under the influence of a widespread tropical sea, the Tethys.

TRIASSIC AND EARLY JURASSIC

The Triassic and Early Jurassic continued with the same temperature conditions of the Late Permian; polar temperature conditions were temperate at their lowest. New Zealand and Australia showed tropical conditions as far south as 40°S latitude.

LATE JURASSIC

The climate of the world oceans was tropical, and there was no polar stratification during the Late Jurassic. Temperatures in northern Europe ranged between 17 and 23°C according to Schwarzbach (1961).

CRETACEOUS

During the Cretaceous the Arctic showed a coldest temperature of 15°C according to Lowenstam and Epstein (1959) (Figure 14-17); northern Europe was temperate to

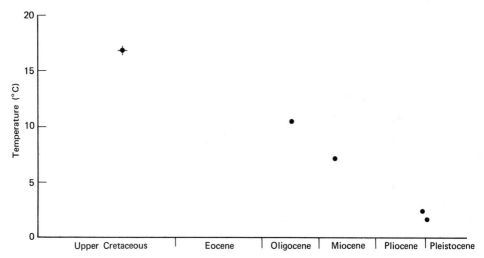

Figure 14-17. Temperature of East Greenland of Late Mid-Cretaceous compared with deep-sea temperature of the Tertiary. [Cretaceous datum from Greenland belemnite; Tertiary data after Emiliani (1954) from deep-sea benthonic Foraminifera.] From Lowenstam and Epstein (1959).

tropical, and New Zealand was tropical. Antarctica appeared to be temperate, according to King (1961).

PALEOCENE TO EOCENE

During the Paleocene and Eocene periods the Arctic and northern Europe were decidedly tropical (Craig, 1961), whereas New Zealand showed boreal conditions (ca. 10°C) according to Gill (1961). We may infer that Antarctica displayed boreal subpolar or polar conditions.

OLIGOCENE

By the Oligocene time the climate at the North Pole and in northern Europe was subtropical, as shown by Craig (1961). The Antarctic inferentially was subpolar at around 6°C, because New Zealand and Australia were boreal–temperate between 15 and 17°C.

MIOCENE

The Arctic Miocene temperature and that of northern Europe may be classified as temperate at 16°C (Craig, 1961), being quite similar to the climate of New Zealand (Gill, 1961). The equatorial deep-sea temperature was 7°C (Lowenstam and Epstein, 1959), and hence that of the Antarctic was probably subpolar, near 3°C, with glaciation a strong likelihood in places. Thus between 36 and 25 million years ago, glaciation commenced in Antarctica, amassing sufficient strength to form cold bottom water that was transported to the equatorial depths.

PLIOCENE

The Arctic water during the Pliocene was of a temperate nature, near 14°C, as was the climate of northern Europe (according to Craig, 1961). New Zealand (Gill, 1961) ranged between 9 and 15°C. Antarctica was glaciated and had a polar climate.

The Pleistocene period marks the commencement of an Arctic polar ice cap some 2 million years ago with a single-step climatological jump from temperate to polar conditions. The deep-sea equatorial bottom climate similarly dropped from 4°C to the +2°C condition, which it retains today. Antarctica continued to cool. Northern Europe and New Zealand showed temperate conditions between 8 and 13°C.

The changes in surface marine climate at the poles, in northern Europe, at the equator, in New Zealand and Australia, and in the sea bottom marine climate at the equatorial deep sea are shown in Table 14-1 and in Figure 14-16. The data are derived from the sources cited, and blanks are inferred from one or more known temperatures during any given geologic period. The likely marine climates on the surface of the earth during each geologic period are also indicated in Figure 14-16.

This condensed version of marine climate through time, although probably not precise to a given unit of temperature in degrees Celsius, does show a revealing pattern that strongly suggests the following conclusions.

The Most Persistent Marine Climates

Two marine climates have apparently persisted on the surface of the earth throughout geological time. These are the tropical (including subtropical) and the temperate, but their position on the earth has shifted — sometimes being at a pole and other times being at the equator. Additionally, the extent of the area covered has varied. The equator appears always to have had a tropical climate, yet the poles have witnessed all types of marine surface climate, depending on the geologic era.

NORTH POLAR SURFACE MARINE CLIMATES

At the North Pole throughout most of geologic time the temperature may be considered as temperate. The two glacial times are in the Pre-Cambrian some 600 million years ago and in the Pleistocene only 2 million years ago. There is a possibility of cool conditions in the Devonian. Tropical conditions prevailed during the Lower Jurassic, the Paleocene, and the Eocene. The transition from the temperate conditions of the Pliocene to the glacial conditions of the Pleistocene appears to have been abrupt, our evidence suggests that there were striking catastrophic consequences as far as the Arctic fauna is concerned.

SOUTH POLAR SURFACE MARINE CLIMATES

Three or four glacial times occurred in the South Pole, one in the Pre-Cambrian, possibly one between the Devonian, and the Penn–Permian, and finally one commencing in the Miocene. In contrast to the Arctic, the transition from the tropical conditions of the late Jurassic to the glacial Miocene appears to have been a gradual process, lasting some 84 million years.

We believe that this strikingly different marine climatological history of the North and South poles accounts for the differences in the faunal composition between the two polar regions (see also Chapters 6 and 7).

THE DEEP–SEA MARINE CLIMATE EVOLUTION
AND EXPECTED FAUNAL CHANGES

This synopsis is based on the surface marine climate history insofar as it is known or suspected. It is assumed that the water volume was constant and that the continents were approximately as they are today, even though proof for these two assumptions is lacking. It is further assumed that the deep-sea temperatures can be no colder than the coldest surface temperature at any given time. This assumption has been developed by Lowenstam and Epstein (1959) and is consistent with what we know about deep-water formation. Deep water is formed only when the surface water sinks because its density has exceeded that of the water below it. Increased density can be caused by increasing salinity or by decreasing temperature, and it is logical to conclude that at no time can the deep and bottom water be colder than the coldest surface temperature. It can be warmer or the same as the coldest surface temperature but no colder. Thus when the earth was uniformly hot as in the Late Jurassic a tropical deep-sea temperature must have existed.

Our knowledge of deep-water formation reminds us that the major source of bottom water is the interconnected southern oceans, and that the enclosed Arctic sea plays no important part in the formation of bottom water in the Pacific, Atlantic, or Indian oceans. The −1.8°C water of Antarctica today yields an average temperature

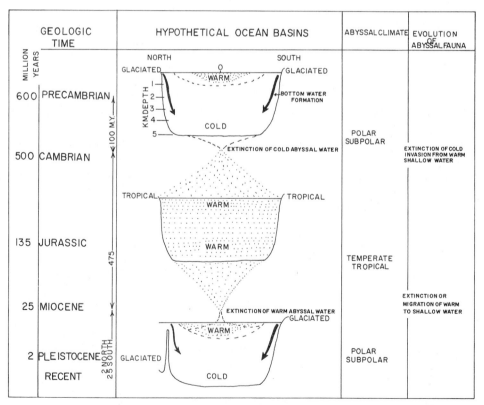

Figure 14-18. Evolution of deep-sea marine climates and associated faunal change. Unipolar Paleozoic glaciation of Antarctica omitted.

of 2.0°C for the Atlantic and Pacific bottom water, and it is imagined that this 4°C gradient was maintained in past eras.

Changes in marine climate from cold to warm on the bottom of an ocean affect the bottom fauna more than changes at the surface affect the shallow-water or pelagic fauna. Surface fauna faced with a climate change can migrate to a "best-liked" climate, providing this climate is present somewhere at the surface. When faced with a warming trend, surface fauna can migrate to deeper water to seek the colder climate. Deep-sea faunas, which are adapted to hypopsychral or psychral climates, have nowhere to go when subjected to a change from a cold to temperate or tropical bottom temperature. If they migrate toward the surface they will meet warmer temperatures rather than colder ones; hence they will become extinct unless they can either evolve or adapt to become thermophiles. When deep-sea faunas previously adapted to a temperate climate are faced with the appearance of cold bottom water, they can migrate to a shallower depth to find warm water, they can evolve into cold adapted species, or they can become extinct. These are the options.

The magnitude of deep-sea bottom temperature change throughout the geologic column is shown in Figure 14-18. It is apparent that this bottom temperature has likely changed throughout time, being first cold and at times tropical in climate and then finally cold again.

Pre-Cambrian

During the Pre-Cambrian glaciation, it is likely that deep-sea temperatures were cold, as they are today.

Cambrian to Silurian

From the Cambrian to the Silurian the bottom temperatures were probably quite warm and any previously cold-adapted fauna would have been forced either to evolve into a warm-water type or to become extinct. The fauna could not seek a cold-water climate because there was none on earth. Any Cambrian invasion of a deep-sea floor must have come from warm-water types of organisms.

Devonian to Permian

The Antarctic glaciation of the Devonian to Permian eras must have resulted in replacement of the warm bottom waters of the Cambro-Silurian with cold water similar to, but perhaps slightly warmer than, the sea of today. The fauna of a deep sea theoretically would be classified as a cold-water fauna and the marine bottom climate as polar or boreal. This means that the previous warm-water fauna of the Cambro-Silurian would have to seek shallower water in order to avoid the cold temperatures; otherwise, it would have had to evolve into a cold-water fauna from its previous warm-water state.

Late Permian

The change to the temperate conditions of the Permian in its late stages must have been catastrophic to the previously existing cold-water abyssal fauna, which could not find cold water anywhere on earth. Evolution or extinction must have been mandatory for the cold-water forms of the early Permian. The abyssal fauna likely experienced a temperate marine climate close to 15°C.

Triassic to Eocene

Between the Triassic and Eocene eras the surface marine climate was warm over most of the earth. The Arctic was temperate in character, as was the Antarctic. The bottom temperature of the deep sea was Mediterranean in character, somewhere between 13 and 15°C. The Eocene Arctic was tropical; the South Pole, showed a continuing cooling trend, culminating in the glaciation of the Antarctic continent, perhaps by the Miocene and certainly by the Pliocene. Thus the deep-sea temperatures also cooled gradually from the warm Cretaceous and Eocene conditions. This implies that if any relict forms exist in the seas today, they probably have temperature or tropical temperature requirements. Such temperatures prevail in the intertidal, the shelf, and the archibenthal zone of transition today, and it is also here that most of the relict marine species are found.

Eocene to Recent

Between the tropical conditions of the Eocene and the Pleistocene glacial times, the history is one of gradual cooling at the poles and hence also in the deep sea. There is little reason to doubt that bottom temperature cooling was accelerated during the Pleistocene when the North Pole took a climate character comparable to that of the South Pole. The abrupt arrival of the Arctic cold-water conditions is a comparatively recent event geologically, and those connections which exist between the Arctic deep-sea fauna and the deep-sea fauna elsewhere can only be viewed in this light.

This climatologic history implies that certain deep-sea species of today likely originated in the southern hemisphere through a gradual process of cold acclimatization.

PERMANENCE VERSUS CONSTANCY IN THE DEEP-SEA

The foregoing treatment of ancient deep-sea marine climates and ancient sediment types and their distribution throughout geological time tells one bold story: neither the marine sediments nor the deep-sea bottom temperatures have been constant throughout geological time, and in contrast to the view of Zenkevitch, they can not be considered as geo-physical constants.

The uniformly hot tropical seas of the Late Jurassic set the stage for interpretation of the existing marine fauna, shallow or deep which, since that warm period, must have evolved into the various climatologic marine faunal types seen on earth today. It follows that all existing marine fauna must have come from warm-water types since the Tethyian Jurassic seas. This statement applies equally well to the fauna of the deep sea which we now view as having, in the main, a Jurassic or Post-Jurassic ancestry. The first occurrence of the cold +2°C water in the ocean deeps since the Jurassic was between the tertiary Miocene and Eocene, and it seems likely that it was between these times that the deep-sea fauna accumulated the few archaic taxa that it has today.

ORIGIN AND AGE OF THE DEEP–SEA FAUNA

Over the past 100 years there have been conflicting views regarding the origin and antiquity of the abyssal fauna.

Author and Date	View
Wallich (vide Agassiz, 1888)	Deep sea has its own special fauna, as it has always had in ages past
Locard (1897)	Deep sea molluscs with affinities to Tertiary
Agassiz (1888)	Deep sea fauna originated at the close of Paleozoic times
Murray, see Herdman (1923)	Freshwater fauna is much more archaic than the deep sea variety
Perrier (1899)	Paleozoic zone in sea located between 400 and 2000 m
Zenkevitch (1966)	Great antiquity because of constancy throughout geologic time
Murray (1895)	Deep sea uninhabitable to higher life during the whole or most of the Paleozoic
Bruun (1957)	Deep sea populated mainly through Pleistocene invasions
Menzies and Imbrie (1958)	Deep sea fauna has fewer Paleozoic types than freshwater or shallow seas

Our study on zonation sheds some light on the problem, especially with reference to site(s) of origin of the contemporary abyssal fauna, and new climatologic evidence aids appreciably in fixing a geologic time for the origin of the abyssal fauna. Previously the abyssal fauna was set at an arbitrary bathymetric limit, in the absence of knowledge that this limit varies in different parts of the ocean, being shallower at the poles and deeper in the low latitudes.

Average Age of the Deep-Sea Fauna

The average age of the deep-sea fauna was determined by Menzies and Imbrie (1958) by calculating the percentage of genera and families in the groups of marine fauna that have a fossil record and for which bathymetric distributional data were available. In general they found that with increasing depth the proportion of Paleozoic types (types having Paleozoic representation in the fossil record) decreased. This was true also for Mesozoic types. Tertiary deep-sea animals showed an increase at depths below 2000 m in all instances except for the Foraminifera.

This evidence was not accepted by Zenkevitch and Birstein (1960), who pointed out that of the million or more living marine species, a very small fraction was treated by Menzies and Imbrie (op. cit.). But this criticism must apply to the entire fossil record, which is admittedly only a small fraction of the fauna of a given geologic epoch or period. Probably the trend is not invalidated because it was repeated for several of the groups of animals. Two major groups of organisms, the Echinodermata and the Mollusca were not treated by Menzies and Imbrie because of a lack of reliable information about those groups at that time. Since then the monographs by

Madsen (1961) on the Porcellanasteridae and by Clarke (1962a, 1962b) on abyssal molluscs of the world have added confirmatory data.

ECHINODERMATA: ASTEROIDS

Madsen's (*op. cit.*) conclusion is that the Porcellanasteridae seem to have originated (probably in the Late Mesozoic) in bathyal depths of the Tethys Sea from astropectinid-like ancestors. All three exclusively bathyal genera (two of which may be the most primitive porcellanasterid genera) are confined to the region of the Indian Ocean which, on the whole, comprises the greatest number of different forms. Some of the world-ranging species of Porcellanasteridae seem to have a rather confined bathymetric distribution. The porcellanasterid genera hitherto recorded from the greatest depths are probably the youngest phylogenetically. Madsen stated:

> This and other considerations support the assumption that in the main the recent abyssal fauna is younger than the bathyal* and littoral faunas, and that probably on the whole the ancestors of the oldest types of abyssal animals today did not invade the abyssal habitat until Early Tertiary.

Madsen's phylogenetic scheme and depth of known young (Y) and ancient (P) genera of the Porcellanasteridae appears in Figure 14-19. Five of the 6 young genera have

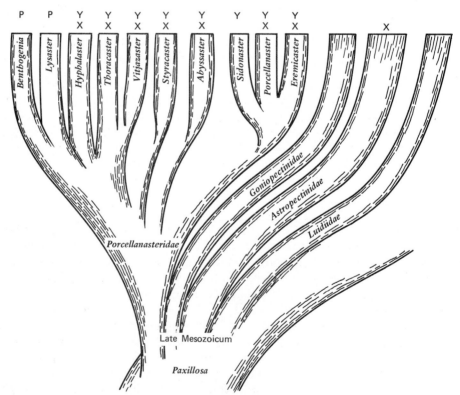

Figure 14-19 Phylogenetic tree, showing the supposed relationships of the recent Porcellanasteridae. (After Madsen, 1961.) × indicates abyssal dwelling, P indicates primitive genus and Y a young genus.

* Bathyal according to Madsen is 1000–2500 m with an upper temperature limit of 12°C. This coincides approximately with our placement of the archibenthal faunal zone of transition.

abyssal representation, whereas only two of the primitive genera live in abyssal depths.

MOLLUSCA

Although Clarke (1961a, 1961b) first concluded that abyssal molluscs indicated a great antiquity, he later (1962b, p. 4) wrote, after completing his monograph on the abyssal world molluscs.

> Moreover the distribution of abyssal molluscs appears to parallel that of the recent shallow-water fauna. Apparently the deep-sea has not, in general, provided a refugium for Paleozoic to Recent survival, and its fauna is probably of relatively recent origin.

DEPTH OF CONSERVATION OF ARCHAIC TYPES

The depth at which archaic types are concentrated in today's oceans is of some interest especially if we derive the more recent taxa from the older ones in the evolutionary process. Of the 44 genera in the following list, none is older geologically than the Jurassic; most date from the Cretaceous onward, and most are found at shelf or archibenthal depths. The Paleozoic relicts such as *Lingula* (shelf–intertidal) and *Xiphosura* (intertidal) are excluded. One illuminating group, the Cidaroidae, whose 80 living species date from the Carboniferous, have 90% of the species confined to shelf or archibenthal depths (Madsen, *op. cit.*, from Mortenson, 1928) (Table 14-2). *Neopilina*, which remains the sole Paleozoic relict species in the abyssal fauna, now constitutes less than 0.01% of the total abyssal species.

PROPORTIONAL SURVIVAL OF ANCIENT FORMS

Ancient forms, in this case genera, are considered to be those with a continuous geologic range from past geologic periods to the Recent. The proportion of surviving genera differs among different groups of animals, and this may reflect differential extinction rates. The fact of paleontology remains — the closer we come in time to one geologic period from another, the higher the number of surviving genera and species. Thus 80 to 95% of the fossil Pleistocene species may be identified with Recent species, whereas only 5 to 6% of the Mesozoic echinoid genera may be identified with Recent genera. But 94 to 95% of the Tertiary genera of echinoids find their

TABLE 14-2.   DEPTH DISTRIBUTION OF CIDAROIDEA RANGING FROM CARBONIFEROUS TO RECENT TIME (10 RECENT GENERA WITH 80 SPECIES)

Numbers of Species	Depth (m)	Zone
60	100–1000	Shelf–archibenthal
13	1000–2000	Archibenthal
4	2000–3000	Abyssal
3	3000–4000	Abyssal

counterparts in the Recent, 36% of the hermatypic corals range from the Cretaceous to the Recent, and 64% range from the Tertiary to the Recent (Durham and Allison, 1960).

Extinction and evolution are the dominant features of the changes taking place in the marine fauna throughout geologic time. Thus it is not astonishing to find that the Recent marine fauna bears a closer relationship to the Tertiary than to the Cretaceous and hardly any to the Paleozoic. What is astonishing (although it was suspected long ago) is that the majority of those marine genera and families which have a long fossil record are conserved in the oceans in shallow water, the shelf, and the archibenthal and not in abyssal depths (Table 14-3).

SITES OF ORIGIN OF THE ABYSSAL FAUNA

It is believed by many that the abyssal fauna originated from a shallow-water fauna by adaptation to the less rigorous abyssal conditions. The intertidal fauna today is strongly separated from the shelf fauna, and the shelf fauna is strongly separated from the archibenthal fauna. It is the archibenthal fauna which shows the closest affinity to the abyssal fauna. However, in any single geographic locality the archibenthal constitutes a small and diminishing proportion of the abyssal fauna. The abyssal fauna is so strongly isolated at low latitudes from the archibenthal, the shelf and the intertidal faunas in any single geographic locality that it must be considered a distinctive faunal unit, quite independent from the Recent shallow-water fauna of the low latitudes.

The suggestion of Beurlen regarding decapods (see Madsen, 1961) is that major periods of regression of ancient seas forced the flourishing fauna of the epicontinental seas onto a constricted and now overpopulated shallow shelf. As a consequence, animals were forced to live in deeper zones, and they were lost from the fossil record. This idea is an attractive, interesting, and reasonable hypothesis which deserves the most careful consideration.

The epicontinental seas of the landmasses of each major geologic period were once flooded and alternately dried. The greatest of these was the Mid-Cretaceous transgression or Tethys Sea, covering more than one-third of the earth's land surfaces. How this came about is not certain, but the etiology of the phenomenon has little relation to the question of abyssal faunal evolution. The fact of existence of the seas and their subsequent shrinkage and loss is the matter of importance. Even Pleistocene sea level has shown a range between-200 to +18 m of the present sea level (Table 14-4).

If there was a deep ocean at the time of the Cretaceous, this would have been an ideal time for a penetration into deeper water because the surface temperatures and bottom temperatures must have been nearly uniform top to bottom (except perhaps at the tropics, where warmer surface conditions were likely to prevail). At this time conditions were probably temperate even in Antarctica, and a bottom temperature close to 15 to 17°C (temperate) may be imagined. Animals occupying the temperate seas of Antarctica should have had no more difficulty in penetrating the homothermal deeps than species do in the Mediterranean today. It is assumed that the earlier Paleozoic cold-water bottom forms had by now become extinct (because they had encountered the Late Permian warm period and could seek cold water nowhere on earth).

Species of Animals	Geological Range	Usual Depth (m)
Crustaceans		
Euryonicus	Jurassic	Archibenthal
Thaumastocheles	Jurassic	Shelf
Bathynomus	Cretaceous	600
Polycheles	Jurassic	Archibenthal
Penaeidae	Jurassic	Littoral shelf–slope
Echinoids		
Brissopsis	Oligocene	Shelf, 600
Cassidulus	Early Cretaceous	600
Periaster	Middle Cretaceous	600
Salenia	Late Cretaceous	600
Echinocyamus	Late Cretaceous	600
Hemiaster	Middle Cretaceous	600
Spatangoidea	Cretaceous	To hadal depth
Asteroids		
Hyphalaster	Late Mesozoic	2275–5415
Thoracaster	Late Mesozoic	2600–5000
Styracaster	Late Mesozoic	2550–5610
Eremicaster	Late Mesozoic	1570–7200
Porcellanaster	Late Mesozoic	1160–6035
Lysaster	Late Mesozoic	1000–
Benthogenia	Late Mesozoic	905–925
Sidonaster	Late Mesozoic	1150–2300
Abyssaster	Late Mesozoic	3200–6280
Vitjazaster		7000 [+]
Hermatypic corals		
Pleurorchium	Cretaceous	Shelf
Tretorete	Cretaceous	Shelf
Phoberus	Jurassic	Shelf
Astrocoenia	Cretaceous	Shelf
Cladocora	Cretaceous	Shelf
Diploria	Cretaceous	Shelf
Favia	Cretaceous	Shelf
Favites	Cretaceous	Shelf
Hydnophora	Cretaceous	Shelf
Madracis	Cretaceous	Shelf
Sidastrea	Cretaceous	Shelf
Stephanocoenia	Cretaceous	Shelf
Solitary corals		
Bathycyathus	Cretaceous	
Caryophyllia	Cretaceous	Shelf
Ceratotrochus	Cretaceous	
Desmophyllium	Cretaceous	Shelf
Microbracia	Cretaceous	
Oculina	Cretaceous	
Platycyathus	Cretaceous	
Platytrochus	Cretaceous	
Sphenotrochus	Cretaceous	
Trochocyathus	Cretaceous	

TABLE 14-4. PLEISTOCENE CLIMATES
AND SEA LEVELS*

General Nomenclature	Sea Levels (m)
Postglacial temperate	Present
Last glacial age	
Phase 3, cool	−30 (ca.)
Interstadial, less cool	−10 (ca.)
Phase 2, cool	−70 (ca.)
Interstadial, temperate	+3
Phase 1, cool	−100 (ca.)
Last interglacial age	
Temperate	+7.5
Cool	Lower
Temperate	+18
Penultimate glacial age	−200 (ca.)

* Modified from Zeuner, 1954, *Eiszeitalter und Gegenwart,* Vol. 4/5, pp. 98–105. (After Ericson et al., 1956.)

It is possible that the Cretaceous deep-sea bed was uninhabitable for reasons of stagnation (again, a situation comparable to the Mediterranean). It is postulated, therefore, that the Cretaceous relict forms entered the deep ocean principally from the Antarctic continent and spread from there along the 15-17°C isotherm in shelf and archibenthal depths of equatorial seas; here, from this time onward the Cretaceous relicts could find such warm temperatures somewhere on earth.

Since the Cretaceous, the deep-sea bottom waters have cooled gradually by around 3°C for each 10 to 16 million years; this pace should have been sufficient to allow for a gradual evolution of abyssal cold-adapted forms from the Cretaceous and Tertiary ancestors in the archibenthal depths of the Antarctic.

As geologic time passed and the Antarctic continued to cool, its evolving fauna could send immigrants into the deep sea with greater ease. This was possible because from the Eocene to the Pleistocene the Antarctic controlled the cold bottom temperatures of the world oceans, with ever-increasing force.

ESTABLISHING ANTIQUITY

The basic tenet of paleoecology is that thermal tolerance is a conservative feature of a species. Applying this useful concept to marine distribution, we can arrive at various distributional patterns that are suggestive of origins.

1. Temperate shelf and archibenthal animals are most likely to be relict descendents of the Tethys Sea, especially if found in temperate water conditions. Examples of such relicts from the shelf and the archibenthal are *Astropecten, Polycheles, Euryonicus,* the cidaroids, and various nonstalked crinoids as well as the stalked *Rhizocrinus.*

2. Endemic genera and species found in hypopsychral temperatures can only be

viewed as having evolved from the Antarctic (Miocene to Recent) and the Arctic (Pleistocene to Recent); the Antarctic forms are likely to be Miocene or Pliocene, whereas the Arctic forms must have come into being only since the Pleistocene.

3. Endemic deep-sea genera living at a positive 2°C temperature fall into three types:

   (a) Worldwide genera, which have likely come from ancestors found world-wide in the Tethys Sea; otherwise it is impossible to account for worldwide distribution.

   (b) Southern hemisphere endemics, which are likely to have come from southern hemisphere subpolar areas and to have followed the 1 to 2°C isotherms since the Pliocene or Miocene.

   (c) Northern hemisphere endemics, which likely originated in the deep sea only since the Pleistocene.

4. Endemic abyssal species have likely evolved in the deep-sea environment between the Miocene and the present.

Geologic Age of a Given Topographic Feature

The age of certain topographic features, such as deep trenches and the Gulf of Aden and perhaps the entire Atlantic Ocean, establish within wide limits the time during which a deep-sea topographic feature could have become populated by benthic marine organisms. It does not tell how old the animals are but only how long in geologic time such population of a deep was possible. Thus *Neopilina* could have invaded the Gulf of Aden only since the Miocene, since the Gulf of Aden is no older than the Miocene. We still are not certain where the *Neopilina* came from. Most likely it was once worldwide in distribution.

Postglacial Invasions

The record of postglacial invasions may be studied in isolated basins such as the Mediterranean where events are tagged by stagnant periods between periods when the sea floor was oxidized. Thus Ninkovitch and Heezen (1967) (Figure 14-20) and Menzies et al. (1961) have identified periods of stagnation at abyssal depths in the eastern basin of the Mediterranean. The extinction and repopulation of Mediterranean sediments by benthos in association with anoxic and oxygenated layers is shown in Figure 14-21.

   Any contemporary invasion of the deep sea would have as the most likely source the isothermal polar regions, where the cold temperatures predominate today, or an area within the Mediterranean and the Red Sea where isothermal temperate conditions prevail from the surface to the seabed.

THE IMPORTANCE OF THE TETHYS WARM–WATER FAUNA

As Ekman (1953) emphasized and explained lucidly, the Tethys Sea provides the main clue to the origin and evolution of the modern marine fauna. It existed, with its tropical–subtropical marine climate, continuously from the Lower Cambrian to the later Tertiary period. During the entire Mesozoic and the early Tertiary it was of con-

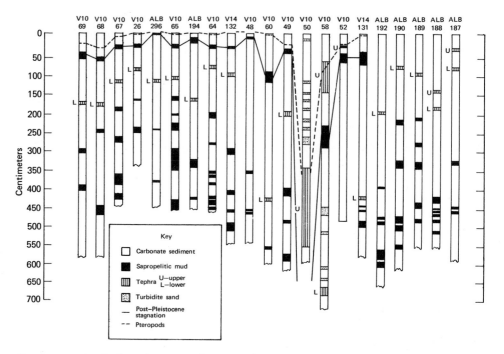

Figure 14-20. Graphic logs of Eastern Mediterranean deep-sea cores containing volcanic ash layers. (After Ninkovitch and Heezen, 1967.)

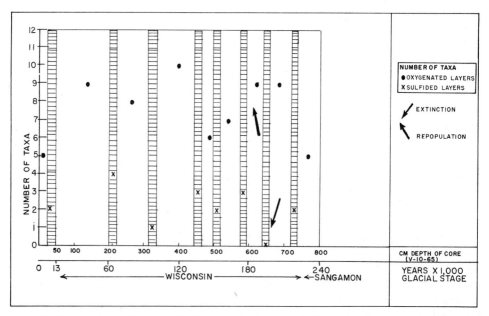

Figure 14-21. Abundance (diversity) of major invertebrate taxa in the Eastern Mediterranean abyssal core.

siderable size and flooded much of the continents as a worldwide phenomenon. The absence of an Atlantic Ocean prior to the Jurassic implies that the waters of the Mediterranean and the West Indies were in complete shallow-water connection at that time and only gradually separated as time passed. Separation became complete enough to influence divergent faunal pathways of evolution only by the later Tertiary, because at that time the Mediterranean and the West Indian faunas took on special characteristics. The Tethys fauna was first split in two by the Miocene orogeny, which formed the Central American Isthmus, thus isolating the tropical Atlantic from the tropical Pacific. At the same time the Isthmus of Suez isolated the tropical Atlantic and Mediterranean from the Indian Ocean and Indo-Pacific (Figure 14-22).

Simultaneously the appearance of cold water in the Southern Ocean (Miocene) prohibited any communication between tropical Atlantic faunas by establishing a cold-water barrier of continuing volume and width.

Post-Mesozoic events that affected the warm-water Atlantic shelf fauna are characterized by a depauperization of the Atlantic fauna, and Ekman has emphasized and documented this in considerable detail.

Ekman explained the ''impoverished'' nature of the Atlantic as being due to climatic deterioration; he believed that the richness of the Indo-Malayan fauna was attributable to the absence of climate deterioration there. He derived the warm-water tropical fauna of the earth from the Tethys rather than from any ''center of evolution'' in the Indo-Pacific. All facts known to us regarding climate, history, and Atlantic Ocean formation are in agreement with his concept.

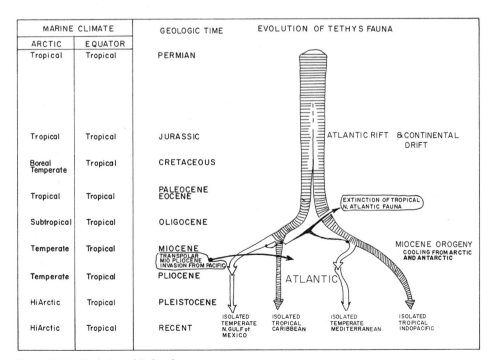

Figure 14-22. Evolution of Tethys fauna.

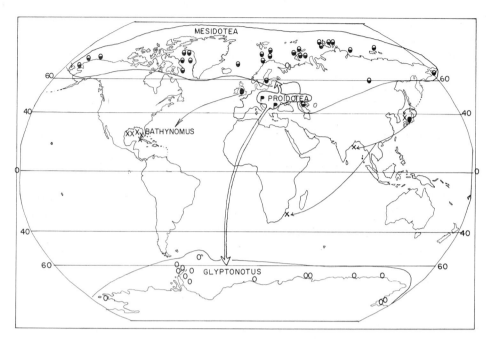

Figure 14-23. Modern relict Tethys distributions of *Mesidothea* ◐ and *Glyptonotus* ○ from Oligocene *Proidotea* ꟼ and relict archibenthal distribution of *Bathynomus* ✕ from Eocene *Bathynomus* ●. (Sources: Zenkevitch, 1963; Kussakin, 1967; and others.)

We can add two examples from the isopods *Bathynomus* and *Glyptonotus-Mesidothea* (Figure 14-23). These two are relict descendents from the Tethys. *Bathynomus* once occurred in the Eocene of England in the Atlantic Ocean. Today it is not found at all in the Atlantic, being restricted to the Gulf of Mexico, the Indian Ocean, and the Sea of Japan and vicinity. It is an archibenthal Tertiary relict genus.

Glyptonotus-Mesidothea tell a different but similar story. They were represented by *Proidotea,* a Tethys genus found in the Atlantic. This genus apparently gave rise to *Mesidothea* of the north polar seas and to *Glyptonotus,* an Antarctic endemic. These two genera look alike superficially, and both are allied to *Proidotea.*

The proponents of a constant deep-sea environment will no doubt view the geologic evidence for a changing deep-sea environment with skepticism. Yet where is the evidence for a constant deep-sea environment? We have not been able to find it. The proponents of high deep-sea community diversity as evidence or argument for a constant deep-sea environment have failed to demonstrate that the fauna of the red clay environment is as diverse as that of the mesoabyssal zone. All existing evidence points to a lesser community diversity in the lower abyssal red clay zone, and also in the upper abyssal zone. To our minds the high species diversity of communities in the mesoabyssal zone is not proof of a time stability of that environment. Factors other than constancy of environment must be sought to explain the high diversity. An enhanced food supply is one factor that should be considered. It is important that an explanation of the high diversity of the mesoabyssal zone be found.

We derive deep-sea endemic genera and families from warm-water ancestors of the Mesozoic and Tertiary. Nonendemic deep-sea genera and families are viewed as secondary and more recent invaders of the deep sea.

Deep-sea animals have likely been derived from shallow-water ancestors in the past. For low, warm-water latitudes, it is improbable that the intertidal or the shelf is making any contribution to deep-sea populations today. It is most likely, instead, that the cold polar climates, and especially the Antarctic, represent the site of origin of modern deep-sea species.

The intertidal fauna is the least likely source of deep-sea species everywhere because of the special adaptations required for life in the intertidal. Similarly, we doubt that the fauna of the sandy shelf has generated deep-sea species to any extent because taxa adapted to a sandy shelf are lacking from the deep sea today.

Deep-sea species have most likely come from level-bottom or calm-water communities, having a preference for fine muds. It is this kind of animal that dominates the deep-sea populations.

In contrast to popular opinion, we believe that it is unlikely that deep-sea invasions are occurring today except from those communities in the archibenthal zone of transition and from polar regions.

Whether the foregoing statements are entirely valid must be determined by future investigations. At the moment they seem to be the most reasonable conclusions that are suggested by the evidence available.

Bibliography

~~~~~~

Abbott, R. T. 1954. *American seashells*. Van Nostrand Co., New York, N.Y., 541 p.

Agassiz, A. 1888. Three cruises of the United States Coast and Geodetic Survey Steamer Blake, Vol. I. *Bull. Mus. Comp. Zool. Harvard College*, **14**:1–314.

Agatep, C. P. 1967. Some elasipodid holothurians of Antarctic and Subantarctic seas, pp. 49–71. *In* G. A. Llano and W. L. Schmitt [eds.], Biology of the Antarctic Seas III. *Antarctic Res. Ser.*, **2**.

Allard, G. O. and V. J. Hurst. 1969. Brazil–Gabon geologic link supports continental drift. *Science*, **163**(3867):528–532.

Allen, J. A. and H. L. Sanders. 1966. Adaptations to abyssal life as shown by the bivalve *Abra profundorum* (Smith). *Deep-Sea Res.*, **13**(6):1175–1184.

Arkell, W. J. 1956. *Jurassic geology of the world*. Hafner Pub. Co., New York, 806 pp.

Ashworth, J. H. 1915. On larvae of *Lingula* and *Pelagodiscus*. *Trans. Roy. Soc. Edinburth*, **51**(1):45–70.

Băcescu, M. C. 1957. The grab "Sonda" for the quantitative study of bottom organisms—a mixed apparatus for simultaneous collecting of macro- and microbenthos. *Bull. Inst. Rech. Pisc.*, **2**:69–77.

_____. 1962. Contribution a la Connaissance du Genre *Macrokylindrus* Stebbing (Crustacea, Cumacea), p. 207–222. *In* J. L. Barnard, R. J. Menzies, and M. C. Băcescu, *Abyssal Crustacea*. Vema Res. Ser. 1, Columbia University Press, New York.

_____. 1971. Contributions to the mysid Crustacea from the Peru—Chile trench (Pacific Ocean). *Anton Bruun Report No. 7* pp. 3–24. Texas A & M Press, College Station, Texas, *in* Sci. Res. Southeast Pac. Exped.

Bandy, O. L. and K. S. Rodolfo. 1964. Distribution of Foraminifera and sediments, Peru—Chile trench area. *Deep-Sea Res.*, **11**:817–837.

Barham, E. G. 1968. A window in the sea. *Oceans Mag.*, **1**:55–60.

_____, N. J. Ayer, Jr., and R. E. Boyce. 1967. Macrobenthos of the San Diego trough: Photographic census and observations from bathyscaphe, *Trieste*. *Deep-Sea Res.*, **14**(6):773–784.

Barnard, J. L. 1959. Epipelagic and under-ice Amphipoda of the central Arctic basin. *Geophysical Research Papers No. 63*, Scientific studies at Fletcher's Ice Island, T-3 (1952–1955), **1**:115–152.

_____. 1962. South Atlantic abyssal amphipods collected by R/V *Vema*, pp. 1–78. *In* J. L. Barnard, R. J. Menzies, and M. C. Băcescu, *Abyssal Crustacea*. Vema Res. Ser. 1, Columbia University Press, New York.

_____, R. J. Menzies, and M. C. Băcescu. 1962. *Abyssal Crustacea*. Vema Res. Ser. 1, Columbia University Press, New York, 222 p.

Barrett, J. R. 1965. Subsurface currents off Cape Hatteras. *Deep-Sea Res.*, **12**(2):173–184.

Beddard, F. 1886. Report on the Isopoda: Report on the scientific results of the voyage of the H.M.S. *Challenger*, 1873–1876, *Zool.*, **17**:1–175.

Beloussov, V. V. 1960. Development of the earth and tectogenesis. *J. Geophys. Res.*, **65**(12):4127–4146.

Belyaev, G. M. 1966. Bottom fauna of the ultra-abyssal depths of the world ocean (in Russian). *Akad. Nauk SSSR, Tr. Inst. Okeanol., 591*(9):1–248.

_____ and M. N. Sokolova. 1960. About a quantitative method for studying deep-sea benthos (in Russian). *Akad. Nauk SSSR, Tr. Inst. Okeanol., 39*:96–100.

Bertelsen, E. 1951. The ceratioid fishes; ontogeny, taxonomy, distribution and biology. Dana Report No. 39, 276 p.

Birstein, Ya. A. 1957. Certain peculiarities of the ultra-abyssal fauna at the example of the genus *Storthyngura* (Crustacea, Isopoda, Asellota) (in Russian). *Zool. Zh., 36*:961–985.

_____. 1963. *Deep-sea isopods of the northwestern Pacific.* Akad. Nauk SSSR, Inst. Okeanol., Moscow, 213 p.

Boschma, H. and T. P. Lowe. 1969. Stylasterina, pp. 14–15. In J. W. Hedgpeth [ed.], *Antarctic map folio ser. 11*, American Geographical Society, New York.

Briggs, J. C. 1970. A faunal history of the North Atlantic Ocean. *Syst. Zool. 19*(1):19–34.

Broch, H. 1961. Benthonic problems in Antarctic and Arctic waters. I. Oslo (Scientific Research of the Norwegian Antarctic Expeditions of 1927–1928 et seq.), instituted and financed by Consul Lars Christensen, **III**(38):1–32.

Brodie, J. W. and T. Hatherton. 1958. The morphology of Kermadec and Hikurangi trenches. *Deep-Sea Res., 5*:18–28.

Brundage, W. L., Jr., C. L. Buchanan, and R. B. Patterson. 1967. Search and serendipity, pp. 75–87. In J. B. Hersey [ed.], *Deep-sea photography.* The Johns Hopkins Press, Baltimore.

Bruun, A. F. 1956a. The abyssal fauna: Its ecology, distribution and origin. *Nature,* **177** (4520):1105–1108.

_____. 1956b. *The Galathea deep-sea expedition.* Macmillan, New York, 296 p.

_____. 1957. Deep-sea and abyssal depths, Chapter 22, pp. 641–672. In J. W. Hedgpeth [ed.], Treatise on marine ecology and paleoecology, Vol. I. *Mem. Geol. Soc. Amer., 67.*

_____. 1959. General introduction to the reports and list of deep-sea stations. *Galathea Rept., 1*:7–48.

_____, S. Greve, and R. Spärck. 1956. *The Galathea deep-sea expedition* 1950–1952, described by members of the expedition. Translated from the Danish by Reginald Spink. George Allen and Unwin Ltd., London, 296 p.

_____ and T. Wolff. 1961. Abyssal benthic organisms: Nature, origin, distribution and influence on sedimentation, pp. 391–397. In M. Sears [ed.], *Oceanography.* American Association for the Advancement of Science Publ. 67.

Bullivant, J. S. 1959. An oceanographic survey of the Ross Sea. *Nature,* **184,** Pt. 1(4684):422–423.

_____. 1967. Ecology of the Ross Sea Benthos. The fauna of the Ross Sea, Part V. *N.Z. Oceanogr. Inst. Mem., 32*:49–75.

_____. 1969. Bryozoa, pp. 22–23. In J. W. Hedgpeth [ed.], *Antarctic map folio ser. 11.* American Geographical Society, New York.

Bunt, J. S. 1964. Primary productivity under sea ice in Antarctic waters. 1. Concentrations and photosynthetic activities of microalgae in the waters of McMurdo Sound, Antarctica, pp. 13–26. In M. O. Lee [ed.], *Antarctic. Res. Ser. 1,* American Geophysical Union.

Bullard, E. 1969. The origin of the oceans. *Sci. Amer., 221*(3):66–75.

Buzas, M. A. and T. G. Gibson. 1969. Species diversity: benthonic Foraminifera in Western north Atlantic. *Science, 163*(3862):72–75.

Caddy, J. F. 1970. A method of surveying scallop populations from a submersible. *J. Fish. Res., 27*:535–549.

Caspers, H. 1948. Okologische Untersuchungen uber die Wattentierwelt im Elbe-Estuar: Verh. Deutschen Zool. Kiel, pp. 350–359.

Cerame-Vivas, M. J. and I. E. Gray. 1966. The distributional pattern of benthic invertebrates of the continental shelf off North Carolina. *Ecology, 47*(2):260–270.

Churkin, J., Jr. 1969. Paleozoic tectonic history of the Arctic basin north of Alaska. *Science,* **165**(3893)549–555.

Clapham, A. R. 1936. Over-dispersion in grassland communities and the use of statistical methods in plant ecology. *J. Ecol., 24*:232–251.

Clarke, A., Jr. 1961a. Structure, zoogeography and evolution of the abyssal mollusk fauna. *Bull. Amer. Malacol. U. (Ann. Rept.)*, **27,** one-page abstract.

———. 1961b. Abyssal mollusks from the south Atlantic Ocean. *Bull. Mus. Comp. Zool.,* **125** (12):343–387.

———. 1962a. On the composition, zoogeography, origin, and age of the deep-sea mollusk fauna. *Deep-Sea Res.,* **9:**291–306.

———. 1962b. Annotated list and bibliography of the abyssal marine molluscs of the world. *Nat. Mus. Can. Bull.,* **181:**1–114.

Cloud, P. E. 1968. Atmospheric and hydrospheric evolution on the primitive earth. *Science,* **160**(3829):729–736.

Craig, G. Y. 1961. Paleozoological evidence of climate. (2) Invertebrates, pp. 207–226. *In* A. E. M. Nairn [ed.], *Descriptive paleoclimatology.* Interscience, New York.

Cushman, J. A. 1948. *Foraminifera, their classification and economic use.* Harvard University Press., Cambridge, 478 p., pls. 1–55.

Czihak, G. and M. Zei. 1960. Photography, television, and the use of bottom-sampler, compared as methods for quantitative analyses of benthic populations. *Extr. Rapp. Proc.-verb. Reunion* C.I.E.S.M.M. 15, pp. 81–83.

Dahl, E. 1954. The distribution of deep-sea Crustacea. On the distribution and origin of the deep-sea bottom fauna. *Int. Union Biol. Sci. (B),* **16:**43–48.

Davis, F. M. 1925. Quantitative studies on the fauna of the seabottom, 2. Southern North Sea. *Gt. Brit. Fish. Invest. Ser. II,* **8**(4):1–50.

Dayton, P. K., G. A. Robilliard, and R. T. Paine. 1970. *Benthic faunal zonation as a result of anchor ice at McMurdo Sound, Antarctica.* Presented at Second SCAR Symposium on Antarctic biology, Cambridge, 1968. Academic Press, New York.

———, ———, ———, and A. L. DeVries. 1969. Anchor ice formation in McMurdo Sound, Antarctica, and its biological effects. *Science,* **163**(3864):273–274.

Deacon, G. E. R. 1937. The hydrology of the southern ocean. *Discovery Rep.,* **15:**1–124.

———. 1963. The southern ocean, pp. 281–296. *In* M. N. Hill [ed.], *The sea,* Vol. 2. Interscience, New York.

Dearborn, J. H. 1967. Stanford University invertebrate studies in the Ross Sea 1958–1961: General account and station list. The fauna of the Ross Sea, pt. 5, *N.Z. Oceanogr. Inst. Mem.,* **32:**31–47.

——— and J. A. Rommel. 1969. Crinoidea, pp. 35–36. *In* J. W. Hedgpeth [ed.], *Antarctic map folio ser. 11.* American Geographical Society, New York.

Dell, R. K. 1969. Benthic mollusca, pp. 25–28. *In* J. W. Hedgpeth [ed.], *Antarctic map folio ser. 11,* American Geographical Society, New York.

Dietz, R. S. 1961. Continent and ocean basin evolution by spreading of the sea floor, *Nature,* **190**(4779):854–857.

———. 1962. Ocean basin evolution by sea floor spreading, pp. 11–22. *In* G. A. Macdonald and H. Kuno [eds.], *The crust of the Pacific basin.* American Geophysical Union Publ. 1035.

——— and J. C. Holden. 1966. Deep-sea deposits in but not on continents. *Bull. Amer. Assoc. Petrol. Geol.,* **50**(2):351–362.

Djakanow, A. 1945. On the interrelation between the Arctic and the northern Pacific marine fauna based on the zoogeographical analysis of the Echinodermata. *J. Gen. Biol.,* **6**(2):125–155. (Not seen by authors; see Zenkevitch 1963.)

Donahue, J. G. 1967. Diatoms as indicators of Pleistocene climatic fluctuations in the Pacific sector of the southern ocean, pp. 133–140. *In* M. Sears [ed.], *Progress in oceanography,* Vol. 4. Pergamon Press, New York.

Donn, W. L., B. D. Donn, and W. G. Valentine. 1965. On the early history of the earth. *Bull. Geol. Soc. Amer.,* **76:**287–306.

Dunbar, M. J. 1957. The determinants of production in northern seas: A study of the biology of *Themisto libellula* Mandt. *Can. J. Zool.,* **35**(6):797–819.

———. 1968. *Ecological development in polar regions: A study in evolution.* Prentice-Hall, Inc., Englewood Cliffs, N.J., 119 p.

Durham, J. W. and E. C. Allison. 1960. The geologic history of Baja California and its marine faunas. *Syst. Zool.*, **9:**47–91.

DuToit, A. L. 1937. *Our wandering continents and hypothesis of continental drift.* Oliver and Boyd, Edinburgh, 366 pp.

Eardley, A. J. 1948. Ancient Arctica. *J. Geol.*, **56:**409–436.

Edmonds, S. J. 1969. Sipuncula and Echiura, pp. 23–24. *In* J. W. Hedgpeth [ed.], *Antarctic map folio ser. 11.* American Geographical Society, New York.

Ekman, S. 1953. *Zoogeography of the sea.* Sidgwick and Jackson, Ltd., London, 417 p.

El-Sayed, S. Z. 1968. On the productivity of the southwest Atlantic Ocean and the waters west of the Antarctic peninsula, pp. 15–47. *In* G. A. Llano and W. L. Schmitt, *Biology of the Antarctic Seas,* Vol. III. Antarctic Res. Ser. 11, American Geophysical Union.

_____ and E. F. Mandelli. 1965. Primary production and standing crop of phytoplankton in the Weddell Sea and Drake Passage. *Antarctic Res. Ser.,* **5:**87–124.

_____, R. D. Ballard, and R. L. Wigley. 1970. A dive aboard *Ben-Franklin* off West Palm Beach, Florida. *Mar. Tech. Soc. J.,* **4**(2):7–16.

_____, A. S. Merrill, and J. V. A. Trumbull. 1965. Geology and biology of the sea floor as deduced from simultaneous photographs and samples. *Limnol. Oceanogr.,* **10**(1):1–21.

Emiliani, C. 1954. Temperatures of Pacific bottom waters and polar superficial waters during the Tertiary. *Science,* **119**(3103):853–855.

_____. 1955. Pleistocene temperatures. *J. Geol.,* **63:**538–578.

_____. 1958. Ancient temperatures. *Sci. Amer.* **198**(2):54–63.

English, T. S. 1961. Some biological oceanographical observations in the central North Polar Sea: Drift station Alpha, 1957–1958. *Arct. Inst. N. Amer., Res. Pap.* 13, 79 p.

Ericson, D. B., W. S. Broecker, J. L. Kulp, and G. Wollin. 1956. Late-pleistocene climates and deep-sea sediments. *Science,* **124**(3218):385–389.

_____, M. Ewing, and B. C. Heezen. 1952. Turbidity currents and sediments in the North Atlantic. *Bull. Amer. Assoc. Petrol. Geol.,* **36**(3):489–511.

_____, _____, and G. Wollin. 1963. Pliocene–Pleistocene boundary in deep-sea sediments. *Science,* **139**(3556):727–737.

_____, _____, _____, and B. C. Heezen. 1961. Atlantic deep-sea sediment cores. *Geol. Soc. Amer.,* **72:**193–286.

_____ and G. Wollin. 1956. Micropleontological and isotopic determinations of Pleistocene climates. *Micropalaeontology,* **2**(3):257–270.

_____ and _____. 1968. Pleistocene climates and chronology in deep-sea sediments. *Science,* **162**(3859):1227–1234.

Ewing, J. and M. Ewing. 1962. Reflection profiling in and around Puerto Rico trench. *J. Geophys. Res.,* **67**(12):4729–4739.

_____ and _____. 1967. Sediment distribution on the Mid-ocean ridge with respect to spreading of the sea floor. *Science,* **156**(3782):1590–1592.

_____, K. Hunkins, and E. M. Thorndike. 1969. Some unusual photographs in the Arctic Ocean. *Mar. Technol. Soc. J.,* **3**(1):41–44.

Ewing, M. and B. C. Heezen. 1955. Puerto Rico trench topographic and geophysical data. *Spec. Pap. Geol. Soc. Amer.,* **62:**255–267.

_____, T. Saito, J. I. Ewing, and L. H. Burckle. 1966. Lower Cretaceous sediments from the northwest Pacific. *Science,* **152**(3723):751–755.

_____ and E. Thorndike. 1965. Suspended matter in deep ocean water. *Science,* **147**(3663):1291–1294.

Fell, H. B. 1961. The fauna of the Ross Sea. Pt. 1., Ophiuroidea. *N.Z. Oceanogr. Inst. Mem.,* **18:**79 p.

_____. 1967. Biological applications of sea-floor photography, pp. 207–221. *In* J. B. Hersey [ed.], *Deep-sea photography.* The Johns Hopkins Press, Baltimore.

_____ and S. Dawsey. 1969. Asteroidea, p. 41. *In* J. W. Hedgpeth [ed.], *Antarctic map folio ser. 11.,* American Geographical Society, New York.

_____, T. Holzinger, and M. Sherraden. 1969. Ophiuroidea, pp. 42–43. *In* J. W. Hedgpeth [ed.], *Antarctic map folio ser. 11,* American Geographical Society, New York.

Filatova, Z. A. 1960. On the quantitative distribution of the bottom fauna in the central Pacific (in Russian). *Tr. Inst. Okeanol., Akad. Nauk SSSR,* **41:**85–97.

—— and N. G. Barsanova. 1964. The communities of bottom fauna of western part of the Bering Sea (in Russian). *Tr. Inst. Okeanol. Akad. Nauk SSSR,* **69:**6–98.

—— and R. J. Levenstein. 1961. The quantitative distribution of the deep-sea bottom fauna in the northeastern Pacific (in Russian). *Tr. Inst. Okeanol., Akad. Nauk SSSR,* **45:**190–213.

—— and L. A. Zenkevitch. 1960. Quantitative distribution of bottom fauna in the abyssal area of the world ocean (in Russian). *Dokl. Akad. Nauk SSSR,* **133**(2):451–453.

Fischer, A. G. 1960. Latitudinal variations in organic diversity. *Evolution,* **14:**64–81.

Fisher, R. A., A. S. Corbet, and C. B. Williams. 1943. The relation between the number of species and the number of individuals in a random sample of an animal population. *J. Anim. Ecol.,* **12:**42–58.

Fisher, R. L. 1961. Middle-America trench: Topography and structure. *Bull. Geol. Soc. Amer.,* **72:**703–720.

—— and H. H. Hess. 1963. Trenches, pp. 411–436. *In* M. N. Hill [ed.], *The sea,* Vol. III. Wiley, New York.

——, B. C. Heezen, R. E. Boyce, D. Bukry, R. G. Douglas, R. E. Garrison, S. A. Kling, V. Krasheninnikov, A. P. Lisitzin, and A. Pimm. 1970. Geological history of the western north Pacific. *Science,* **168**(3936):1210–1214.

Forbes, E. 1844. Report on the Mollusca and Radiata of the Aegean Sea, and their distribution, considered as bearing on geology. *Report (1843) to the 13th meeting of the British Association for the Advancement of Science,* pp. 30–193.

——. 1846. On the connection between the distribution of existing fauna and flora of the British Isles and the geological changes which have affected their area, especially during the epoch of the northern drift. *Mem. Geol. Surv. Gt. Brit.,* Vol. 1, London.

——. 1854. Distribution of marine life. *In* K. Johnston [ed.], *Physical atlas of natural phenomena.* Edinburgh.

——. 1859. The natural history of the European seas. *In* R. G. Austen, F.R.S. [ed.], London, 306 p.

Foster, M. W. 1969. Brachiopoda, pp. 21–22. *In* J. W. Hedgpeth [ed.], *Antarctic map folio ser. 11.* American Geographic Society, New York.

Frankenberg, D. and R. J. Menzies. 1968. Some quantitative analyses of deep-sea benthos off Peru. *Deep-sea Res.,* **15:**623–626.

Gallardo, A. 1963. Notas sobre la densidad de la fauna bentónica en el sublitoral del norte de Chile. *Gayana Zool.,* **8:**3–15.

Garth, J. S. and J. Haig. 1971. The decapod Crustacea of *Anton Bruun* Cruise XI. *Anton Bruun Report* **6,** pp. 1–20. Texas A & M Press, College Station, Texas, *in* Sci. Res. Southeast Pac. Exped.

George, R. Y. and R. J. Menzies. 1967. Indication of cyclic reproductive activity in abyssal organisms. *Nature,* **215**(5103):878.

—— and ——. 1968a. Further evidence for seasonal breeding cycles in deep sea. *Nature,* **220**(5162):80–81.

—— and ——. 1968b. Distribution and probable origin of the species in the deep-sea isopod genus *Storthyngura. Crustaceana,* **15**(2):171–187.

—— and ——. 1968c. Species of *Storthyngura* (Isopoda) from the Antarctic with descriptions of six new species. *Crustaceana,* **14**(3):275–301.

Gilat, E. 1964. The marobenthonic invertebrate communities on the Mediterranean continental shelf of Israel. *Bull. Inst. Oceanogr., Monaco,* **62**(1290):1–46.

Gill, E. D. 1961. The climates of Gondwanaland in Kainozoic times, pp. 332–353. *In* A. E. M. Nairn [ed.], *Descriptive paleoclimatology.* Interscience, New York.

Gleason, H. A. 1922. On the relation between species and area. *Ecology,* **3:**158–162.

Goodell, H. G. 1963. *Marine geology of the Drake Passage, Scotia Sea and South Sandwich Trench* (mimeo). Contribution 10. Sedimentology Research Laboratory, The Florida State University, 263 p.

——. 1965. *The marine geology of the southern ocean: I. Pacific–Antarctic and Scotia Basins* (mimeo). Contribution 11. Sedimentology Research Laboratory, The Florida State University, 196 p.

Gorbunov, G. 1946. Bottom life of the Novosiberian shoal waters and the central part of the Arctic Ocean (in Russian). *3rd Compendium of Results of Drifting Expedition of Ice-Breaker Cedov 1937–1940.* Chief Office of North Road, Moscow, 138 p.

Gordon, A. L. 1966. Potential temperature, oxygen and circulation of bottom water in the southern ocean. *Deep-Sea Res.,* **13**(6):1125–1138.

———. 1967. Structure of Antarctic waters between 20°W and 170°W. *Antarctic map folio ser., folio 6,* American Geographical Society, New York, 10 p., 14 pls.

Green, K. E. 1959. Ecology of some Arctic Foraminifera. *In* V. Bushnell [ed.], *Scientific studies at Fletcher's ice island, T-3 (1952–1955).* U.S. Air Force Cambridge Research Center, Bedford, Mass., *Geophys. Res. Pap.,* **63**(l):59–81

Grieg-Smith, P. 1964. Quantitative plant ecology, 2nd ed. Butterworths, London, 255 p.

Griggs, G. B., A. G. Carey, and L. D. Kulm. 1969. Deep-Sea sedimentation and sediment-fauna interaction in Cascadia Channel and on Cascadia Abyssal Plain. Deep-Sea Res., **16**(6):157–170.

Guille, A. 1965. Observations faites en soucoupe plongeante à la limite inferieure d'un fond a *Ophiothrix quinquemalculata* D. Ch. au large de la côte du Roussillon. *Rapp. Comm. Int. Mer Medit.,* **18**:115–118.

Gurjanova, E. 1933. Die marinen Isopoden der Arktis, pp. 391–470. *In* Romer and Schaudinn [eds.], *Fauna Artica,* **6.**

———. 1934. Crustacean fauna of the Kara Sea and routes of penetration of the Atlantic sea fauna into the Arctic. *Dokl. Akad. Nauk SSSR,* **1**(2):91–96.

———. 1938. On the question of the composition and origin of the abyssal fauna of the polar basin. C. R. Acad. Sci. SSSR, **20**(4):333–336.

Haeckel, E. 1887. Report on the Radiolaria collected by H.M.S. *Challenger* during the years 1873–1876. *Rept. Sci. Results H.M.S. Challenger, Zoology,* **18.** 1803 pp.

Habib, D. 1968. Spores, pollen, and microplankton from the horizon beta outcrop. *Science,* **162**(3861):1480–1481.

Hansen, H. J. 1913. Crustacea Malacostraca II. *Danish Ingolf Exped.,* **3**(3):1–145.

———. 1916. Crustacea Malacostraca III., *Danish Ingolf Exped.,* **3**(5):1–262.

Hartman, O. 1965. Deep-water benthic polychaetous annelids off New England to Bermuda and other north Atlantic areas. Allan Hancock Foundation Publication, *Occasional Pap.,* **28**:1–378.

———. 1967. Larval development of benthic invertebrates in Antarctic seas: Early development of *Nothria notialis* (Monro) and *Paronuphis antarctica* (Monro) in Bransfield Strait, Antarctic Peninsula. *Proc. Symp. Pacific-Antarctic Sci., Sci. Rept. Spec. Issue* **1**:205–208.

——— and J. Barnard. 1958. The benthic fauna of the deep basins off Southern California. *Allan Hancock Pa. Exped.,* **22**(1):1–67.

Hayes, J. D. 1967. Quaternary sediments of the Antarctic Ocean. *Progr. Oceanogr.,* **4**:119–131.

——— and N. D. Opdyke. 1967. Antarctic Radiolaria, magnetic reversals, and climatic change. *Science,* **158**(3804):1001–1011.

Hedgpeth, J. W. 1957. Classification of marine environments. Treatise of marine ecology and paleoecology. *Geol. Soc. Amer. Mem.,* **67**(1):17–27.

———. 1969a. Preliminary observations of life between tidemarks at Palmer Station, 64°45′S. 64°05′W. *Antarct. J. U.S.,* **IV**(4):106–107.

——— [ed.]. 1969b. Distribution of selected groups of marine invertebrates in waters south of 35°S latitude. *Antarctic map folio ser. and maps. Amer. Geogr. Soc. N.Y.,* **11**:1–44.

Heezen, B. C. and C. Hollister. 1964. Deep-sea current evidence from abyssal sediments. *Mar. Geol.,* **1**:141–174.

———, ———, and W. F. Ruddiman. 1966. Shaping of the continental rise by deep geostrophic contour currents. *Science,* **152**(3721):502–508.

——— and G. L. Johnson. 1965. The South Sandwich trench. *Deep-Sea Res.,* **12**:185–197.

Helmcke, J. G. 1940. Die Brachiopoden der Deutschen Tiefsee Expedition. *Wiss. Ergebn. Deutsch. Tiefsee Exped.* (*"Valdivia" 1898–1899),* **24**(3)215–316.

Herdman, W. A. 1911. The life and work of Edward Forbes. *Ann. Rept. Liverpool Mar. Biol. Comm.,* **29**:17–44.

———. 1923. *Founders of oceanography and their work.* Edward Arnold and Co., London, 340 p.

Hersey, J. B. [ed.]. 1967. *Deep-sea photography.* The Johns Hopkins Press, Baltimore, 310 p.

Hessler, R. R. and H. L. Sanders. 1967. Faunal diversity in the deep-sea. *Deep-Sea Res.,* **14**(1):65–78.

Holme, N. A. 1964. Methods of sampling the benthos. *Advan. Mar. Biol.,* **2**:171–260.

Hopkins, T. 1964. A survey of marine bottom samplers, pp. 213–256. *In* M. Sears [ed.], *Progress in oceanography,* Vol. 2. Pergamon Press, New York.

———. 1969. Zooplankton standing crop in the Arctic basin. *Limnol. Oceanogr.,* **14**(1):80–85.

Horikoshi, M. 1970. Quantitative studies on the smaller macrobenthos inhabiting various topographical environments around the Sagami Bank in the deep-sea system of Sagami Bay. *J. Oceanogr. Soc. Jap.,* **26**(3):159–182.

Hult, J. 1941. On the soft-bottom isopods of the Skager Rak. *Zool. Bidr. Uppsala,* **21** (1):1–234.

Huxley, J. 1955. Morphism and evolution. *Heredity,* **9**(1):1–52.

Hyman, L. H. 1955. *The invertebrates: Echinodermata,* Vol. 4. McGraw-Hill, New York.

Imlay, R. W. 1965. Jurassic marine faunal differentiation in North America. *Jour. Paleont.* **39**:1023–1038.

Ivanov, A. V. 1963. *Pogonophora.* Academic Press, London, 479 p.

Jones, N. S. 1950. Marine bottom communities. *Biol. Rev.,* **25**(3):283–313.

———. 1952. The bottom fauna and the food of flatfish off the Cumberland Coast. *J. Anim. Ecol.,* **21**(2):182–205.

Jousé, A. P., G. S. Koroleva, and G. A. Nagaieva. 1963. Stratigraphic and palaeogeographic investigations in the Indian sector of the southern ocean (in Russian). *Mezhd Geofiz, Komitet, Prezidiume, Akad. Nauk SSSR, Res. Issl. Program Mezhd Geofiz. God. Okeanol. Issl.,* **8**:137–161.

King, L. C. 1961. The paleoclimatology of Gondwanaland during the Paleozoic and Mesozoic eras, pp. 307–331. *In* A. E. M. Nairn [ed.], *Descriptive paleoclimatology.* Interscience, New York.

Kinney, P., M. E. Arhelger, and D. C. Burrell. 1970. Chemical characteristics of water masses in the Amerasian basin of the Arctic Ocean. *J. Geophys. Res.* **75**:4097–4104.

Klopfer, P. H. 1959. Environmental determinants of faunal diversity. *Amer. Natur.* **93**:337–342.

Kobayashi, T. and T. Shikama. 1961. The Climatic History of the Far East, pp. 292–306. *In* A. E. M. Nairn [ed.] *Descriptive Paleoclimatology.* Interscience, New York.

Koltun, V. M. 1969. Porifera, pp. 13–14. *In* J. W. Hedgpeth [ed.], *Antarctic map folio ser.* 11. American Geographical Society, New York.

Kort, E. G. 1962. The Antarctic Ocean. *Scientific American,* **207**(3):113–128.

Kott, P. 1969. Ascidiacea, pp. 43–45. *In* J. W. Hedgpeth [ed.], *Antarctic map folio ser.* 11. American Geographical Society, New York.

Kriss, A. E. 1962. *Marine microbiology (Deep-Sea).* Translated by J. M. Shewan and Z. Kabata. Oliver and Boyd. Edinburgh and London, 536 p.

Ku, T. L. and W. S. Broecker. 1967. Rates of sedimentation in the Arctic Ocean, p. 95–104. *In* M. Sears [ed.], *Progress in oceanography,* Vol. 4. Pergamon Press, New York.

Kuenen, Ph. 1950. *Marine geology.* Wiley, New York, 568 p.

Kullenberg, B. 1956. The technique of trawling, pp. 112–118. *In* A. Bruun [ed.], *The Galathea deep-sea expedition* 1950–1952. Macmillan, New York.

Kulp, J. L. 1961. Geologic time scale. *Science,* **133**(3459):1105–1114.

Kussakin, O. G. 1967. Fauna of Isopoda and Tanaidacea in the coastal zones of the Antarctic and Subantarctic. *Biol. Rept. Sov. Antarct. Exped.* **3**:220–389.

Kusnetzov, A. 1960. Data on the quantitative distribution of bottom fauna on the floor of the Atlantic Ocean. Dokl. Akad. Nauk SSSR, **130**(6):1345–1348 [in Russian].

Laborel, J., J. M. Peres, J. Picard, and J. Vacelet. 1961. Etude directe des fonds des parages de Marseille de 30 a 300 m avec la soucoupe plongeante Cousteau. *Bull. Inst. Oceanogr., Monaco,* **58**(1206): 1–16.

Ladd, H. S., G. Gunter, K. E. Lohman, and R. Revelle. 1949. Report of the committee on a treatise of marine ecology and paleoecology. Washington, D.C., National Research Council, *Rept. 9,* 1948–1949, 121 p.

————, ————, ————, and ————. 1951. Report of the committee on a treatise of marine ecology and paleoecology. Washington, D.C., National Research Council, *Rept. 11,* 1950–1951, 82 p.

Lamarck, J. B. 1809. *Philosophie zoologique.* Reprinted 1960. Hafner, New York, 475 p.

Laughton, A. S. 1966. The Gulf of Aden. *Phil Trans. Roy. Soc., London,* **259A:**150–171.

Le Danois, E. [ed.]. 1948. *Les profondeurs de la mer. Trente ans de recherches sur la fauna sousmarine au large des côtes de France.* Payot, Paris, 303, p.

Lemche, H. and K. G. Wingstrand. 1959. The anatomy of *Neopilina galatheae* Lemche, 1957 (Mollusca Tryblidiacea). *Galathea Reports,* Danish Science Press, Copenhagen, **3**(71):9–71.

Lie, U. 1968. A quantitative study of benthic infauna in Puget Sound, Washington, USA, in 1963–1964. *Fiskeridirektorat. Skrifter, Ser. Havundersoek,* **14:**231–550.

———— and M. M. Pamatmat. 1965. Digging characteristics and sampling efficiency of the 0.1 m² Van Veen Grab. *Limnol. Oceanogr.,* **10:**379–383.

Locard, A. 1897. *Mollusques testaces, II. Expeditions Scientifiques Travailleur et du Talisman 1880–1883.* Paris, **1:**1–561.

Loeblich, A. R. and H. Tappan. 1955. Revision of some recent foraminiferal genera. *Smithsonian Misc. Collect.,* 128(5):1–37.

Longhurst, A. R. 1964. A review of the present situation in benthic synecology. *Bull. Inst. Oceanogr., Monaco,* **63**(1317):1–54.

Lowenstam, H. A. 1964. Paleotemperatures of the Permian and Cretaceous periods, pp. 227–252. *In* A. E. M. Nairn [ed.], *Problems in paleoclimatology.* Interscience, New York.

———— and S. Epstein. 1959. Cretaceous paleotemperatures as determined by the oxygen isotope method, their relations to and the nature of rudisid reefs. *Proc. 20th Sess. Congr. Geol. Int., Mexico City, D. F.,* 67–76.

Maack, R. 1964. Characteristic features of paleogeography and stratigraphy of the Devonian of Brazil and South Africa, pp. 285–293. *In* A. E. M. Nairn [ed.], *Problems in paleoclimatology.* Interscience, New York.

MacArthur, R. H. 1965. Patterns of species diversity. *Biol. Rev.,* **40:**510–533.

MacGinitie, G. E. 1955. Distribution and ecology of the marine invertebrates of point Barrow, Alaska. *Smithsonian Misc. Collect.,* **128**(9):1–201.

Madsen, F. J. 1961. On the zoogeography and origin of the abyssal fauna, in view of the knowledge of the Porcellanasteridae. *Galathea Rept.,* **4:**177–215.

Margalef, R. 1957. La teoria de la informacion en ecolgia. *Memo. R. Acad. cie. art. (Barcelona),* **33:**373–449.

————. 1958. Temporal succession and spatial heterogeneity in natural phytoplankton, pp. 323–349. *In Perspectives in marine biology.* University of California Press, Berkeley and Los Angeles.

————. 1968. *Perspectives in ecological theory.* University of Chicago Press, Chicago, 111 p.

Marshall, N. B. 1953. Egg size in Arctic, Antarctic and deep-sea fishes. *Evolution,* 7(4):328–341.

————. 1954. Aspects of deep sea biology. Hutchinson, London, 380 p.

McIntyre, A. D. 1956. The use of trawl, grab, and camera in estimating marine benthos. *J. Mar. Biol. Assoc. U.K.,* **35**(2):419–429.

McIntyre A. 1967. Coccoliths as paleoclimatic indicators of Pleistocene glaciation. *Science,* **158**(3806):1314–1317.

McNulty, J., R. C. Work, and H. B. Moore. 1962. Level sea bottom communities in Biscayne Bay and neighboring areas. *Bull. Mar. Sci. Gulf Carib.,* **12**(2):204–233.

Mead, G. W., E. Bertelsen, and D. M. Cohen. 1964. Reproduction among deep-sea fishes. *Deep-Sea Res.,* **11**(4):569–596.

Menzies, R. J. 1962a. The zoogeography, ecology and systematics of the Chilean marine isopods. *Rep. Lund Univ. Chile Exped. 1948–1949,* **42:**1–162.

————. 1962b. The isopods of abyssal depths in the Atlantic Ocean, pp. 79–206. *In* J. L. Barnard, R. J.

Menzies, and M. C. Bačescu, *Abyssal Crustacea. Vema* Res. Ser. 1, Columbia University Press, New York.

_____. 1962c. On the food and feeding habits of abyssal organisms as exemplified by the Isopoda. *Int. Rev. Ges. Hydrobiol.* **47**(3):339–358.

_____. 1963. *The abyssal fauna of the sea floor of the Arctic Ocean.* Proceedings of Arctic Basin Symposium, 1962, Arctic Institute of North America, pp. 46–66.

_____. 1964. Improved techniques for benthic trawling at depths greater than 2000 meters. *Biol. Antarct. Seas, Antarct. Res. Ser., Amer. Geophys. Union,* **1**:93–109.

_____. 1965. Conditions for the existence of life on the abyssal sea floor. *Oceanogr. Mar. Biol. Ann. Rev.,* **3**:195–210.

_____. 1968. New species of *Neopilina* of the Cambro-Devonian class Monoplacophora from the Milne-Edwards deep of the Peru-Chile trench, R/V *Anton Bruun. Proc. Symp. Mollusca (India),* **1**:1–9.

_____. 1972. Biological history of the Mediterranean Sea with reference to the abyssal benthos. *Rapp. Proc. Verb. C.I.E.S.M.* (in press) (Manuscript submitted for publication; publication not seen by authors.)

_____ and J. L. Barnard. 1959. Marine Isopoda on coastal shelf bottoms of southern California: Systematics and ecology. *Pacif. Natur.,* **1**(11):1–35.

_____ and E. Chin. 1966. *Cruise report, research vessel Anton Bruun cruise 11.* Spec. Rept. Marine Laboratory, Texas A. & M University (unpublished manuscript).

_____ and D. Frankenberg. 1966. *Handbook on the common marine isopod Crustacea of Georgia.* University Georgia Press, Athens, 93 p., 27 figs., 4 pls.

_____ and _____. 1968. Systematics and distribution of the bathyal-abyssal genus *Mesosignum* (Crustacea: Isopoda), pp. 113–140. *In* L. Schmitt and G. A. Llano [eds.], Antarctic Res. Ser. Vol. 11, *Biology of the Antarctic Seas III.* American Geophysical Union.

_____ and R. Y. George. 1967. A re-evaluation of the concept of hadal or ultra-abyssal fauna. *Deep-Sea Res.,* **14**(6):703–723.

_____ and _____. 1972. Isopod Crustacea of the Peru—Chile trench. *Anton Bruun Report No. 9* pp. 1–124 Texas A & M Press, College Station, Texas *in*Sci. Res. Southeast Pac. Exped.

_____, _____, and G. T. Rowe. 1968. Vision index for isopod Crustacea related to latitude and depth. *Nature,* **217**(5123):93–95.

_____ and P. W. Glynn. 1968. The common marine isopod Crustacea of Puerto Rico, pp. 1–133. *In* Studies on the fauna of Curacao and other Caribbean islands (P. W. Hummelinck), Nijhoff, Hague.

_____ and J. Imbrie. 1958. On the antiquity of the deep-sea bottom fauna. *Oikos,* **9**(2):192–210.

_____, _____, and B. C. Heezen. 1961. Further considerations regarding the antiquity of the abyssal fauna with evidence for a changing abyssal environment. *Deep-Sea Res.,* **8**(2):79–94.

_____ and W. Layton. 1963. A new species of monoplacophoran mollusc *Neopilina (Neopilina) veleronis* from the slope of the Cedros trench, Mexico. *Ann. Mag. Nat. Hist. Ser. 13,* **5**:401–406.

_____ and J. L. Mohr. 1962. Benthic Tanaidacea and Isopoda from the Alaskan Arctic and the polar basin. *Crustaceana,* **3**(3):191–202.

_____, O. H. Pilkey, B. W. Blackwelder, D. Dexter, P. Huling, and L. McCloskey. 1966. A submerged reef off North Carolina. *Int. Rev. Ges. Hydrobiol.,* **51**(3):393–431.

_____ and G. T. Rowe. 1968. The LUBS, a large undisturbed-bottom sampler. *Limnol. Oceanogr.,* **13**(4):708–714.

_____ and _____. 1969. The distribution and significance of detrital turtle grass, *Thallassia testudinata,* on the deep-sea floor off North Carolina. *Int. Rev. Ges. Hydrobiol.,* **54**(2):217–222.

_____, L. Smith, and K. O. Emery. 1963. A combined underwater camera and bottom grab: A new tool for investigation of deep-sea benthos. *Int. Rev. Ges. Hydrobiol.,* **48**(4):529–545.

_____ and J. B. Wilson. 1961. Preliminary field experiments on the relative importance of pressure and temperature on the penetration of marine invertebrates into the deep sea. *Oikos,* **12**(2):302–309.

_____, J. S. Zaneveld, and R. M. Pratt. 1967. Transported turtle grass as a source of organic enrichment of abyssal sediments off North Carolina. *Deep-Sea Res.,* **14**(1):111–112.

Mohr, J. L. 1959. Marine biological work, pp. 82–103. *In* V. Bushnell [ed.], *Scientific studies at Fletcher's Ice Island, T-3 (1952–1955).* U.S. Air Force, Cambridge Research Center, Bedford, Mass., Geophys. Res. Pap. 63, **1**.

Mortensen, T. 1921. *Studies of the development and larval forms of echinoderms.* G. E. C. Gad., Copenhagen, 261 p.

———. 1928. A monograph of the Echinoidea, *Vol. I. Cidaroidea.* Reitzel, Copenhagen, 551 p.

———. 1933. Ophiuoroidea. *Danish Ingolf Exped.,* **4**(8):121 p.

Mosby, H. 1963a. Interaction between the polar basin and peripheral seas, particularly the Atlantic approach. *Proc. Arct. Basin Symp.,* 1962, 109–116.

———. *1963b. Water, salt and heat balance in the north polar sea.* Proc. Arct. Basin Symp., 1962: 69–89.

Murray, J. 1895. A summary of the scientific results obtained at the sounding, dredging and trawling stations of H. M. S. *Challenger. Challenger Rept., Summ. Res.* **2**:797–1608.

——— and J. Hjort. 1912. *The depths of the ocean.* Macmillan London, 821 p., 575 figs., 4 maps, 9 pls.

Nairn, A. E. M. [ed.]. 1961. *Descriptive paleoclimatology.* Interscience, New York, 380 p.

———. 1964. *Problems in paleoclimatology.* Interscience, New York, 705 p.

Neiman, A. A. 1966. La repartition des groupes trophiques du benthos sur le plateau continental. *Abstr. 2nd Int. Oceanogr. Congr.,* Moscow, pp. 270–271.

Ninkovitch, D. and B. C. Heezen. 1967. Physical and chemical properties of volcanic glass shards from Pozzuolana Ash, Thera Island, and from upper and lower ash layers in eastern Mediterranean deep-sea sediments. *Nature,* **213**(5076):582–584.

Odhner, N. H. 1966. Brachiopoda, Reports of the Swedish deep-sea expedition, Vol. II. *Zoology,* **23**:401–406.

Odum, H. T., J. E. Cantlon, and L. S. Kornicker. 1960. An organization hierarchy postulate for the interpretation of species individual distributions, species entropy, ecosystem evolution and the meaning of a species variety index. *Ecology,* **41**:395–399.

Orton, J. H. 1920. Sea temperature, breeding and distribution in marine animals. *J. Mar. Biol. Assoc., U.K.,* n.s., **12**:339–366.

Ostenso, N. A. 1963. Geomagnetism and gravity of the Arctic basin. *Proc. Arct. Basin Symp.,* 1962:9–40.

Owen, D. M., H. L. Sanders and R. R. Hessler. 1967. Bottom photography as a tool for estimating benthic population, pp. 229–234, J. B. Hersey [ed.], Deep Sea Photography. The Johns Hopkins Press, Baltimore.

Parker, R. H. 1961. Speculations on the origin of the invertebrate faunas of the lower continental slope. *Deep-Sea Res.,* **8**(3/4):286–293.

———. 1963. Zoogeography and ecology of some macroinvertebrates, particularly mollusks, in the Gulf of California and the continental slope off Mexico. *Vidensk. Medd. Dansk. Naturh. Foren.* **126**:1–178.

Pasternak, F. A. 1958. *Die Tiefsee-Antipatharien des Kurilen–Kamtschatka Geabens. Tr. Inst. Okeanol. Acad. Nauk SSSR,* **27**:180–191.

Pawson, D. L. 1969. Holothuroidea and Echinoidea, pp. 36–41. In J. W. Hedgpeth [ed.], *Antarctic map folio ser. 11.* American Geographical Society, New York.

Pérès, J. 1957. Le probleme de 'etagement des formations benthiques, pp. 4–21. In Université d' Aix-Marseille facultie des Sciences, Rec. Trav. Station Mar. d' endoume Bull. 21(12).

———. 1961. *Oceanographie biologique et biologie marine I. La vie benthique.* Presses Universitaires de France, Paris, 542 p.

Perrier E. 1899. Les Exploration sous-marínes (not seen by authors).

Petersen, C. G. 1914. On the distribution of the animal communities of the sea bottom. *Rept. Danish Biol. Stat.,* **22**:1–7, 2 charts.

Pettersson, H. [ed.]. 1966. pp. 1–124. Reports of the Swedish deep-sea expedition 1947–1948, vol. 1 Göteborg. Elanders Boktrycheri.

——— and B. Kullenberg. 1940. A vacuum core-sampler for deep-sea sediments. *Nature,* **145** (3669):306.

Pettibone, M. H. 1951. A new species of polychaete worm of the family Polynoidae from Point Barrow, Alaska. *J. Washington Acad. Sci.,* **41**(1):44–45.

———. 1954. Marine polychaete worms from Point Barrow, Alaska, with additional records from the north Atlantic and north Pacific. *Proc. U.S. Nat. Mus.,* **103**(3324):203–356.

Pianka, E. R. 1966. Latitudinal gradients in species diversity: A review of concepts. *Amer. Nat.,* **100**(910):33–46.

Pielou, E. C. 1969. *An introduction to mathematical ecology.* Wiley–Interscience, New York, 286 p.

Pitman, W. C. and J. R. Hertzler. 1966. Magnetic anomalies over the Pacific–Antarctic ridge. *Science,* **154**(3753):1164–1171.

Powell, A. W. B. 1937. Animal communities of the sea-bottom in Aukland and Manukau Harbors. *Trans. Roy. Soc. N.Z.,* **66**:354–401.

Purdy, E. G. 1964. Sediments as substrates, pp. 238–271. *In* J. Imbrie and N. Newell [eds.], *Approaches to paleoecology.* Wiley, New York.

Rankin, J. S. Jr., K. B. Clark, and B. W. Found. 1968. Zonation of the Weddell Sea benthos. *Antarct. J. U.S.,* **3**(4):85–86.

Rass, R. S. 1941. Analagous or parallel variations in structure and development of fishes in northern and Arctic seas, pp. 1–60 (in Russian). *Moscow Society of Naturalists, Jubilee Publication, 1805–1940.*

Reyss, D. 1964a. Contribution à l'étude du rech Lacaze–Duthiers, vallée sousmarine des cotes du Roussillon. *Vie et milieu,* **15**(1):1–46.

———. 1964b. Observations faites en soucoupe plongeante dan deux vallées sousmarines de la mer Catalane: le rech du Cap et le rech Lacaze–Duthiers. *Bull. Inst. Océanogr. Monaco,* **63**(1308):1–8.

——— and J. Soyer. 1965. Étude de deux vallees sousmarines de la mer Catalane (Compte rendu de plongees en soucoupe plongeante SP300). *Bull. Inst. Océanogr. Monaco,* **65**(1356):1–27.

Richardson, H. 1905. A monograph on the isopods of North America. *Bull. U.S. Nat. Mus.* **54**:1–727.

Rona, P. A., E. D. Schneider, and B. C. Heezen. 1967. Bathymetry of the continental rise off Cape Hatteras. *Deep-Sea Res.,* **14**(5):625–633.

Ross, A. and W. A. Newman. 1969. Cirripedia, pp. 30–32. *In* J. W. Hedgpeth [ed.], *Antarctic map folio ser. 11.* American Geographical Society, New York.

Ross, D. A. 1968. Current action in a submarine canyon. *Nature,* **218**(5148)1242–1245.

Ross, R. 1954. III. Algae: Planktonic. The cryptogamic flora of the Arctic. *Bot. Rev.,* **20**(6/7):400–416.

Rowe, G. T. 1968. Distribution patterns in populations of large, deep-sea benthic invertebrates off North Carolina. Doctoral thesis. Duke University, Durham, N.C., 296 p.

———. 1971. Observations on bottom currents and epibenthic populations in Hatteras submarine canyon. *Deep-Sea Res.,* **18**(6):569–581.

Rowe, G. T. 1972. Benthic biomass and surface productivity. *Fertility of the Sea.* Vol. II. [J. Costlow, ed.], pp. 441–454. Gordon and Breach, New York.

——— and R. J. Menzies. 1967. Use of sonic techniques and tension recordings as improvement in abyssal trawling. *Deep-Sea Res.,* **14**(2):271–274.

——— and ———. 1968. Deep bottom currents off the coast of North Carolina. *Deep-Sea Res.,* **15**(6):711–719.

——— and ———. 1969. Zonation of large benthic invertebrates in the deep sea off the Carolinas. *Deep-Sea Res.,* **16**(5):531–537.

Rudwick, M. J. S. 1964. The Infra-Cambrian glaciation and the origin of the Cambrian fauna, pp. 150–155. *In* A. E. M. Nairn [ed.], *Problems in paleoclimatology.* Interscience, New York.

Runcorn, S. K. 1962. Paleomagnetic evidence for continental drift and its geophysical cause. *Continental drift,* Vol. 3. Academic Press, New York.

Ryther, J. H. 1963. Geographic variations in productivity, pp. 347–380. *In* M. N. Hill [ed.], *The sea,* Vol. 11. Interscience, New York.

Saito, T., L. H. Burckle, and M. Ewing. 1966. Lithology and paleontology of the reflective layer horizon A. *Science,* **154**(3753):1173–1176.

———, M. Ewing, and L. H. Burckle. 1966. Tertiary sediment from the Mid-Atlantic ridge. *Science,* **151**(3714):1075–1079.

Sanders, H. L. 1956. Oceanography of Long Island Sound, 1952–1954, X: The biology of marine bottom communities. *Bull. Bingham Oceanogr. Collect.,* **15**:345–414.

———. 1958. Benthic studies in Buzzards Bay, I: Animal–sediment relationships. *Limnol. Oceanogr.,* **3**:245–258.

_____. 1960. Benthic studies in Buzzards Bay. III. The structure of the soft-bottom community. *Limnol. Oceanogr.*, **5**:138–153.

_____. 1963. Some observations on the benthonic fauna of the deep-sea. *Proc. Int. Congr. Zool.*, **4**:311.

_____. 1968. Marine benthic diversity: A comparative study. *Amer. Nat.*, **102**(925):243–282.

_____, E. M. Goudsmit, E. L. Mills, and G. R. Hampson. 1962. A study on the intertidal fauna of Barnstable Harbor, Mass., *Limnol. Oceanogr.*, **7**(1):63–79.

_____ and R. R. Hessler. 1969. Ecology of the deep-sea benthos. *Science* **163**(3874):1419–1424.

_____, _____, and G. R. Hampson. 1965. An introduction to the study of deep-sea benthic faunal assemblages along the Gay Head–Bermuda Transect, *Deep-Sea Res.*, **12**(6):845–867.

Savage, J. M. 1960. Evolution of a peninsular Herptofauna. *Syst. Zool.*, **9**:184–212.

Schlieper, C. 1968. High pressure effects on marine invertebrates and fishes. *Mar. Biol.* **2**:5–12.

Schneider, E. D. and B. C. Heezen. 1966. Sediments of the Caicos Outer Ridge, the Bahamas. *J. Geol. Soc. Amer.*, **77**:1381–1398.

Schoener, A. 1967. Post-larval development of five deep-sea ophiuroids. *Deep-Sea Res.*, **14**(6):645–660.

_____. 1968. Evidence for reproductive periodicity in the deep-sea. *Ecology*, **49**:81–87.

_____ and G. T. Rowe. 1970. Pelagic *Sargassum* and its presence among the deep-sea benthos. *Deep-Sea Res.*, **17**:923–925.

Schorygin, A. 1945. Changes in the quantity and composition of the benthos in the northern part of the Caspian Sea in the course of the years 1935–1940. *Zool. Zh.* **24**(3). (Not seen by writers, see L. A. Zenkevitch 1963.)

Schwarzbach, M. 1961. The climatic history of Europe and North America, pp. 255–291. *In* A. E. M. Nairn [ed.], *Descriptive paleoclimatology*. Interscience, New York.

Shannon, C. E. and W. Weaver. 1963. *The mathematical theory of communication.* University of Illinois press, Urbana. 125 p.

Shirley, J. 1964. The distribution of lower Devonian faunas, pp. 255–261. *In* A. E. M. Nairn [ed.], *Problems in paleoclimatology*. Interscience, New York.

Shoemaker, C. R. 1955. Amphipoda collected at the Arctic Laboratory, Office of Naval Research, Point Barrow, Alaska by G. E. MacGinitie. *Smithsonian Misc. Collect.*, **128**(1):1–78.

Sokolova, M. N. 1959. On the distribution of deep-water bottom animals in relation to their feeding habits and the character of sedimentation. *Deep-Sea Res.*, **6**(1):1–4.

_____. 1966. La structure trophique du benthos de l'abysse. *Proc. 2nd Int. Oceanogr. Congr., Moscow, pp. 345–346.*

Southward, E. C. and T. Brattegard. 1968. Pogonophora of the northwest Atlantic: North Carolina region. *Bull. Mar. Sci.*, **18**(4):836–875.

Spjeldnaes, N. 1964. Climatically induced faunal migrations: Examples from the littoral fauna of the late Pleistocene of Norway, pp. 353–356. *In* A. E. M. Nairn [ed.], *Problems in paleoclimatology*. Interscience, New York.

Squires, D. F. 1969. Scleratinia, pp. 15–18. *In* J. W. Hedgpeth [ed.], *Antarctic map folio ser. 11*, American Geographical Society, New York.

Stanley, D. J. and G. Kelling. 1968. Photographic investigation of sediment texture, bottom current activity, and benthonic organisms in the Wilmington Submarine Canyon. *U.S.C.G.C. Rockaway*—Smithsonian Institution cruise, *U.S.C.G. Oceanogr. Rept.*, **22**, 95 p.

Stephenson, T. A. and A. Stephenson. 1949. The universal features of zonation between tide-marks on rocky coasts. *J. Ecol.*, **37**:289–305.

Stuxberg, A. 1882. Evertebratfauna i Sibiriens. Ishaf. *Vega Expedit,* Vetensk. Jaktagesker. I. (vide Zenkevitch, 1963).

Strelnikov, I. D. 1929. La faune de la mer de Kara et ses conditions écologiques. *C.R. Ac. Sci. Paris*, 188.

Sverdrup, H. U., M. W. Johnson, and R. H. Fleming. 1942. *The oceans: Their physics, chemistry and general biology.* Prentice-Hall, Englewood Cliffs, N.J., 1087 p.

Suyehiro, Y. and others. 1960. Notes on the sampling gears and animals collected by the second cruise of the Japanese Expedition of Deep-Sea (JEDS-2). *Oceanogr. Mag.* **11**(2):187–198.

Thomson, C. W. 1874. *The depths of the sea.* Macmillan, London, 527 p.

————. 1885. *Reports on the Scientific Results of the Voyage of H.M.S. Challenger during the years 1873–76, under the command of Capt. George S. Nares and the late Capt. Frank Tourle Thomson. Narrative,* Vol. 1, Pt. 2. London, Edinburgh and Dublin, 1110 p. (See Appendix)

Thorson, G. 1933. Investigations on shallow-water animal communities in the Franz Joseph Fjord (East Greenland) and adjacent waters. *Medd. Grønland,* **100**(2):1–68.

————. 1934. Contributions to the animal ecology of the Scoresby Sound Fjord complex (East Greenland). *Medd. Grønland,* **100**(3):1–67.

————. 1936. The larval development, growth and metabolism of Arctic marine bottom invertebrates. *Medd. Grønland,* **100**(6):1–55.

————. 1946. Reproduction and larval development of Danish marine bottom invertebrates. *Medd. Komn. Danm. Fisk. og Havunders, Plank.* **4**:1–523.

————. 1950. Reproductive and larval ecology of marine bottom invertebrates. *Biol. Rev.,* **25**:1–45.

————. 1956. Marine level-bottom communities of recent seas, their temperature adaptation and their "balance" between predators and food animals. *Trans. N.Y. Acad. Sci., Ser. 2,* **18**(8):693–700.

————. 1957a. Sampling the benthos, p. 61–73. *In* J. W. Hedgpeth [ed.], Treatise on marine ecology and paleoecology. *Geol. Soc. Amer., Mem.* **67**(1).

————. 1957b. Bottom communities (sublittoral or shallow shelf), pp. 461–534. *In* J. W. Hedgpeth [ed.], Treatise on marine ecology and paleoecology. *Geol. Soc. Amer. Mem.,* **67**(1).

Tizard, T., H. N. Moseley, J. Y. Buchanan, and J. Murray. 1885. *Narrative of the cruise of H.M.S. Challenger with a general account of the scientific results of the expedition.* H.M. Stationery Office, London, **1**(1):1–509.

Townsend, C. H. 1901. Dredging and other records of the United States Fisheries Commission Steamer Albatross, with bibliography relative to the work on the vessel. Report of the Commissioner for Fish and Fisheries for the Year ending June 30, 1900, Pt. 26, p. 387–562.

Uchupi, E. and K. O. Emery. 1963. The continental slope between San Francisco, California, and Cedros Island, Mexico. *Deep-Sea Res.,* **10**(4):397–447.

Ushakov P. V. 1963. Quelques particularités de la bionomie benthique de l'Antarctique de l'Est. *Cahiers Biol. Mar.* **4**:81–89.

Vaissiére, R. and C. Capine. 1964. Compte rendu de plongees en soucoupe plongeante SP 300(region A-1). *Bull. Inst. Oceanogr., Monaco* **63**(1314):1–36.

Vaughn, T. W. and J. W. Wells. 1943. Revision of the suborders, families, and genera of the Scleractinia. *Geol. Soc. Amer., Spec. Pap.,* **44**:363 p.

Vine, F. J. 1966. Spreading of the ocean floor: New evidence. *Science,* **154**(3755):1405–1415.

Vinogradov, M. E. 1961. The feeding of deep-sea zooplankton, *In* J. H. Fraser and J. Corlett, *Rapp. Proc. Verb. Reunion Cons. Perm. Int. 'Explor. Mer.* **153**:114–120.

Vinogradova, N. G. 1956a. Some regularity in the vertical distribution of the abyssal bottom fauna of the world ocean. *Dokl. Akad. Nauk SSSR,* **110**(4):684–687.

————. 1956b. Zoogeographical subdivision of the abyss of the world ocean. *Dokl. Akad. Nauk SSSR.,* **111**(2):195–198.

————. 1958. The vertical distribution of the deep-sea bottom fauna of the ocean (in Russian, title in German). *Tr. Inst. Okeanol. Akad. Nauk SSSR,* **27**:87–122.

————. 1959. The zoogeographical distribution of the deep-water bottom fauna in the abyssal zone of the ocean. *Deep-Sea Res.,* **5**(3):205–208.

————. 1962a. Vertical zonation in the distribution of the deep-sea benthic fauna in the ocean. *Deep-Sea Res.,* **8**(3/4):245–250.

————. 1962b. Some problems of the study of deep-sea bottom fauna. *J. Oceanogr. Soc. Jap., 20th Ann. Vol.,* pp. 724–741.

Wall, R. E. and M. Ewing. 1967. Tension recorder for deep-sea winches. *Deep-Sea Res.,* **14**(3):321–324.

Walther, J. 1926. Die Methoden der Geologie als Historischer und Biologischer Wissenschaft, pp. 529–649. *In* E. Abderholden [ed.], *Handbuch der biologischen Arbeismethoden.* Urban & Schwarzenberg, Berlin, 1930, Abt. 10.

Wells, H. W. and I. E. Gray. 1960. The seasonal occurrence of *Mytilus edulis* on the Carolina coast as a result of transport around Cape Hatteras. *Biol. Bull.*, **119**:550–559.

Whittaker, R. H. 1965. Dominance and diversity in land plant communities. *Science*, **147**:249–260.

Wiebe, P. H. 1970. Small-scale spatial distribution in oceanic zooplankton. *Limnol. Oceanogr.*, **15**(2):205–217.

Wigley, R. L. and K. O. Emery. 1967. Benthic animals, particularly *Hyalinoecia* (Annelida) and *Ophiomusium* (Echinodermata), in sea-bottom photographs from the continental slope, pp. 235–250. *In* J. B. Hersey [ed.], *Deep-sea photography*. The Johns Hopkins Press, Baltimore.

⸻ and A. D. McIntyre. 1964. Some quantitative comparisons of offshore microbenthos and macrobenthos south of Martha's Vineyard. *Limnol. Oceanogr.*, **9**:485–493.

Williams, C. B. 1964. *Patterns in the balance of nature and related problems in quantitative ecology.* Academic Press, New York. 324 p.

Wilson, G. and A. Geike. 1861. *Memoir of Edward Forbes, F.R.S.* Macmillan, London, 589 p.

Wilson, J. T. 1963. Continental drift. *Sci. Amer.* **208**(4):86–100.

Windisch, C. C., R. J. Leyden, J. L. Worzel, T. Saito, and J. Ewing. 1968. Investigation of Horizon Beta. *Science*, **162**(3861):1473–1479.

Wiseman, J. D. H. and C. D. Ovey. 1953. Definitions of features on the deep-sea floor. *Deep-Sea Res.*, **1**(1):11–16.

Wolff, T. 1956. Isopoda from depths exceeding 6000 meters. *Galathea Rept.* **2**:85–157.

⸻. 1960. The hadal community, an introduction. *Deep-Sea Res.*, **6**(2):95–124.

⸻. 1961. Animal life from a single abyssal trawling. *Galathea Rept.*, **5**:129–162.

⸻. 1962. The systematics and biology of bathyal and abyssal Isopoda Asellota. *Galathea Rept.*, **6**:1–358.

⸻. 1970. The concept of the hadal or ultra-abyssal fauna. *Deep-Sea Res.*, **17**(6):983–1003.

Worzel, J. L. and M. Ewing. 1954. Gravity anomalies and structure of the West Indies, Part II, *Bull. Geol. Soc. Amer.*, **65**:195–200.

Wust, G. 1964. The major deep-sea expeditions and research vessels 1873–1960, pp. 1–52. *In* M. Sears [ed.], *Progress in oceanography*, Vol. 2. Pergamon Press, London.

Wyrtki, K. 1962. The oxygen minima in relation to ocean circulation. *Deep-Sea Res.*, **9**(1):11–23.

⸻. 1966. Oceanography of the eastern equatorial Pacific Ocean. *Ann. Rev. Oceanogr. Mar. Biol.*, **4**:33–68.

Zaneveld, J. S. 1966. Vertical zonation of Antarctic and sub-Antarctic benthic marine algae. *Antarct. J. U.S.*, **1**:211–213.

⸻. 1968. Benthic marine algae, Ross Island to Balleny Islands. *Antarctic map folio ser., Amer. Geogr. Soc., N.Y.*, **10**:10–12, pl. 13.

Zarenkov, N. A. 1968. Crustacea Decapoda collected in the Antarctic and antiboreal regions by the Soviet Antarctic expeditions. *Rez. Biol. Issl. Sov. Antarct. Eksped.*, **4**:153–199.

Zeigler, J. M., W. D. Athearn, and H. Small. 1957. Profiles across the Peru–Chile trench. *Deep-Sea Res.*, **4**(4):238–249.

Zenkevitch, L. A. 1930. A quantitative evaluation of the bottom-fauna in the sea region about the Kanin Peninsula. *Tr. Morsk. Nauk. Inst., Moscow*, **4**(3):5–23.

⸻. 1954. Erforschungen der Tiefseefauna im nordwestlichen Teil des stillen Ozeans. *Int. Zool. Congr., Copenhagen, Unon Int. Sci. Biol. (Ser. B.)*, **(16)**:72–85.

⸻. 1959. Certain zoological problems connected with the study of the abyssal and ultra-abyssal zones of the ocean. *Proc. 15th Int. Congr. Zool., London*, pp. 215–218.

⸻. 1961. Certain quantitative characteristics of the pelagic and bottom life of the ocean, pp. 323–336. *In* M. Sears [ed.], *Oceanography.* American Association for the Advancement of Science, Vol. 67.

⸻. 1963. *Biology of the seas of the USSR.* Wiley–Interscience, New York, 955 p.

⸻. 1966. On the antiquity of the ocean and the role of marine fauna history in the solution of this problem (in Russian; summary in English). *Dokl. Akad. Nauk Okeanol. SSSR*, **6**(2):195–207.

⸻ [ed.-in-Chief]. 1969. *Biology of the Pacific Ocean*, Vol. 2. *The deep-sea bottom fauna*, Pt I. P. P.

Shirshov Institute of Oceanology Publishing House "Nauka," Moscow, 353 p.

_____, N. G. Barsanova, and G. M. Belyaev. 1960. Quantitative distribution of bottom fauna in the abyssal area of the world ocean. *Dokl. Akad. Nauk SSSR,* **130**(1):183–186.

_____ and Ya. A. Birstein. 1956. Studies of the deep-water fauna and related problems. *Deep-Sea Res.,* **4**(1):54–64.

_____ and _____. 1960. On the problem of the antiquity of the deep-sea fauna. *Deep-Sea Res.,* **7**(1):10–23.

_____, _____, and G. M. Belyaev. 1954. Studies of the fauna of the Kurile–Kamchatka trench (in Russian). *Priroda, Moscow* **2**:61–74.

_____, _____, and _____. 1955. Investigations of the bottom fauna of the Kurile–Kamchatka trench (in Russian). *Tr. Inst. Okeanol. Akad. Nauk SSSR,* **12**:345–381.

_____ and Z. A. Filatova. 1958. General characteristics of the quantitative distribution of the bottom fauna in the northwestern part of the Pacific Ocean (in Russian; title in German). *Tr. Inst. Okeanol. Akad. Nauk SSSR,* **27**:154–160.

_____ and _____. 1960. Quantitative distribution of bottom fauna in the northern part of the Pacific Ocean at a depth below 2000 meters (in Russian). *Dokl. Akad. Nauk SSSR,* **133**(2):451–453.

Zharkova, I. S. 1966. Changes in size and number of cells with increase of body dimensions at different depth as illustrated by isopod. *2nd Int. Oceanogr. Congr., Moscow, Abstr.,* **482**:408–409.

ZoBell, C. E. 1946. Marine microbiology, a monograph on hydrobacteriology. *Chron. Bot., n. s., Waltham, Mass.,* Vol. 17, xvi+240 p., 11 figs.

_____. 1953. *The occurrence of bacteria in the deep sea and their significance for animal life.* 14th Intern. Zool. Congr. Intern. Un. of Biol. Sci. Deep-Sea Colloquium, Copenhagen, (abstr, 1 page).

Scientific Publications of the Great Men in Deep-Sea Ecology

LIST OF MARINE AND GEOLOGICAL WORKS OF EDWARD FORBES (1831–1859)*

1834.	On British Species of Patella	University Journal
1835.	Natural History Tour in Norway (four Papers)	London's Mag. of Nat. Hist., Ser. 1, VIII, 65, 249, 305; IX 169
	Records of Dredging (three papers)	Mag. Nat. Hist. Ser. 1, VIII 68, 591; IX. 191
1836.	Note on Blennius Ocellaris,	Do., Ser. 1, IX. 203
1837.	On the Comparative Elevation of Testacea in the Alps	Mag. Zool. and Bot., 1837, I. 257
	On a New British Viola	Trans. Bot. Soc. Edin.
	On a New British Polygala	do.
	On New and Rare Forms of British Plants and Animals (read before the British Association, Liverpool)	Rep. Brit. Ass. Sect. p. 102
1838.	Malacologia Monensis	1 vol. 12 mo, 70 pp.
	On the Distribution of Pulmoniferous Mollusca in Europe (read before Brit. Association)	Brit. Ass. Rep. Sect. p. 112
	On the Land and Fresh-water Mollusca of Algiers and Bougia.	Ann. Nat. Hist. II. 250
1839.	Review of Johnston's British Zoophytes	Ann. Nat. Hist. III. 46
	On Two British Species of Cydippe	Ann. Nat. Hist. III. 145
	Notice of Researches in Shetland	Brit. Ass. Rep. 1839
	On the British Ciliograda (in conjunction with Mr. Goodsir)	do.
	Report on the Distribution of Pulmoniferous Mollusca in the British Islands	do p. 127
	The Dudley Expedition, an Association Medley	Lit. Gazette, Sept. 7
	On the Asteriadae of the Irish Sea.	Wernerian Mem. VIII
	On the Association of Mollusca on the British Coasts, considered with reference to Pleistocene Geology	Edin. Acad. Ann. 1840. See Mem. Geol. Survey Great Brit. I. 371
	On a Shell-bank in the Irish Sea	Ann. Nat. Hist. IV. 217
1840.	On the Botany of Trieste (read before Botanical Society, 14th November 1839	Ann. Nat. Hist. IV. 307
	On some New and Rare British Mollusca.	Ann. Nat. Hist. V. 102
	On the British Actiniadae.	Ann. Nat. Hist. V. 180

* From Wilson and Geike, 1861.

On Corymorpha Nutans (along with Mr. Goodsir) . . . . .   do.   V. 309
Note on Animalcules . . . .   do.   V. 363
On the British Ciliograda⎤ in conjunction with Mr. Goodsir
On Pelonaia. . . . .⎦   Brit. Assoc, Rep. Sect. pp. 137, 141

Dredging Report . . . .   Brit. Ass. Rep. Sect. p. 444
On a Pleistocene Tract in the Isle of Man. . . . .   do.   p. 104
Zoo-Geological Considerations on the Fresh-water Mollusca   Ann. Nat. Hist. VI. 241
On Lottia Pulchella . . . .   do.   VI. 316
On the Genus Euplocamus . . . .   do.   VI 317
On the Blood of Nudibranchia. . . . .   do.   do.

1841.   Note on the Cause of Ciliary Motion . . . .   Edin. Monthly Medical Journal, I

Abstract of Classification of the Mollusca. . . . .   Jameson's Jour. 1841
Contributions to British Actinology . . . .   Ann. and Mag. Nat. Hist. VII. 81

Note on the Appendages of the Anthers in the Genus Viola .   Do., p. 157
On a New Genus of Ascidian Molluscs . . . .   Do., p. 345
On Two Remarkable Marine Invertebrata, inhabiting the Aegean Sea. . . . .   Rep. Brit. Ass., 1841, Sect. p. 72

On Thalassema and Echiurus, in conjunction with Mr. Goodsir . . . .   Edin. New Phil. Jour. XXX. 369

On Pelonaia, in conjunction with Mr. Goodsir . . . . . .   Do., XXXI. 29
A History of British Star-fishes. 1 vol. 8vo.

1842.   Letters on Travels in Lycia, etc. . . . .   Ann. Nat. Hist. IX. 239; X. 59, 124, 205, 348

1843.   On a New Star-fish . . . .   Ann. Nat. Hist. XI. 280
On Pectinura, and the Species of Ophiura inhabiting the Eastern Mediterranean . . . .   Do., XI. 463; Trans. Linn. Soc. XIX

Retrospective Comments (Natural History). . . . .   Ann. Nat. Hist. XII. 40
Note in Reply to Mr. Hassal . . . .   do.   XII 188
On the Radiata of the Eastern Mediterranean . . . .   Trans. Linn. Soc. XIX 143
Abstract of Papers read at the British Association. . . . . .   Athenaeum and Literary Gazette

Report on the Mollusca and Radiata of the Aegean Sea . .   Rep. Brit. Ass. 1843, 130
On the Species of Neaera inhabiting the Aegean Sea. . . .   Ann. Nat. Hist. XIII. 306

1844.   Lecture on the Bearing of Marine Researches on Geology . .   Jameson's Jour. XXXVI. 318

Paper on British Fossil Ophiuridae . . . .   Proc. Geol. Soc. IV
Abstract of Papers read before the British Association, York. . . . .   Athenaeum
On the Morphology of the Reproductive System of the Sertularian Zoophyte . . . .   Ann. Nat. Hist. XIV. 385
On some Additions to the British Fauna . . . .   do.   do.   410
On the Medusa Proboscidalis . . . .   Proc. Linn. Soc. Nov. 1844
Report on the Tertiary Fossils of Malta and Gozo,. . . . . .   Proc. Geol. Soc. IV

1845.   On some Cretaceous Fossils from New Jersey . . . .   Quart. Jour. Geol. Society, I. 61

Report on Fossils from Southern India (read January 31, 1844 . . . .   Do., I. p. 79
Report on the Lower Greensand Fossils in the possession of the Geological Society.   Do., I. 78
Account of Two Fossil Species of Creseis (read March 1844) .   Do., I. p. 145
On the Fossils of the Fresh-water Tertiary Formation of the Gulf of Smyrna (read April 17, 1844) . . . .   Do., I. p. 162

Report on the Fossils from Santa Fe de Bogota (read May 1, 1844)	Do., I. p. 174
On some remarkable Analogies between the Animal and Vegetable Kingdoms (Lecture at Royal Institution).	Literary Gazette, Feb. 14
1845. Catalogue of the Lower Greensand Fossils in the possession of the Geological Society	Quart. Jour. Geol. Society, I. 237, 345
On a New Species of Cardium from the Greensand of Devonshire	Do., I. p. 408
On the Section between Black Gang Chinc and Atherfield Point (in conjunction with Capt. Ibbetson)	Do., I. p. 190
On some Echinodermata from the Miocene Strata of North America	Do., I. 425
On the Distribution of Endemic Plants	Brit. Ass. Rep. 1845
Notice of Additions to the Marine Fauna of Britain.	do.
On the Fresh-water Fossils of Cos	do.
On Preserving Medusae.	do.
1846. On the Geology of Lycia (in conjunction with Lieutenant Spratt, R. N.)	Quart. Jour. Geol. Society, II. 8
List of Pleistocene Fossils from the Isle of Man	Do., 346
On the Connexion between the Existing Fauna and Flora of of the British Isles, and the Geological Changes which have affected their Area	Mem. Geol. Sur. I. 336
On the Pulmograde Medusae of the British Seas	Ann. Nat. Hist. XVIII 284; Brit. Ass. Rep. 1846
Notices of Natural History Observations bearing upon Geology.	Brit. Ass. Rep. 1846
Note on Sections of Isle of Wight	do. do.
Travels in Lycia (in conjunction with Lieut. Spratt), two vols.	Van Voorst
Monograph on the Cretaceous Fossils of Southern India . . .	Trans. Geol. Soc. 2d Series, VII
On Palaeozoic and Secondary Fossil Molluscs of South America	App. to Darwin's Geology of South America
On some Fossils from Samos and Euboea (read 18th November)	Jour. Geol. Soc. III. 73
1847. New and Rare British Animals.	Ann. Nat. Hist. XIX 96, 390
Notice of Munby's Flora of Algiers	Do., p. 398
1847. On Orbitolites Mantelli (in Lyell's paper)	Quart. Jour. Geol. Society, IV. 11
On the Families of British Lamellibranchiate Mollusca . . .	Brit. Ass. Rep. 1847, Sect. I. 75
On Dredging Researches in progress	Do., 77
List of Shells in the Temple of Serapis (an Appendix to Mr. Babbage's paper)	Quart. Journ. Geol. Society, III. 216
1848. History of British Mollusca (with Mr. Hanley, begun in January).	Van Voorst
Palaeontological Map of the British Isles, with a Summary of British Palaeontology and Geology, ...	Keith Johnston's Physical Atlas
On some New Fossil Shells from Barbadoes	Ann. and Mag. Nat. Hist. 2d Series, I. 347
Notice of Discoveries among the British Cystideae	Brit. Ass. Rep. for 1848

On some Marine Animals from the British Channel	do.	do.
Monograph of the Naked-Eyed Medusae.	1 vol. 4to (Ray Society)	
Notice of some Peloria Varieties of Viola Canina (read before Linnaean Society, 6th June)	Ann. and Mag. Nat. Hist. 2d Series, II. 352	
Monograph of the British Fossil Asteriadae	Mem. Geol. Sur. II. Part 2	
Monograph of the Silurian Cystideae of Britain	do.	do.
Memorandum respecting some Fossiliferous Localities alluded to in Mr. Ramsay's and Mr. Aveline's Paper	Quart. Jour. Geol. Society, IV. 297	

1849.

Note on a Letter from Mr. Macandrew about the Mollusca of Vigo Bay	Ann. Nat. Hist., 2d Series, III. 507
British Organic Remains, Decade 1; and paper on Ampyx, in Decade 2.	Memoirs of the Geological Survey

On a Monstrosity of a Vinca	Brit. Ass. Rep. for 1849	
On the Reproduction of Beroe Cucumis	do.	do.
On the British Patellaceae.	do.	do.
British Mollusca, vol. ii. (with Mr. Hanley).	Van Voorst	

1850.

Note on Fossiliferous Deposits in the Middle Island of New Zealand.	Quart. Jour. Geol. Society, VI. 343

On the Succession of Organic Remains in the Dorsetshire Purbecks	Brit. Ass. Rep. for 1850	
On the Distribution of European Echini	do.	do.
Note of Discoveries in Hebrides.	Jameson's Jour. October	
British Organic Remains, Decade 3, Echinoderms.	Memoirs of the Geological Survey	
Description of Fossil Echinidae from Portugal (a Note to Mr. Sharpe's Paper, read 21st November 1849)	Quart. Jour. Geol. Society, VI. 195	
On Cardiaster.	Ann. Nat. Hist. 2d Series, VI. 442	
On the Species of Mollusca collected during the Surveying Voyages of the "Herald" and "Pandora."	Proc. Zool. Soc. 1850, pp. 53, 270	
Lecture on Recent Researches into the Natural History of the British Seas	Proc. Royal. Inst.; Ann. and Mag. Nat. Hist. 2d Series, VII. 232	
On the Discovery by Dr. Overweg of Devonian Rocks in North Africa	Brit. Ass. Rep. for 1851	
On the Echinodermata of the Crag	do.	do.
On a New Species of Maclurea.	do.	do.
On some Indications of the Molluscous Fauna of the Azores and St. Helena	do.	do.
On a New Testacean discovered during the Voyage of H. M. S. Rattlesnake.	do.	do.
On the Estuary Beds and the Oxford Clay at Loch Staffin .	Quart. Jour. Geol. Society, VII. 104	
On the Vegetable Remains from Ardtun Head (Note to the Duke of Argyle's Paper)	Do., 103	
Essay on the Vegetable Works contributed to the Great Exhibition	Art Journal	
Notes on various Vegetable and Animal Products in the Great Exhibition.	Hunt's Guide; Illustrated Catalogue of Great Exhibition	

British Mollusca, vol. iii. (with Mr. Hanley.)	Van Voorst
On Australian Mollusca	Voyage of the "Rattle-snake," vol. II
1851. Lecture on Natural History applied to Geology and the Arts.	Records of the School of Mines, vol. I. Part I
1851. On a Species of Aequorea inhabiting the British Seas . . .	Ann. and Mag. Nat Hist. 2d Series, XIV 294
1852. Note on Eocene Echinodermata (Lyell's Paper on Belgian Tertiary Formations)	Quart. Journ. Geol. Society, VIII
On the Supposed Analogy between the Life of an Individual and the Duration of Species	Lecture at Royal Institution; Ann. and Mag. Nat. Hist., 2d Series, X. 59
On the Extinct Land Shells of St. Helena	Quart. Jour. Geol. Society, VIII. 197
On Arctic Echinoderms	Append. to Dr. Sutherland's Arctic Voyage
Monograph of British Tertiary Echinoderms	Palaeontographical Society, vol. for 1852
On the Fossils of the Yellow Sandstone of the South of Ireland	Brit. Ass. Rep. for 1852
On a Species of Sepiola new to Britain	do. do.
On a New Map of the Geological Distribution of Marine Life, and on the Homoiozoic Belts	do. do.
British Mollusca, vol. iv	Van Voorst
1853. Lecture on some New Points in British Geology	Brit. Inst. Rep. for 1853, Part III. 316
On the Fluvio-Marine Tertiaries of the Isle of Wight, ...	Quart. Jour. Geol. Society, IX. 259
1854. Note on Spadix Purpurea	Ann. and Mag. Nat. Hist. 2d Series, XIII. 31
On the Manifestation of Polarity in the Distribution of Organized Beings in Time (April 28)	Royal Inst. Rep. Part IV. 428
Note on an Indication of Depth of Primeval Seas afforded by the Remains of Colour in Fossil Testacea.	Proc. Royal Soc. Mar. 23
Map of Homoiozoic Belts, etc.	Johnston's Physical Atlas, April
Presidential Address at the Geological Society.	Quart. Jour. Geol. Soc. X
Inaugural Address to the Natural History Class at Edinburgh	Edinburg Monthly Jour. of Science
On Foliation of Metamorphic Rocks in Scotland.	Brit. Ass. Rep. for 1854
On some Points connected with the Natural History of the Azores.	do. do.
1854. Article on Murchison's Siluria	Quarterly Review. Oct.
1855. Introductory Lecture to Natural History Class at Edinburgh, Session 1854–55	Edin. New Phil. Journal, January 1855
1858. On the Fluvio-Marine Tertiary Strata of the Isle of Wight (completed by Mr. Austen, Professor Ramsay, and Mr. Bristow).	Memoirs of the Geological Survey
1859. Natural History of the European Seas (continued and completed by Mr. Austen)	One vol. 12 mo, Van Voorst

LIST OF SELECTED CONTRIBUTIONS OF ALEXANDER AGASSIZ (1861–1909)*

1. Agassiz, Alexander. 1861. Acalephan fauna of the southern coast of Massachusetts (Buzzard's Bay). Proc. Boston Soc. Nat. Hist., Vol. 8, pp. 224–225.
2. Agassiz, Alexander. 1861. Notes on the described species of Holconoti found on the western coast of North America. Boston. In Proc. Boston Soc. Nat. Hist., Vol. 8, pp. 122–133.
3. Agassiz, Alexander. 1862. The mode of development of the marginal tentacles of the free Medusae of some hydroids. Boston. Proc. Boston. Soc. Nat. Hist., Vol. 9, pp. 88–101.
4. Agassiz, Alexander. 1862. On alternate generation in annelids and on the embryology of *Autolytus cornutus*. Boston J. Nat. Hist., Vol. VII, pp. 384–409.
5. Agassiz, Alexander. 1862. On *Arachnactis brachiolata,* a species of floating actinia found at Nahant, Massachusetts. Boston J. Nat. Hist., Vol. VII, pp. 525–531.
6. Agassiz, Alexander. 1863. Berenicidae (*Halopsis ocellata*). Proc. Boston Soc. Nat. Hist., Vol. IX, pp. 219–220.
7. Agassiz, Alexander. 1863. List of the echinoderms sent to different institutions in exchange for other specimens, with annotations. Bull. Mus. Comp. Zool., Vol. 1, pp. 17–28.
8. Agassiz, Alexander. 1863. *Nanomia cara,* gen. sp. nov. Proc. Boston Soc. Nat. Hist., Vol. 9. pp. 180–181.
9. Agassiz, Alexander. 1863. On the embryology of *Asteracanthion berylinus* Ag., and a species allied to *A. rubens* M. T., *Asteracanthion pallidus* Ag., Proc. Amer. Acad. Sci., Vol. VI. pp. 106–112.
10. Agassiz, Alexander. 1863. Synopsis of the echinoids collected by Dr. Simpson on the North Pacific Exploring Expedition under the command of Captains Ringgold and Rogers. Proc. Acad. Nat. Sci. Philadelphia), pp. 352–361.
11. Agassiz, Alexander. 1863. *Toxopneustes drobachiensis.* Proc. Boston Soc. Nat. Hist., Vol. IX, pp. 191–193.
12. Agassiz, Alexander. 1864. Embryology of the starfish. Cambridge. Mem. Mus. Comp. Zool., 1877, Vol. 5, pp. 1–83. Also in Agassiz, L., contrib. Nat. Hist. U.S., 1877, Vol. 5, pt. 1.
13. Agassiz, Alexander. 1865. Function of the Pedicellariae. Proc. Boston Soc. Nat. Hist., Vol. 9, p. 329.
14. Agassiz, Alexander. 1864. On the embryology of echinoderms. Mem. Amer. Acad., Vol. 9, pp. 1–30.
15. Agassiz, Alexander. 1865. Habits of *Spinalis flemingii?* Proc. Boston Soc. Nat. Hist., Vol. 10, pp. 14–15.
16. Agassiz, Alexander. 1865. North American Acalephae. Cambridge. Museum of Comparative Zoology, Illustrated Catalogue No. 2—Memoirs Vol. I, No. 2, 234 pp.
17. Agassiz, Alexander. 1865. Recherches sur l'embryologie des echinides, des ophiures, des holothuries et des asteries. Ann. Sci. Nat. Vol. III, No. 5, pp. 367–377.
18. Agassiz, Alexander. 1865. Young in echinoderms. Proc. Boston Soc. Nat. Hist., Vol. 9, p. 326.
19. Agassiz, Alexander. 1866. Description of *Salpa Cabotti* Desor. Proc. Boston Soc. Nat. Hist., Vol. 11, pp. 17–23.
20. Agassiz, Alexander. 1866. Notes on the embryology of starfish (*Tornaria*). Ann. Lyc. Nat. N.Y. Hist., Vol. VIII, pp. 240–246.
21. Agassiz, Alexander. 1866. On the young stages of a few annelids. Ann. Lyc. Nat. Hist. N.Y., Vol. VIII, pp. 303–343.
22. Agassiz, Alexander. 1869. Note on Lovéns' article on "Leskia Mirabilis Gray." Ann. Lyc. Nat. Hist. N.Y., Vol. 9, pp. 242–248.
23. Agassiz, Alexander. 1869. On the habits of a few echinoderms. Proc. Boston Soc. Nat. Hist., Vol. 13, pp. 104–107.
24. Agassiz, Alexander. 1869. Preliminary report on the Echini and starfishes dredged in deep water between Cuba and the Florida reef, by L. F. de Pourtalés. Bull. Mus. Comp. Zool., Vol. I, No. 9, pp. 253–308.
25. Agassiz, Alexander. 1870. Principal results arrived at by the American Dredging Expedition. Nature, Vol. I, p. 168.

* Courtesy holdings of Museum of Comparative Zoology, Harvard College, Cambridge, Mass.

26. Agassiz, Alexander. 1871. Appendix to the preliminary report on the Echini collected by L. F. de Pourtalés. Bull. Mus. Comp. Zool., Vol. 2, pp. 455–457.

27. Agassiz, Alexander. 1871. Systematic zoology and nomenclature. Salem. Amer. Nat., Vol. 5, pp. 353–356.

28. Agassiz, Alexander. 1872. Geographical distribution of the Echini. Mem. Mus. Comp. Zool. Vol. 3, pp. 213–220.

29. Agassiz, Alexander. 1872. Preliminary notice of a few species of Echini. Bull. Mus. Comp. Zool., Vol. III, No. 4, pp. 55–58.

30. Agassiz, Alexander. 1872–1874. Revision of the Echini. Cambridge. Illustrated Catalog, Museum of Comparative Zoology, Harvard College, No. 7.

31. Agassiz, Alexander. 1873. Agassiz and Forbes. Nature, Vol. 8, pp. 222–223.

32. Agassiz, Alexander. 1873. The echini collected on the Hassler Expedition. Bull Mus Comp. Zool., Vol. III, No. 8, pp. 187–190.

33. Agassiz, Alexander. 1873. The history of *Balanoglossus* and *Tornaria*. Mem. Amer. Acad. Arts Sci., Vol. IX, pp. 421–436.

34. Agassiz, Alexander. 1873. The homologies of Pedicellariae. Am. Nat., Vol. 7, pp. 398–406.

35. Agassiz, Alexander. 1873. Originators of glacial theories. Nature, Vol. 8, pp. 24–25.

36. Agassiz, Alexander. 1873. Revision of the Echini. Pt. 3, pp. 379–628, 45 plates. Notice. N. Neues Jahrbuch für Mineralogie, 1875, pp. 105–106.

37. Agassiz, Alexander. 1874. Echini. (*In* Agassiz, Alexander and L. F. de Pourtalés, Zoological results of the Hassler Expedition, Vol. 1, pp. 1–23.) Cambridge. Museum of Comparative Zoology Illustrated Cataglogue No. 8 (Mem. Vol. 4).

38. Agassiz, Alexander. 1874. Embryology of the Ctenophorae. Mem. Amer. Acad. Sci. X (Supplement), pp. 357–398. (This paper published and distributed in 1874, was not issued by the Academy until 1885.)

39. Agassiz, Alexander. 1874. Note sur la fertilisation artificielle de deux espèces d'étoiles de mer. Arch. Zool. Exp. Gen., notes et revue, Vol. 3, p. xlvi.

40. Agassiz, Alexander and others. 1874–1875. Zoological results of the Hassler Expedition. Cambridge. Mem. Mus. Comp. Zool., Vol. 4, No. 8.

41. Agassiz, Alexander. 1875. Instinct? in hermit crabs. Amer. J. Sci. (3), Vol. X pp. 290–291.

42. Agassiz, Alexander. 1875. Notice of papers on embryology by A. Kowalevsky. Ann. Mag. Nat. Hist. (4), 15, pp. 92–93.

43. Derby, Orville Adelbert, and Agassiz, Alexander. 1876. Exploration of Lake Titicaca, by Alexander Agassiz and S. W. Garman. 2. Notice of the Palaeozoic fossils. Bull. Mus. Comp. Zool., Vol. 3, pp. 279–286.

44. Agassiz, Alexander. 1876. The development of flounders. Am. Nat., Vol. 10, pp. 705–708.

45. Agassiz, Alexander. 1876. Haeckel's Gastraea theory. Am. Nat., Vol. X, 73–75.

46. Agassiz, Alexander. 1876. Hydrographic sketch of Lake Titicaca. Proc. Am. Acad. Sci. Arts, Vol. XI, pp. 283–292.

47. Agassiz, Alexander. 1876. On viviparous Echini from the Kerguelen Islands. Proc. Am. Acad., Vol. XI, pp. 231–236.

48. Agassiz, Alexander. and Pourtalés, L. F. Recent corals from Tilibiche, Peru. Bull. Mus. Comp. Zool., Vol. 3, pp. 287–290.

49. Agassiz, Alexander. 1877. Le développement des pleuronettes. Rev. Sci. Nat. Montpellier, Vol. 6, pp. 129–139.

50. Agassiz, Alexander. 1877. North American starfishes. In Mus. Comp. Zool. (Cambridge, Mass.), Mem., Vol. 1. Also forms Vol. V of "Contributions to the natural history of the U.S." by Louis Agassiz.

51. Agassiz, Alexander. 1877. Observations sur des echinides vivipares provenent des îles Kerguelen. Ann. Sci. Nat. (6), Vol. 5, art. no. 6.

52. Agassiz, Alexander. 1877. Sir Wyville Thomson, and the working up the "*Challenger*" collections. New Haven, Am. J. Sci. (3), Vol. 14, pp. 161–162.

53. Agassiz, Alexander. 1877–1882. On the young stages of some osseous fishes. Proc. Am. Acad. Arts Sci., Vol. 13, 1877; Vol. 17, 1882.

54. Agassiz, Alexander. 1878. The development of *Lepidosteus I*. Proc. Am. Acad., Vol. XIV, pp. 65–76.

55. Agassiz, Alexander. 1878. Note on the habits of young *Limulus*. Am. J. Sci. (3), Vol. 15, pp. 75–76.

56. Agassiz, Alexander. 1878. On the young stages of bony fishes. Boston. In Proc. Am. Acad., Vol. XIV, pp. 1–25. (Part 2 of "On the young stages of some osseous fishes, 1877–1882.".)

57. Agassiz, Alexander and others. 1878. Report on the Echini, crinoids and corals, and ophiurans (*Blake* Expedition. Bull. Mus. Comp. Zool., Vol. 5, pp. 181–238.

58. Agassiz, Alexander. 1878. Der Zoologischer Anzeiger of J. V. Carus. Am. J. Sci. (3), Vol. 16, p. 405.

59. Agassiz, Alexander, 1878–1881. Letters to Carlile P. Patterson, Superintendent United States Coast Survey on the dredging operations of the U.S.C.S. Steamer "*Blake*." Bull. Mus. Comp. Zool., Vol. I, No. 1, 6, 14, 1878–1879; Vol. VI, No. 8, 9, 1879–1880; Vol. IX, No. 3, 1881.

60. Agassiz, Alexander. 1879. List of dredging stations occupied by the United States Coast Survey streamers "*Corwin*," "*Bibb*," "*Hassler*," and "*Blake*," from 1867 to 1879. Bull. Mus. Comp. Zool., Vol. 6, pp. 1–15.

61. Agassiz, Alexander. 1879. Preliminary report on the "Challenger" Echini. Proc. Am. Acad. Arts Sci., Vol. XIV, pp. 190–212.

62. Agassiz, Alexander. 1879. Sketches of West Indies Islands as seen from the "*Blake*."

63. Agassiz, Alexander. 1879. A zoological laboratory. Nature, Vol. XIX, pp. 317–319.

64. Agassiz, Alexander. 1880. Note on some points in the history of the synonymy of Echini. Proc. Zool. Soc. London, pp. 33–38.

65. Agassiz, Alexander. 1880. Paleontological and embryological development. Cambridge.

66. Agassiz, Alexander. 1880. Preliminary report on the Echini. Reports on the results of dredging by the U.S.C.S. Steamer "*Blake*." Vol. IX. Bull. Mus. Comp. Zool., Vol. VIII, No. 2, pp. 69–84.

67. Agassiz, Alexander. 1880–1881. Das System der Medusen (1e Theil. 1 & 2e Halfte) von Ernst Haeckel. Am. J. Sci. (3), Vol. XIX, pp. 245–248, 1880; Vol. XXII, pp. 150–162, 1881.

68. Agassiz, Alexander. 1881. Biographical sketch of Louis Francois de Pourtalés. Cambridge. In Proc. Am. Acad. Arts Sci., Vol. XVI, pp. 435–443.

69. Agassiz, Alexander. 1881. Ctenophorae, by Dr. C. Chun. (A review.) Am. J. Sci. (3), Vol. 21, pp. 81–83.

70. Agassiz, Alexander. 1881. Etude sur le développment paléontologique et embryologique. (Translated by P. de Loriol.) Arch. Sci. Phys. Nat. (3), Vol. V, pp. 516–558.

71. Agassiz, Alexander. 1881. Letter No. 5 to C. P. Patterson, on the explorations in the vicinity of the Tortugas during March and April 1881. Bull. Mus. Comp. Zool., Vol. IX, No. 3, pp. 145–149.

72. Agassiz, Alexander. 1881. List of dredging stations occupied during the year 1880 by the U.S. Coast Survey Steamer "*Blake*," Commander J. R. Bartlett, U.S.N., commanding. Bull. Mus. Comp. Zool., Vol. 8, pp. 95–98.

73. Agassiz, Alexander. 1881. Paleontological and embryological development. Proc. Am. Soc. Adv. Sci., Vol. 29, pp. 389–414.

74. Agassiz, Alexander. 1881. *Polydonia frondosa*. Nature, Vol. 24, p. 509.

75. Agassiz, Alexander. 1881. Report on the Echinoidea dredged by H.M.S. *Challenger* during the years 1873–1876. London. Zool. Challenger Exped., Vol. 3, pt. IX.

76. Agassiz, Alexander. 1882. Bibliography to accompany "Selections from embryological monographs, etc. II. Echinodermata." Cambridge. Bull. Mus. Comp. Zool., Vol. X, pp. 109–134.

77. Agassiz, Alexander. 1881. Treasurer. The Darwin Memorial.

78. Agassiz, Alexander and Fewkes, J. W. 1882–1883. Exploration of the surface fauna of the Gulf Stream. Cambridge. Bull. Mus. Comp. Zool., 1882, Vol. 9, pp. 251–289. Mem. Am. Acad. Arts Sci., 1885, Vol. 11, pp. 107–134. Mem. Mus. Comp. Zool., 1883, Vol. 8, No. 2. Bull. Mus. Comp. Zool., 1883, Vol. 11, pp. 79–90.

79. Agassiz, Alexander and others. 1882–1884. Selections from embryological monographs and bibliography. Cambridge.

80. Agassiz, Alexander. 1883. A chapter in the history of the Gulf Stream. Cambridge.

81. Agassiz, Alexander. 1883. Echinodermata. Mem. Mus. Comp. Zool., Vol. IX, No. 2.

82. Agassiz, Alexander. 1883. The Porpitidae and Velellidae. Cambridge. Mem. Mus. Comp. Zool., Vol. 8, No. 2.

83. Agassiz, Alexander. 1883. Report on the Echini (of the *Blake*). Cambridge. In Mem. Mus. Comp. Zool., Vol. 10, No. 1.

84. Agassiz, Alexander. 1883. The Tortugas and Florida reefs. Cambridge. Mem. Am. Acad. Arts Sci., Vol. 11, pp. 107–134.

85. Agassiz, Alexander and Whitman, Charles Otis. 1884. On the development of some pelagic fish eggs. Proc. Am. Acad. Arts Sci., Vol. 20, pp. 23–75.
86. Agassiz, Alexander. 1885. The coast survey and "political scientists." Nation, pp. 235–236.
87. Agassiz, Alexander and Whitman, Charles Otis. 1885–1915. The development of osseous fishes. Cambridge. Mem. Mus. Comp. Zool., 1885–1889, Vol. 14, No. 1, 1915, Vol. 40, No. 9.
88. Agassiz, Alexander. 1886. Three cruises of the United States Coast and Geodetic survey steamer "Blake" in the Gulf of Mexico, in the Caribbean Sea, and along the Atlantic coast of the United States, from 1877 to 1880, by Alexander Agassiz. Cambridge, John Wilson and Son. Bull. Mus. Comp. Zool., Vols. 14 and 15.
89. Agassiz, Alexander. 1886. The work of the congressional commission of the surveys. Nation, No. 1094, p. 502.
90. Agassiz, Alexander. 1889. The coral reefs of the Hawaiian Islands. Bull. Mus. Comp. Zool., Vol. XVII, pp. 121–170.
91. Agassiz, Alexander. 1890. Notice of *Calamocrinus Diomedae*, a new stalked crinoid from the Galapagos, dredged by the U.S. Fish Commission Steamer "Albatross," Cambridge. In Mus. Comp. Zool. Bull. xx, 6.
92. Agassiz, Alexander. 1890. On the rate of growth in corals. Bull. Mus. Comp. Zool., Vol. 20, pp. 61–64.
93. Agassiz, Alexander. 1890. Ueber einen neuen Tiefsee-Crinoiden aus der Familie der Apiocriniden. Neues Jahrbuch f. Min. Geol. Pal., Vol. I, pp. 94–95.
94. Agassiz, Alexander. 1891. "Albatross" notebooks; February 10–April 15, February 22–April 11, April 15–23.
95. Agassiz, Alexander. 1891. From Hai-Phong in Tong-King to Canton, overland. London. In Proc. Roy. Geogr. Soc. and Monthly Rec. Geogr.
96. Agassiz, Alexander. *Calamocrinus diomedae*, a new stalked crinoid, with notes on the apical system and the homologies of echinoderms. Cambridge. In Mem. Mus. Comp. Zool., Vol. XVII, No. 2, pp. 95.
97. Agassiz, Alexander. 1892. General sketch of the expedition of the "Albatross," from February to May, 1891. Cambridge. In Bull. Mus. Comp. Zool., Vol. XXIII, 1.
98. Agassiz, Alexander. 1892. Preliminary note on some modifications of the chromatophores of fishes and crustaceans. Cambridge. In Bull. Mus. Comp. Zool., Vol. XXIII, 4.
99. Agassiz, Alexander. 1893. The Gulf Stream. Washington. Bull. Mus. Comp. Zool., Vol. XIV, pp. 241–259; Smithsonian Rept., 1891, pp. 189–206.
100. Agassiz, Alexander. 1893. Observations in the West Indies. Am. J. Sci., Vol. XLV, pp. 358–362.
101. Agassiz, Alexander. Notes from the Bermudas. Am. J. Sci. (3), Vol. 47, pp. 411–416.
102. Agassiz, Alexander. 1894. A reconnaissance of the Bahamas and of the elevated reefs of Cuba in the stream yacht "Wild Duck," Jan. to April, 1893. Cambridge. In Bull. Mus. Comp. Zool., Vol. XXVI, 1.
103. Agassiz, Alexander. 1895. Notes on the Florida reef. Am. J. Sci. (3), Vol. 49, pp. 154–155.
104. Agassiz, Alexander. 1895. A visit to the Bermudas in March, 1894. Cambridge. In Bull. Mus. Comp. Zool., Vol. XXVI, 2.
105. Agassiz, Alexander. 1896. The elevated reef of Florida; by Alexander Agassiz. With notes on the geology of southern Florida; by L. S. Griswold. Cambridge. In Bull. Mus. Comp. Zool. Vol. XXVIII, No. 2, pp. 27–62.
106. Agassiz, Alexander and Woodworth, W. 1896. Some variations in the genus *Eucope*. Bull. Mus. Comp. Zool., Vol. 30, pp. 119–150.
107. Agassiz, Alexander. 1896. A visit to the Great Barrier reef of Australia. Am. J. Sci. (4), Vol. II, pp. 240–244.
108. Agassiz, Alexander. 1898. The islands and coral reefs of the Fiji group. Am. J. Sci. (4), Vol. 5, pp. 113–123.
109. Agassiz, Alexander and Mayer, Alfred Goldsborough. 1898. On Dactylometra. (Studies from the Newport Marine Laboratory, No. 41.) Bull. Mus. Comp. Zool., Vol. 32, pp. 1–11.
110. Agassiz, Alexander and Mayer, Alfred Goldsborough. 1898. On some Medusae from Australia. Bull. Mus. Comp. Zool., Vol. 32, pp. 15–19. (*Desmonema rosea* n. sp. *Crambessa mosaica*.)
111. Agassiz, Alexander. 1898. Preliminary report on the Echini. Cambridge. Bull. Mus. Comp. Zool., Vol. 32, pp. 69–86.
112. Agassiz, Alexander. 1898. The Tertiary elevated limestone reefs on Fiji. Am. J. Sci. (4), Vol. 6, pp. 165–167.

113. Agassiz, Alexander. 1898. A visit to the Great Barrier Reef of Australia in the Steamer "*Croydon,*" during April and May, 1896. Bull Mus. Comp. Zool. Vol. 28, No. 4, pp. 95–148.

114. Agassiz, Alexander and Mayer, Alfred Goldsborough. 1899. Acalephs from the Fiji Islands. Bull. Mus. Comp. Zool., Vol. 32, pp. 155–189.

115. Agassiz, Alexander. 1899. The Islands and Coral Reefs of Fiji. Bull. Mus. Comp. Zool., Vol. 33, pp. 1–167.

116. Agassiz, Alexander. 1899–1900. Cruise of the *Albatross* 1–4. (4 letters to the U.S. Fish Commission on the voyage of the *Albatross.*) Science, 1899, Vol. 10, pp. 833–841; 1900, Vol. 11, pp. 92–98, 288–292, 574–578.

117. Agassiz, Alexander. 1900. Explorations of the "*Albatross*" in the Pacific Ocean. Am. J. Sci. (4), Vol. 9, pp. 33–43, 109–116, 193–198, 369–374.

118. Agassiz, Alexander. 1900. Introductory note (to) notes on the limestones and general geology of the Fiji Islands by E. C. Andrews. Bull. Mus. Comp. Zool., Vol. 38, pp. 3–4.

119. Agassiz, Alexander. 1901. Nouvelle exploration des îles à coraux de l'Océanie . . . du 20 août 1899 au mois de mars 1900 . . . Translated by Aug. Mayor., Bull. Soc. Sci. Nat. Neuchatel, Vol. 29, pp. 415–428.

120. Agassiz, Alexander. 1902. An expedition to the Maldives. Am. J. Sci. (4), Vol. 13, pp. 297–308.

121. Agassiz, Alexander. 1902. Reports on the scientific results of the expedition to the tropical Pacific. I. Preliminary report and list of stations, with remarks on the deep-sea deposits by Sir John Murray. Mem. Mus. Comp. Zool., Vol. 26, pp. 1–114.

122. Agassiz, Alexander and Mayer, Alfred Goldsborough. 1902. Reports on the scientific results of the expedition to the tropical Pacific. Mem. Mus. Comp. Zool., Vol 26, pp. 137–176.

123. Agassiz, Alexander. 1903. Biographical memoir of Louis Francois de Pourtalés 1824–1880. Biogr. Mem. Nat. Acad., pp. 79–89. Portr.

124. Agassiz, Alexander. 1903. The coral reefs of the Maldives. Cambridge. Mem. Mus. Comp. Zool., Vol. 29.

125. Agassiz, Alexander. 1903. The coral reefs of the tropical Pacific. Cambridge. Mem. Mus. Comp. Zool., Vol. 28.

126. Agassiz, Alexander. 1903. Exploration and study of the tropical Pacific Ocean. Yearbook No. 1, 1902, Carnegie Institution of Washington, Jan. 1903, pp. 272–274.

127. Agassiz, Alexander. 1903. On the formation of barrier reefs and of the different types of atolls. Proc. Roy. Soc. London, Vol. 71, pp. 412–414.

128. Agassiz, Alexander. 1904. The Panamic deep-sea Echini. Cambridge. Mem. Mus. Comp. Zool., Vol. 31.

129. Agassiz, Alexander. 1905. On the progress of the *Albatross* Expedition to the Eastern Pacific. Am. J. Sci. (4), Vol. 19, pp. 143–148, 274–276, 367–376.

130. Agassiz, Alexander. 1905. Three letters from Alexander Agassiz to the Hon. George M. Bowers . . . on the cruise in the eastern Pacific, of the U.S. Fish Commission Steamer "*Albatross.*" Bull. Mus. Comp. Zool., Vol. 46, pp. 63–84.

131. Agassiz, Alexander. 1906. General report of the expedition (eastern tropical pacific). Cambridge. Mem. Mus. Comp. Zool., Vol. 33.

132. Agassiz, Alexander and Clark, Humbert Lyman. 1907. Hawaiian and other Pacific Echini, *Cidaridae.* Mem. Mus. Comp. Zool., Vol. 34, pp. 1–4, 1–42.

133. Agassiz, Alexander and Clark, Hubert Lyman. 1907. Preliminary report on the Echini collected in 1902 among the Hawaiian Islands. Bull. Mus. Comp. Zool., Vol. 50, pp. 229–260.

134. Agassiz, Alexander and Clark, Hubert Lyman. 1907. Preliminary report on the Echini collected in 1906, from May to December, among the Aleutian Islands, Bering Sea. Bull. Mus. Comp. Zool., Vol. 51, pp. 107–140.

135. Agassiz, Alexander. 1907. The museum which Agassiz founded. Harvard Graduates Mag., Vol. 15, pp. 595–603.

136. Agassiz, Alexander. 1908. *Camelopardalus.* Harvard Coll. Observ. Circular *146*, p. 2.

137. Agassiz, Alexander and Clark, Hubert Lyman. 1908. Hawaiian and other Pacific Echini. The Salenidae, Arbacidae, Aspidodiadematidae, and Diadematidae. Mem. Mus. Comp. Zool., Vol. 34, pp. 43–132.

138. Agassiz, Alexander. 1908. Reports . . . on the tropical Pacific. Mem. Mus. Comp. Zool., Vol. 39, pp. 1–8, 1–34.

139. Agassiz, Alexander. 1909. Address. Cambridge. In Proc. 7th Int. Zool. Congr. 1907.

140. Agassiz, Alexander. 1909. "Golobiferen" and "Cystacanths." Zool. Anzeiger, Vol. 34, p. 623.

141. Agassiz, Alexander and Clark, Hubert Lyman. 1909. Hawaiian and other Pacific Echini, the Echinothuridae. Mem. Mus. Comp. Zool., Vol. 34, pp. 133–204.
142. Agassiz, Alexander. 1909. On the existence of teeth and of a lantern in the genus *Echinoneus van Phels*. Am. J. Sci. (4), pp. 490–492.

LIST OF SCIENTIFIC WORKS OF LEV. A. ZENKEVITCH (1916–1972)*

1916 Nephridium of Sipunculida (*Phascolion spitsbergense* and *Phascolosoma eremita*). Preliminary report. [Russ.]. Dnev. zool. Otd. imp. Obsch. Lyub. Estestvozn. Antrop. Etnogr. (N.S.) **3** (5). 197–220.

1922 The Floating Marine Scientific Institute. [Russ.]. Russk. gidrobiol. Zh. **1** (9–10), 301–302.

1922 *Fabricia sabella* subsp. *caspica*. subsp. *nova*. from the Caspian Sea. [Russ.]. Russk. gidrobiol. Zh. **1** (11–12), 320–322.

1923 Nephridial system of *Polycirrus albicans*. [Russ.]. Russk. zool. Zh. **3** (3–4), 408–426.

1924 Report on the first cruise of 1924 of the Floating Marine Scientific Institute on the research vessel "Persey." [Russ.]. Russk. gidrobiol. Zh. **3** (8–10), 225–230.

1925 Polychaeta of the Belushaya Guba (Novaya Zemlja). [Russ.]. Trudȳ plav. morsk. nauch. Inst. **1** (6), 1–12.

1926 Fourth expedition of the Floating Marine Scientific Institute (1924). [Russ.]. Trudȳ plav. morsk. nauch. Inst. **1** (1), 21–22.

1926 Sixth expedition of the Floating Marine Scientific Institute (1925). [Russ.]. Trudȳ plav. morsk. nauch. Inst. **1** (1), 27–28.

1926 Seventh expedition of the Floating Marine Scientific Institute (1925). [Russ.]. Trudȳ plav. morsk. nauch. Inst. **1** (1), 29–30.

1927 Materialien zur quantitativen Untersuchung der Bodenfauna des Barents- und des Weissen Meeres. [Russ., German Summary]. Trudy plav. morsk. nauch. Inst. **2** (4), 39–64.

1928 (With V. A. Brozkaya and M. S. Idelson) Materials for the study of the productivity of the sea-bottom in the White, Barents and Kara Seas. J. Cons. perm. int. Explor. Mer **3** (3), 371–379.

1929 The further aims in the studies of our Northern seas. *In:* S. A. Obruchev: On the *Persey* through the Polar seas, [Russ.]. pp 168–221. Moscow: Society of Writers.

1929 Expedition des Wissenschaftlichen Meeresinstituts 13. (1927). [Russ., German Summary]. Trudȳ morsk. nauch. Inst. **4** (1), 38–41, and 92–93.

1929 Expedition des Wissenschaftlichen Meeresinstituts 15. (1929). [Russ., German Summary]. Trudȳ morsk. nauch. Inst. **4** (1), 56–59 and 98–100.

1930 Results of the four years studies of the Marine Scientific Institute on the bottom productivity of the Northern seas. [Russ.]. Trudȳ gos. gidrobiol. Inst. **2** (3), 228–229.

1930 A quantitative evaluation of the bottom fauna in the sea region about the Kanin-peninsula. [Russ., Engl. Summary]. Trudȳ morsk. nauch. Inst. **4** (3), 5–23.

1931 On the aeration of the bottom waters through vertical circulation. J. Cons. perm. Int. Explor. Mer **6** (3), 402–418.

1932 On the aeration of the bottom waters through vertical circulation. [Russ., Engl. Summary]. Byull. gosud. okeanogr. Inst. **5**, 1–22.

1932 (Ed.) Fish food in the Barents Sea, [Russ.]. 60 pp. Vladimir-City: City's Typography.

1932 (With V. A. Brozkaya) A quantitative evaluation of the bottom fauna of the Cheshskaya Guba (Bay). [Russ., Engl. Summary]. Trudȳ gos. okeanogr. Inst. **2** (2), 41–57.

1933 Beiträge zur Zoogeographie des nördlichen Polarbassins in Zusammensetzung mit der Frage über dessen paläogeographischen Vergangenheit. [Russ., Engl. Summary]. Zool. Zh. **12** (4), 17–32.

1934 Historical review and methodics of biological and biostatistical investigations. *In:* Materials

---

* From Belyaev and Mileikovsky, 1971.

for studies of wood-borers in the seas of the USSR and for discovering measures for the protection of the wooden parts of the harbour hydrotechnical constructions. [Russ.]. Trudy tsent. nauchno-issled. Inst. vod. transp. **87,** 3–32.

1934   Productivity of the marine basins of the USSR. (Abstract) [Russ.]. Trudy faunisticheskoy Konferencii Zoologicheskogo Instituta Akademii Nauk SSSR, 3–8 Fevralja 1932, Sekzija-gidrobiol. pp 70–77. Leningrad Academy of Sciences of the USSR Press.

1934   (With Ya. A. Birstein) On the possible measures for the enlarging of the productivity of the Caspian and Aral Seas. [Russ.]. Ryb. Khoz. SSSR **3,** 38–40.

1935   On the question of growth rates during different seasons of the year. [Russ.]. Uchen. Zap. mosk. gos. Univ. (Biol.) **4,** 135–138.

1935   Some observations on fouling in the Ekatherininsky Bay (Kola fjord). [Russ.]. Byull. mosk. Obshch. Ispyt. Prir. **44** (3), 103–112.

1935   Über das Vorkommen der Brackwasserpolychaete *Manayunkia* (*M. polaris* n. sp.) an der Murmanküste. Zool. Anz. **109** (7/8), 195–203.

1936   Conditions of life in the ocean. Conclusion to the Russian translation of the book by W. Beebe "In the depths of the Ocean. A dive in the bathysphere at a depth of 923 m." [Russ.]. pp 70–93. Ed. by L. A. Zenkevitch. Moscow, Leningrad: Biomedgiz Press.

1936   (With V. A. Brozkaya) Biological productivity of the sea basins. [Russ.]. Zool. Zh. **15** (1), 13–25.

1937   The history of the system of the Invertebrata. *In:* Manual of zoology, Vol. 1. Invertebrata: Protozoa through Rotatoria, [Russ.] pp 1–55. Ed. by L. A. Zenkevitch. Moscow, Leningrad: Biomedgiz Press.

1937   Achievements in the sea-fauna studies of the USSR for twenty years. [Russ.]. Zool. Zh. **16** (5), 830–870.

1937   (With Ya. A. Birstein) On the problem relative to the acclimatization of new animal species in the Caspian and Aral Seas. [Russ., Engl. Summary]. Zool. Zh. **16** (3), 443–447.

1937   (With Ya. A. Birstein) Against the proposal about the acclimatization of the "chinese crab". [Russ.]. Ryb. Khoz. **6,** 33–34.

1937   (With V. A. Brozkaya) Materials on the ecology of the leading forms in the Barents Sea benthos. [Russ.]. Uchen. Zap. mosk. gos. Univ. **13** (Zoology), 203–226.

1938   On the weight changes during moulting in *Leander adspersus.* [Russ.]. Zool. Zh. **17** (3), 505–508.

1938   The influence of Caspian and Black Sea waters of different concentrations upon some common Black Sea invertebrates. Part I. Survivorship and body weight changes. [Russ., Engl. Summary]. Zool. Zh. **17** (5), 845–876.

1938   The influence of Caspian and Black Sea waters of different concentrations upon some common Black Sea invertebrates. Part II. Change of internal salinity. [Russ., Engl. Summary]. Zool. Zh. **17** (6), 976–1002.

1939   System and phylogeny. [Russ., Engl. Summary]. Zool. Zh. **18** (4), 600–611.

1939   (With V. A. Brozkaya) Quantitative evaluation of the bottom fauna of the Barents Sea. [Russ., Engl. Summary]. Trudy vses. nauchno-issled. Inst. morsk. ryb. Khoz. Okeanogr. **4,** 5–126. (Engl. Summary pp 99–126).

1939   (With V. A. Brozkaya) Ecological depth-temperature areas of benthos mass forms of the Barents Sea. Ecology **20** (4), 569–576.

1940   Sur l'acclimatisation dans la mer Caspienne de nouveaux invèrtebres (pour les poissons) et sur les prémisses théorétiques concernant cette acclimatisation. [Russ., French Summary]. Byull. mosk. Obshch. Ispyt. Prir. (Sect. Biol) **49** (1), 19–22.

1940   Sipunculoidea, Priapuloidea. *In:* Manual of zoology, Vol. 2. Invertebrata, Annelida, Mollusca, [Russ.]. pp 258–275 and 276–283. Ed. by L. A. Zenkevitch and V. A. Dogel. Moscow/Leningrad: Academy of Sciences of USSR Press.

1940   The distribution of fresh- and brackish-water Coelenterata. [Russ., Engl. Summary]. Zool. Zh. **19** (4). 580–602.

1941   The work of the Floating Marine Scientific Institute on the R/V "Persey." [Russ.]. Sov. Arkt. **1941** (2), 70–77.

1944   Essays on the evolution of the apparatus of animal locomotion. Part I. Some general aspects of the evolution of animal locomotion. [Russ., Engl. Summary]. Zh. obshch. Biol. **5** (3), 129–171.

1945   In memoriam of Sergei Alexandrivitch Zernov. [Russ.]. Zool. Zh. **24** (4), 201–214.

1945   The evolution of animal locomotion. J. Morphol. **77** (1), 1–52.

1945   (With YA. A. BIRSTEIN and A. F. KARPEVITCH) First successes in the reconstruction of the fauna of the Caspian Sea. [Russ.]. Zool. Zh. **24** (1), 25–31.

1945   (With YA. A. BIRSTEIN and A. F. KARPEVITCH) The first successes of the planned systematic reconstruction of the Caspian Sea fauna. [Russ.]. Rȳb. Khoz. **1945** (1), 40–44.

1946   (With YA. A. BIRSTEIN and N. A. BOBRINSKY) Animal geography, [Russ.]. 455 pp. Moscow: Soviet Science.

1947   The fauna and the biological productivity of the sea. Vol. 2. The seas of the USSR, [Russ.]. 588 pp. Moscow: Soviet Science.

1947   On the aims, object and method of marine biogeography. [Russ.]. Zool. Zh. **26** (3), 201–220.

1948   Biological structure of the ocean. [Russ.]. Zool. Zh. **27** (2), 113–124.

1948   The Russian studies of the fauna of the seas. [Russ.]. Trudȳ Inst. Okeanol. **2**, 170–196.

1948   (With V. A. ZAZEPIN and Z. A. FILATOVA) Material on the quantitative evaluation of the bottom fauna of the Kola fjord's littoral zone. [Russ.]. Trudȳ gos. okeanogr. Inst. **6**, 13–54.

1949   The role of the Sevastopol Biological Station of the Academy of Science of USSR in the development of Russian and Soviet Biology. [Russ.]. Trudȳ sevastopol'. biol. Sta. **7**, 5–10.

1949   Evolution des structures morphologiques chez les animaux. XIII Int. Congr. Zool. **1**, 432–437.

1949   La structure biologique de l'Ocean. XIII Int. Congr. Zool. **1**, 522–529.

1949   Sur l'ancienneté de l'origine de la faune marine d'deau froide. XIII Int. Congr. Zool. **1**, p. 550.

1949   On the antiquity of the origin of the cold water marine fauna and flora. [Russ.]. Trudy Inst. Okeanol. **3**, 191–199.

1949   (With YA. A. BIRSTEIN) In memoriam of A. A. SHORYGIN. [Russ.]. Priroda, Mosk. **1949** (5), 75–76.

1951   Fauna and the biological productivity of the sea. Vol. 1. The world Ocean, [Russ.]. 507 pp. Moscow: Soviet Science.

1951   The seas of the USSR, their fauna and flora, [Russ.]. 368 pp. Moscow: Uchpedgiz Press. (2nd revised edition, 1956, 424 pp. Moscow: Uchpedgiz Press).

1951   On the practice of working with the bottom-sampler at great depths. [Russ.]. Trudȳ Inst. Okeanol. **5**, 63–64.

1951   Achievements and perspectives of the development of Soviet hydrobiology with special reference to marine basins. [Russ.]. Zool. Zh. **30** (2), 111–120.

1951   (Ed.) Manual of zoology, Vol. 3. Invertebrata. Part. 2. Pentastomida through Chaetognatha, [Russ.]. 608 pp. Moscow: Soviet Science.

1951   Some biogeographical problems of the seas as part of general geography. [Russ.]. Vop. Geogr. **24**, 234–250.

1952   The reconstruction of the marine fauna of the USSR. [Russ.]. Priroda, Mosk. **1952** (4), 66–70.

1952   Life in the depths of the Ocean. [Russ.]. Priroda, Mosk. **1952** (6), 60–64.

1953   On the provision of fishes in the marine basins with food resources. *In:* Proceedings of the All-Union Conference on Fisheries problems (problems of the number dynamics of fishes, fisheries prognoses and maintenance of the fish stock), December 17–26, 1951, [Russ.]. pp 529–537. Moscow: Academy of Sciences USSR.

1953   Theoretical basis (of acclimatization). *In:* Acclimatization of *Nereis* in the Caspian Sea, [Russ.]. pp 10–35. Ed. by V. N. NIKITIN. Moscow: Moscow Society of Naturalists publications.

1953   A complex method in the studies of the biological processes in water bodies. [Russ.]. Trudȳ vses. gidrobiol. Obshch. **5**, 212–223.

1954   (With L. A. BIRSTEIN and G. M. BELYAEV) studies of the fauna of the Kurile–Kamchatka trench. [Russ.]. Priroda, Mosk. **1954** (2), 61–74.

1954   Erforschungen der Tiefseefauna im Nordwestlichen Teil des Stillen Ozeans. Un. int. Sci. biol. (Ser. B) **1954** (16), 72–85.

1955   The general characteristics of the Okhotsk and Bering Seas. *In:* Geographical distribution of fishes and other commercial animals of the Okhotsk and Bering Seas, [Russ.]. pp 5–9. Moscow: Academy of Sciences of USSR Press.

1955   The distribution of the sea-floor fauna in the Northwest Pacific [Engl., French Summary]. Proc. UNESCO Symp. phys. Oceanogr., pp 238–245. Tokyo UNESCO & Japanese Soc. for Promotion of Sciences.

1955   Oceanographic research conducted by the USSR in the North-west Pacific. [Engl., French Summary]. Proc. UNESCO Symp. phys. Oceanogr., pp 251–252. Tokyo: UNESCO & Japanese Soc. for Promotion of Sciences.

1955   IVAN ILLARIONOVITCH MESYATZEV. [Russ.]. Trudȳ vses. gidrobiol. Obshch. **6**, 5–16.

1955   (With YA. A. BIRSTEIN) Studies of the deep-water fauna and related problems. [Russ.]. Vest. mosk. gos. Univ. **1955** (4–5), 231–242.

1955   (With YA. A. BIRSTEIN and G. M. BELYAEV) Studies on the bottom fauna of the Kurile–Kamchatka trench. [Russ.]. Trudȳ Inst. Okeanol. **12**, 345–381.

1955   (With YA. A. BIRSTEIN and G. M. BELYAEV) Die Erforschung der Tierwelt des Kurilen-Kamtschatka-Tiefseegrabens. Urania, Lpz. **18** (1), 20–28.

1956   Neue Vertreter der Mittelmeerfauna im Kaspischen Meer. Int. Congr. Zool. **14**, 113–118. (Danish Scientific Press).

1956   The newest oceanological investigations of the Northwestern part of the Pacific Ocean. [Russ.]. Izv. Akad. Nauk SSSR (Ser. Geogr.) **1956** (4), 26–37.

1956   Biological grounds for the commercial use of the ocean. [Russ.]. Priroda, Mosk. **1956** (1), 35–45.

1956   Importance of the studies of the ocean depths. Deep-Sea Res. **4** (1), 67–70.

1956   Oceanology versus oceanography. Deep-Sea Res. **4** (1), p. 70.

1956   (With YA. A. BIRSTEIN) Studies of the deep-water fauna and related problems. Deep-Sea Res. **4** (1), 54–64.

1956   (With YA. A. BIRSTEIN and G. M. BELYAEV) Vertical distribution of the bottom fauna of the Kurile–Kamchatka trench. [Russ.]. Trudȳ probl. temat. Soveshch. zool. Inst. **6**, 15–16.

1957   Program of the researches of marine biology proposals for the International Geophysical Year. Colloques int. Biol. mar. (Station Biologique de Roscoff, 27 June–4 July, 1956). Annee biol. **33**, 317–320.

1957   New genus and two new species of the deep-water echiurids from the Far-eastern Seas and the northwestern part of the Pacific Ocean. [Russ.]. Trudȳ Inst. Okeanol. **23**, 291–295.

1957   Caspian and Aral Seas. Chapter 26. *In:* Treatise on marine ecology and paleoecology, Vol. 1. Ecology, pp 891–916. Ed. by J. W. HEDGPETH. New York: Geological Society of America (Memoir No. 67).

1957   Investigaciones de la URSS en las aguas profundas del Oceano Pacifico. Apropose del Seminario de Oceanografía convocado por la UNESCO, pp 21–25. Lima, Peru, Mayor de San Marcos: Universite Nacional, Instituto de Geografia. Ser. Conferencias No. 1.

1957   Evolution of the apparatus of animal locomotion. Part 3. Extremities of Arthropoda. [Russ., German Summary]. Trudȳ leningr. Obshch. Estest. **73** (4) (Sect. Zoology) 19–31.

1957   Biological taxation of the ocean and the problem of transoceanic acclimatization. *In:* Materials of the International Conference on the preservation of the stock of fishes and other marine animals, [Russ.]. Vol. 1. pp 140–160. Ed. by K. E. BABAYAN. Moscow.

1957   (With B. G. BOGOROV and T. S. RASS) The World Ocean and its natural resources. [Russ.]. Izv. Akad. Nauk SSSR (ser. Geogr.) **1957** (5), 39–49.

1957   (With Z. A. FILATOVA) Quantitative distribution of bottom fauna in the Kara Sea. [Russ.]. Trudȳ vses. gidrobiol. Obshch. **8**, 3–67.

1958   Die Tiefsee-Echiuriden des Nord-Westlichen Teiles des Stillen Ozeans. Part 2 [Russ., German title]. Trudȳ Inst. Okeanol. **27**, 192–203.

1958   On the direction of biological studies of the Institute of Oceanology of the Academy of Sciences of USSR on the Far-eastern Seas. [Russ.]. Trudȳ Okeanogr. Kom. **3**, 66–74.

1958   Immediate problems in the development of marine biology. *In:* Perspectives in marine biology, pp 27–31. Ed. by A. A. BUZZATI-TRAVERSO. Berkeley & Los Angeles: University of California Press.

1958   Investigations of bottom fauna of the Far Eastern Seas and the adjacent parts of the Pacific. Proc. pan-Pacif. Sci. Congr. **16** (Oceanography), (Bangkok, Thailand), p. 208.

1958   Study of the abyssal bottom fauna in the northwest part of the pacific by "Vityaz". Proc. pan-Pacif. Sci. Congr. **16** (Oceanography), (Bangkok, Thailand), p. 209.

1958   (With Z. A. FILATOVA) Allgemeine Charakteristik der quantitativen Verbreitung der Bodenfauna der Fernöstlichen Meere der USSR und des Nord-westlichen Teiles des Stillen Ozeans. [Russ., German title]. Trudȳ Inst. Okeanol. **27**, 154–160.

1959   The classification of brackish-water basins as examplified by the seas of the USSR. Archo. Oceanogr. Limnol. **11** (Suppl.) 53–61.

1959   Certain zoological problems connected with the study of the abyssal and ultra-abyssal zones of the oceans. Int. Congr. Zool. **15** (Sect. 3), (Paper 29), 1–3.

1959   Classifications of the brackish-water basins on the example of the seas of the USSR. [Russ.]. Izv. Akad. Nauk SSSR (Ser. Geogr.) **1959** (2), 3–11.

1959   (With G. M. Belyaev, Ya. A. Birstein, B. G. Bogorov, N. G. Vinogradova and M. E. Vinogradov) On the scheme of vertical biological zonation of the ocean. [Russ.]. Dokl. Akad. Nauk SSSR **129** (3), 658–661.

1959   (With Ya. A. Birstein, G. M. Belyaev, Z. A. Filatova) Qualitative and quantitative characteristics of the deep-water bottom fauna. [Russ.]. Ed. by L. A. Zenkevitch. Itogi Nauki. Dostizh. Okeanol. **1**, 106–147.

1959   (With A. N. Bogojavlensky) Oceanographic investigations of the Kurile–Kamchatka Trench, May–July, 1953. [Russ.]. Trudy Inst. Okeanol. **16**, 24–46.

1959   (With G. B. Udintzev and A. P. Lisitzyn) The depths of the ocean as a subject for studies. [Russ.]. Ed. by L. A. Zenkevitch. Itogi Nauki. Dostizh. Okeanol. **1**, 7–27.

1960   Adsorption and biocirculation in the ocean waters. [Russ., transl. in English, French, Spanish]. Int. atom. Energy Ag. Bull. **1**, 99–103.

1960   Special quantitative estimation of deep-water life in the oceans. [Russ., Engl. title]. Izv. Akad. Nauk SSSR (Ser. Geogr) **1960** (2), 10–16).

1960   (With N. G. Barsanova and G. M. Belyaev) Quantitative distribution of bottom fauna in the abyssal area of the World Ocean. [Russ.]. Dokl. Akad. Nauk SSSR **130** (1), 183–186.

1960   (With Ya. A. Birstein) On the problem of the antiquity of the deep-sea fauna. Deep-Sea Res. **7** (1), 10–23.

1960   (With Z. A. Filatova) Quantitative distribution of bottom fauna in the Northern part of the Pacific, at a depth below 2000 m. [Russ.]. Dokl. Akad. Nauk SSSR **133** (2), 451–453.

1961   Studies of the World Ocean. [Russ.]. Moscow: Znanie 1–48. (Booklet).

1961   Problems involved with deep-sea studies. [Russ.]. Okeanologija **1** (3), 382–398.

1961   Certain quantitative characteristics of the pelagic and bottom life of the ocean. *In:* Oceanography, pp 323–336. Ed. by M. Sears. Washington, D. C.: American Association for the Advancement of Science.

1961   L'étude des profoundeurs de l'océan. Impact Sci. Soc. **11** (2), 139–159.

1963   The biology of the seas of the USSR. [Russ.]. 740 pp. Moscow: Academy of Sciences of USSR Press.

1963   The biology of the seas of the USSR, 955 pp, London: Allen & Unwin.

1963   (With O. G. Reznitchenko) Pyotr Ivanovitch Usachev (1892–1962). [Russ.]. Trudy vses. gidrobiol. Oshch. **13**, 146–147.

1964   Soviet oceanology. Oceanus **11** (1), 14–21.

1964   New deep-sea echiuroids from the Indian Ocean. [Russ.] Trudy Inst. Okeanol. **49**, 178–182.

1964   New representatives of deep-water echiuroids (*Alomasoma belyaevi* Zenk. sp.n. and *Choanostoma filatovae* sp. n.) in the Pacific. [Russ., Engl. Summary]. Zool. Zh. **43** (12), 1863–1864.

1964   (With V. S. Muraveyskaya) Hydraulic method of locomotion in animals. [Russ.]. Priroda, Mosk. **1964** (6), 89–95.

1966   On the antiquity of the ocean and the role of marine fauna history in the solution of this problem. [Russ., Engl. Summary]. Okeanologija **6** (2), 195–207.

1966   The position of botany and zoology in the system of biological science. [Russ.]. Byull. mosk. Obshch. Ispyt. prir. **71** (3), 5–13.

1966   Peculiarities in the biological regime of the Polar Basin and Soviet Northern Seas (Some results and prospects of their scientific and commercial development). [Russ.]. Trudy vses. nauchno-issled. Inst. morsk. rýb. Khoz. Okeanogr. **60**, 19–26.

1966   The systematics and distribution of abyssal and hadal (ultra-abyssal) Echiuroidea. Galathea Rep. **8**, 175–184.

1966   (With B. G. Bogorov) Biological structure of the ocean. *In:* Ecology of the water organisms, [Russ.]. pp 3–14. Moscow: Nauka.

1967   Some data on the comparative biogeocoenology of land and ocean. [Russ., Engl. Summary]. Zh. obshch. Biol. **28** (5), 523–537.

1967   On the history of oceanology in the USSR (some stages). Bottom fauna of the seas. *In:* The development of the sciences of the Earth in the USSR (1917–1967), [Russ.]. pp 455–458 and 547–561. Ed. by A. P. Vinogradov. Moscow: Nauka.

1967   Studies of the fauna of the seas and oceans. *In:* Development of biology in the USSR (1917–1967), pp 323–344, Ed. by B. E. Bychovsky. Moscow: Nauka.

1967   (With S. D. Osokin) Soviet contribution to the science of the Ocean. [Russ.]. Priroda, Mosk. **11,** 90–110.

1968   Introduction. Mezozoa, Temnocephalida, Udonellida, Hyrocotilida, Priapulida, Echiurida, Sipunculida. *In:* The life of the animals in six volumes, Vol. 1. Invertebrata, [Russ.]. pp 7–53, 181, 367, 368, 374, 455, 456, 526–528. Ed. by L. A. Zenkevitch. Moscow: Prosveschenie.

1968   Chaetognatha. *In:* The life of the animals in six volumes, Vol. 2. (Continuation). Invertebrata, [Russ.]. pp 298–302. Ed. by L. A. Zenkevitch. Moscow: Prosveschenie.

1968   (With G. B. Zevina) The reconstruction of the Caspian Sea fauna. [Russ.]. Priroda, Mosk. **1968** (1), 12–22.

1969   Introduction. The taxonomy of the Pacific Ocean deep-sea bottom fauna: Echiuroidea. The origin and the antiquity of the deep-sea fauna. *In:* The Pacific Ocean. Biology of the Pacific Ocean, Book II. The deep-sea bottom fauna. Pleuston. [Russ.]. pp 5–6, 62–63, 235–243. Ed. by L. A. Zenkevitch. Moscow: Nauka.

1969   (With Z. A. Filatova) On the temporary distribution of the ancient primitive molluscs Monoplacophora in the World Ocean and on fossil Pogonophora in the sediments of the Cambrian Seas. [Russ., Engl. Summary]. Okeanologija **9** (1), 162–171.

1969   (With G. B. Zevina) Chapter X. Fauna and flora. *In:* The Caspian Sea, [Russ.]. pp 229–255. Ed. by A. D. Dobrovolsky, A. N. Kosarev and O. K. Leonov. Moscow: University of Moscow Press.

1970   General characteristics of the biogeocoenoses of the ocean and their comparison to those of dry land. 2. Bottom fauna of the ocean. *In:* Program and method of investigation of biogeocoenological water environments. The biogeocoenology of the ocean, [Russ.]. pp 7–27 and 213–227. Ed. by L. A. Zenkevitch. Moscow: Nauka.

1970   The most important problems in the studies of the seas and oceans and in the exploitation of their resources. *In:* Problems of the World Ocean, [Russ.]. pp 3–16. Ed. by A. M. Gusev and V. V. Alekseev. Moscow: University of Moscow Press.

1971   (With Z. A. Filatova) Some interesting zoological records from the region of the Peru–Chile trench. [Russ., Engl. Summary]. Trudy Inst. Okeanol. **89,** 77–80.

1971   (With Z. A. Filatova, G. M. Belyaev, T. S. Luk'janova and I. A. Suetova) Quantitative distribution of zoobenthos in the World Ocean. [Russ., Engl. Summary]. Byull. mosk. Obshch. Ispyt. Prir. **76** (3), 27–33.

1971   On the antiquity of marine fauna as an indicator of the antiquity of the ocean. *In:* The history of the World Ocean, [Russ.]. pp 77–83. Ed. by L. A. Zenkevitch. Moscow: Nauka.

In press   Histoire des recherches biologiques quantitatives dans les mers et les oceans. I-er Congr. Int. d'Histoire d'Océanographie.

In press   Shallow-water Echiuroidea from the Galathea Expedition. Vidensk. Meddr dansk. naturh. Foren. **129.**

In press   General characteristics of oceanic biogeocoenoses and their comparison with terrestrial biogeocoenoses. Int. Revue ges. Hydrobiol.

Scientific Contributions of Important Deep-Sea Expeditions

H.M.S. CHALLENGER*

* From xerox of titles.

REPORT

ON THE

SCIENTIFIC RESULTS

OF THE

VOYAGE OF
H.M.S. CHALLENGER

DURING THE YEARS 1873–76

UNDER THE COMMAND OF

Captain GEORGE S. NARES, R.N., F.R.S.

AND THE LATE

Captain FRANK TOURLE THOMSON, R.N.

PREPARED UNDER THE SUPERINTENDENCE OF
THE LATE

Sir C. Wyville Thomson, Knt., F.R.S., &c.

REGIUS PROFESSOR OF NATURAL HISTORY IN THE UNIVERSITY OF EDINBURGH
DIRECTOR OF THE CIVILIAN SCIENTIFIC STAFF ON BOARD

AND NOW OF

JOHN MURRAY

ONE OF THE NATURALISTS OF THE EXPEDITION

Narrative — Vol. I.

SECOND PART.

Published by Order of Her Majesty's Government

1885

Great Britain and Ireland. [*Voyages, &c.*]

[*Challenger Expedition.*] Report on the Scientific Results of the Voyage of H.M.S. Challenger during . . . 1873–76 . . . prepared under the superintendence of Sir C. W. Thomson (and . . . of J. Murray), &c. 40 vol. [in 50.] 4°. *London, Edinburgh & Dublin*, 1880–95.

NARRATIVE.

Vol. I. Narrative of the Cruise . . . with a general account of the scientific results of the expedition. By . . . T. H. Tizard . . . H. N. Moseley . . . J. Y. Buchanan . . . and . . . J. Murray. *pp. liv, 1110: 75 pls. (14 col.), 44 maps, text illust. (1 col.)*   1885.

" II. [Magnetic Observations by Naval Officers.] *pp. viii, 744.*   1882.

    Append.

       A. The pressure errors of the . . . Thermometers. By Prof. Tait. *pp. 42: 1 pl., text illust.*

       B. Report on the petrology of the rocks of St. Paul (Atlantic). By the Rev. A. Renard. *pp. 29: 1 pl. col.*

SUMMARY.

A Summary of the Scientific Results. By J. Murray. *pp. liii, (xix,) 1608: 22 pls., 54 maps (12 col.), text illust.* 2 Vol.   1895.

    [With the second volume were issued Pt. *lxxxiii* of the "Zoology," and Pt. *viii* of the "Physics and Chemistry."]

BOTANY.

Vol. I.   1885.

    Insular Floras, being an Introduction to the Botany of the Expedition. By W. B. Hemsley. *pp. 75.*

    Pt.   *i.* Botany of the Bermudas and various other Islands of the Atlantic and Southern Oceans. By W. B. Hemsley.

       1. The Bermudas. *pp. 135: 13 pls.*

" *ii.* ——

       2. [Other Islands.] *pp. 299: pls. xiv–liii.*

" *iii.* Botany of Juan Fernandez, the south-eastern Moluccas, and the Admiralty Islands. By W. B. Hemsley. *pp. 333: pls. liv–lxv.*

Vol. II.   1886.

    Pt.   *iv.* Diatomaceæ. By Conte F. Castracane degli Antelminelli. *pp. iii, 178: 30 pls.*

DEEP-SEA DEPOSITS. *pp. xxix, 525, 1 tab.: 51 pls. (15 col.), 43 maps (1 col.), text illust.*   1891.

Report on Deep-Sea Deposits, based on the specimens collected . . . By J. Murray . . . and Rev. A. F. Renard.

    Append.

       II. Report on an analytical examination of Manganese Nodules, with special reference to the presence or absence of the rarer elements. By J. Gibson.

       III. Chemical Analyses.

PHYSICS AND CHEMISTRY.

Vol. I.   1884.

    Pt.   *i.* Composition of Ocean-Water. By W. Dittmar. *pp. 251: 3 pls., text illust.*

" *ii.* Specific Gravity of . . . Ocean-Water. By J. Y. Buchanan. *pp. 46: 11 pls. col., 1 map col., text illust.*

" *iii.* Deep-Sea Temperature Observations of Ocean-Water. *pp. 2, 7 tabs.: 258 pls.*

Vol. II.   1889.

    Pt.   *v.* Physical Properties of Fresh Water and of Sea-Water. By Prof. P. G. Tait. *pp. 76: 2 pls.*

" *v.* Atmospheric Circulation. By A. Buchan. *pp. 78, 263: 2 pls., 52 maps col., text illust.*

" *vi.* Magnetical Results. By E. W. Creak. *pp. 18: 1 pl., 5 maps.*

" *vii.* Petrology of Oceanic Islands. By Prof. A. Renard. *pp. ii, 180: 7 maps, text illust.*

" *viii.* Oceanic Circulation. By A. Buchan. *pp. 38: 16 maps col.*   1895.

       [Pt. *viii* was issued with the second volume of the "Summary."]

ZOOLOGY.
Vol. I.   1880.

               General Introduction. By Sir C. W. Thomson. *pp. xiv, 62: text illust.*
Pt.   *i.* Brachiopoda. By T. Davidson, *pp. 67: text illust., 4 pls.*
"   *ii.* Pennatulida. By A. von Kölliker. *pp. 41: 11 pls.*
"   *iii.* Ostracoda. By G. S. Brady. *pp. 184: 44 pls.*
"   *iv.* Bones of Cetacea. By W. Turner. *pp. 45: 3 pls.*
"   *v.* Green Turtle. By W. K. Parker. *pp. 58: 13 pls. (col.)*
"   *vi.* Shore Fishes. By A. Günther. *pp. 82: 32 pls.*

Vol. II.   1881.
Pt.   *vii.* Hydroid, Alcyonarian, and Madreporarian Corals. By H. N. Moseley. *pp. 248: 32 pls., text illust.*
" *viii.* Birds. By P. L. Sclater. *pp. 166: 30 pls. col., text illust.*
          1. On the Birds collected in the Philippine Islands. By Arthur, Marquis of of Tweeddale.
          2. On the Birds collected in the Admiralty Islands. By P. L. Sclater.
          3. On the Birds collected in Tongatabu, the Fiji Islands, Api (New Hebrides), and Tahiti. By O. Finsch.
          4. On the Birds collected in Ternate, Amboyna, Banda, the Ki Islands, and the Arrou Islands. By T. Salvadori.
          5. On the Birds collected at Cape York, Australia, and on the neighbouring Islands (Raine, Wednesday, and Booby Islands). By W. A. Forbes.
          6. On the Birds collected in the Sandwich Islands. By P. L. Sclater.
          7. On the Birds collected in Antarctic America. By P. L. Sclater and O. Salvin.
          8. On the Birds collected on the Atlantic Islands and Kerguelen Island, and on the miscellaneous collections. By P. L. Sclater.
          9. On the Steganopodes and Impennes collected during the expedition. By P. L. Sclater and O. Salvin.
         10. On the Laridiæ collected during the expedition. By H. Saunders.
         11. On the Procellariidæ collected during the expedition. By O. Salvin.
         Append.
            1. List of the eggs collected during the expedition. By P. L. Sclater.
            2. Note on the gizzard and other organs of *Carpophaga latrans*. By A. H. Garrod.

Vol. III.   1881.
Pt.   *ix.* Echinoidea. By A. Agassiz. *pp. 321: 65 pls.*
"   *x.* Pycnogonida. By P. P. C. Hoek. *pp. 167: 21 pls. (col.), text illust.*

Vol. IV.   1882.
Pt.   *xi.* Anatomy of the Petrels. By W. A. Forbes. *pp. 64: 7 pls. (3 col.), text illust.*
"   *xii.* Deep-Sea Medusæ. By E. Haeckel. *pp. cv, 154: 32 pls. (col.), text illust.* [For German edition, *See* HAECKEL (E. H. P. A.) Monographie der Medusen. Pt. 2.]
" *xiii.* Holothurioidea. Pt. I. By H. Theel. *pp. 176: 46 pls. (col.)*

Vol. V.   1882.
Pt. *xiv.* Ophiuriiodea. By T. Lyman. *pp. 386: 48 pls.*
" *xvi.* Marsupialia. By D. J. Cunningham. *pp. 192: 13 pls., text illust.*

Vol. VI.   1882.
Pt. *xv.* Actiniaria. By R. Hertwig. *pp. 136: 14 pls.*
" *xvii.* Tunicata. Pt. I. By W. A. Herdman. *pp. 296: 37 pls., text illust.*

Vol. VII.   1883
Pt. *xviii.* Anatomy of the Spheniscidæ. By M. Watson. *pp. 244: 19 pls. (col.), text illust.*
" *xix.* Pelagic Hemiptera. By F. B. White. *pp. 82: 3 pls. (col.)*
"   *xx.* Hydroida. By G. J. Allmann. Pt. I. *pp. 54: 20 pls., text illust.*
" *xxi.* Orbitolites. By W. B. Carpenter. *pp. 47: 8 pls., text illust.*

Vol. VIII.   1883.
Pt. *xxiii.* Copepoda. By G. S. Brady. *pp. 142: 55 pls., text illust.*
" *xxiv.* Calcarea. By N. Poléjaeff. *pp. 76: 9 pls. (col.)*
" *xxv.* Cirripedia. By P. P. C. Hoek. (Systematic Part.) *pp. 169: 13 pls. (col.)*

Vol. IX.   1884.
 Pt.   *xxii.* Foraminifera. By H. B. Brady. *pp. xxi, 814: 116 pls. (col.), 2 maps, text illust.*
         (2 Vol.)
Vol. x.   1884.
 Pt.   *xxvi.* Nudibranchiata. By R. Bergh. *pp. 154: 14 pls.*
 ''   *xxvii.* Myzostomida. By L. von Graff. *pp. 82: 16 pls. (col.), text illust. col.*
 ''   *xxviii.* Cirripedia. By P. P. C. Hoek. (Anatomic Part.) *pp. 47: 6 pls. (5 col.)*
 ''   *xxix.* Human Skeletons. By Sir W. Turner. Pt. I. The Crania. *pp. 130: 7 pls., text
         illust.*
 ''   *xxx.* Polyzoa. By G. Busk. Pt. I. The Cheilostomata. *pp. xxiii, [i,] 216: 36 pls., 1
         map, text illust.*
Vol. XI.   1884.
 Pt.   *xxxi.* Keratosa. By N. Poléjaeff. *pp. 88: 10 pls., text illust.*
 ''   *xxxii.* Crinoidea. By P. H. Carpenter. Pt. I. *pp. x, 442: 69 pls., text illust.*
 ''   *xxxiii.* Isopoda. By F. E. Beddard. Pt. I, *pp. 85: 10 pls.*
Vol. XII.   1885.
 Pt.   *xxxiv.* Annelida Polychæta, By W. C. M'Intosh. *pp. xxxvi, 554: 94 pls. (col.), 1 map,
         text illust.*
Vol. XIII.   1885.
 Pt.   *xxxv.* Lamellibranchiata. By E. A. Smith. *pp. 341: 25 pls., text illust.*
 ''   *xxxvi.* Gephyrea. By E. Selenka. *pp. 24 [1]: 4 pls.*
 ''   *xxxvii.* Schizopoda. By G. O. Sars. *pp. 228: 38 pls., text illust.*
Vol. XIV.   1886.
 Pt. *xxxviii.* Tunicata. Pt. II. By W. A. Herdman. *pp. 432: 49 pls. (1 col.), 1 map, text illust.*
 ''   *xxxix.* Holothurioidea. Pt. II. By H. Théel. *pp. 290: 16 pls.*
Vol. xv.   1886.
 Pt.   *xli.* Marseniadæ. By R. Bergh. *pp. 24: 1 pl.*
 ''   *xlii.* Scaphopoda and Gasteropoda. By R. B. Watson. *pp. v, 756: 53 pls. (col.), text
         illust.*
                     Append. B. Cæcidæ. By Leopold, Marquis de Folin.
 ''   *xliii.* Polyplacophora. By A. C. Haddon. *pp. 50: 3 pls. (col.).*
Vol. XVI.   1886.
 Pt.   *xliv.* Cephalopoda. By W. E. Hoyle. *pp. vi, 245: 33 pls., 1 map, text illust.*
 ''   *xlv.* Stomatopoda. By W. K. Brooks. *pp. 116: 16 pls.*
 ''   *xlvi.* Reef Corals. By J. J. Quelch. *pp. 203: 12 pls.*
 ''   *xlvii.* Human Skeletons. By Sir W. Turner. Pt. II. The Bones of the Skeleton. *pp.
         136: 3 pls., text illust.*
Vol. XVII.   1886.
 Pt.   *xlviii.* Isopoda. By F. E. Beddard. Pt. II. *pp. 178: 25 pls., 1 map, text illust.*
 ''   *xlix.* Brachyura. By E. J. Miers. *pp. l, 362: 29 pls.*
 ''   *l.* Polyzoa. By G. Busk. Pt. II. The Cyclostomata, Ctenostomata, and Ped-
         icellinea. *pp. viii, 47: 10 pls., text illust.*
Vol. XVIII.   1887.
 Pt.   *xl.* Radiolaria. By E. Haeckel, *pp. viii, clxxxviii, 1803: 140 pls. (col.), 1 map. (3
         Vol.)*
Vol. XIX.   1887.
 Pt.   *liv.* Nemertea. By A. A. W. Hubrecht. *pp. 150: 16 pls. (col.), text illust.*
 ''   *lv.* Cumacea. By G. O. Sars. *pp. 78: 11 pls.*
 ''   *lvi.* Phyllocarida, By G. O. Sars. *pp. 38: 3 pls., text illust.*
 ''   *lviii.* Pteropoda. By P. Pelseneer. Pt. I. *pp. 74: 3 pls., text illust.*
Vol. xx.   1887.
 Pt.   *lix.* Monaxonida. By S. O. Ridley and A. Dendy. *pp. lxviii, 275: 51 pls. (col.), 1
         map, text illust.*
 ''   *lxi.* Myzostomida. By L. von Graff. (Supplement.) *pp. 16: 4 pls. (col.)*
 ''   *lxii. Cephalodiscus dodecalophus,* a new type of Polyzoa. By W. C. M'Intosh.
         pp. 47: 7 pls. (col.), text illust.
Vol. xxi.   1887.
 Pt.   *liii.* Hexactinellida. By F. E. Schulze. *pp. 513: 104 pls. (col.), 1 map col., text
         illust. (2 Vol.)*

Vol. xxii. 1887.

Pt. *lvii.* Deep-Sea Fishes. By A. Günther. *pp. lxv, 335: 73 pls.*

Append.
A. Structure of the Peculiar Organs on the Head of Ipnops. By H. N. Moseley.
B. Structure of the Phosphorescent Organs of Fishes. By R. von Lendenfeld.

Vol. xxiii. 1888.

Pt. *lxv.* Pteropoda. By P. Pelseneer. Pt. ii. *pp. 132: 2 pls., text illust.*
" *lxvi.* Pteropoda. By P. Pelseneer. Pt. iii. *pp. 97: 5 pls., text illust.*
" *lxx.* Hydroida. By G. J. Allman. Pt. ii. *pp. lxix, 90: 39 pls., 1 map.*
" *lxxi.* Entozoa. By O. von Linstow. *pp. 18: 2 pls., text illust.*
" *lxxii.* Heteropoda. By E. A. Smith. *pp. 51: text illust.*

Vol. xxiv. 1888.

Pt. *lii.* Crustacea Macrura. By C. S. Bate. *pp. xc, 942: 157 pls., text illust.* (2 Vol.)

Appendix A. Description of *Sylon challengeri*, n.sp., a parasitic cirriped. By P. P. C. Hoek.

Vol. xxv. 1888.

Pt. *lxiii.* Tetractinellida. By W. J. Sollas. *pp. clxvi, 458: 44 pls. col., 1 map, text illust.*

Vol. xxvi. 1888.

Pt. *lx.* Crinoidea. By P. H. Carpenter. Pt. ii. *pp. ix, 399: 70 pls., text illust.*
" *lxviii.* Seals. By Sir W. Turner. *pp. 240: 10 pls., text illust.*

Appendix. Myology of the Pinnipedia. By W. C. S. Miller.
" *lxxiii.* Actiniaria. By R. Hertwig. (Supplement.) *pp. 56: 4 pls.*

Vol. xxvii. 1888.

Pt. *lxix.* Anomura. By J. R. Henderson. *pp. xi, 221: 21 pls.*
" *lxxiv.* Anatomy of the Deep-Sea Mollusca. By P. Pelseneer, *pp. 42: 4 pls.*
" *lxxv.* *Phoronis Buskii*, n.sp. By W. C. M'Intosh. *pp. 27: 3 pls.*
" *lxxvi.* Tunicata. Pt. iii. By W. A. Herdman. *pp. 166: 11 pls., text illust.*

Vol. xxviii. 1888.

Pt. *lxxvii.* Siphonophoræ. By E. Haeckel. *pp. ii, 380: 50 pls. (col.)*

Vol. xxix. 1888.

Pt. *lxvii.* Amphipoda. By T. R. R. Stebbing. *pp. xxiv, (xiii,) 1737: 212 pls., 1 map, text illust.* (3 Vol.)

Vol. xxx. 1889.

Pt. *li.* Asteroidea. By W. P. Sladen. *pp. xliii, 893: 118 pls., 1 map.* (2 Vol.)

Vol. xxxi. 1889.

Pt. *lxiv.* Alcyonaria. By Prof. E. P. Wright and Prof. T. Studer. *pp. lxxii, 314: 49 pls., text illust.*
" *lxxviii.* Pelagic Fishes. By A. Günther. *pp. 47: 6 pls.*
" *lxxix.* Polyzoa, Suppl. Report. By A. W. Waters. *pp. 41: 3 pls.*

Vol. xxxii. 1889.

Pt. *lxxx.* Antipatharia. By G. Brook. *pp. iii, 222: 15 pls., text illust.*
" *lxxxi.* Alcyonaria, Suppl. Report. By Prof. T. Studer. *pp. 31: 6 pls.* (2 col.)
" *lxxxii.* Deep-Sea Keratosa. By Prof. E. Haeckel. *pp. 92: 8 pls. (col.)*
Pt. *lxxxiii.* Genus *Spirula*. By the Rt. Hon. T. H. Huxley and Prof. P. Pelseneer. *pp. 32: 6 pls., text illust.* 1895.

[Pt. *lxxxiii* was issued with the second volume of the "Summary."]

Great Britain and Ireland. [*Voyages, &c.*]

[*Challenger Expedition.*] Deep-Sea Fauna of New Zealand. Extracted from the Reports of the . . . Expedition. *See* HAMILTON (A.) 8⁰.1896.

Great Britain and Ireland. [*Voyages, &c.*]

[*Challenger Expedition.*] [For references to other works relating to this expedition] *See* CHALLENGER, *H.M.S.*

Great Britain and Ireland. [*Voyages, &c.*]
[*China.*] Despatch . . . forwarding a Report by Mr. A. Hosie . . . of a
Journey through the Provinces of Kueichow and Yünnan, *&c. pp. 37.*
4^0. *London*, 1883.
> Extracted from the Blue-Book Report: China, No. 1.

Great Britain and Ireland. [*Voyages, &c.*]
[*China.*] Report by Mr. Hosie of a Journey through central Ssu-Ch'uan in . . .
1884. 4^0. *London*, 1885.
> Blue-Book Report: China, No. 2 (1885).

Great Britain and Ireland. [*Voyages, &c.*]
[*Euphrates and Tigris.*] The expedition for the Survey of the Rivers Euphrates
and Tigris, carried on . . . in . . . 1835–37, *&c.* 2 Vol. *See* CHESNEY (F.R.)
8^0. 1850.

Great Britain and Ireland. [*Voyages, &c.*]
[*Herald.*] The Zoology of the Voyage of H.M.S. Herald . . . 1845–51, *&c. See*
FORBES (E.) 4^0. 1852–54.

Great Britain and Ireland. [*Voyages, &c.*]
[*Herald.*] Narrative of the Voyage of H.M.S. Herald during . . . 1845–51, *&c.*
See SEEMAN (B. C.) 8^0. 1853.

GREAT BRITAIN & IRELAND. — ADMIRALTY.
A Manual of Scientific Enquiry: prepared for the use of Her Majesty's Navy
and . . . Travellers in general. Edited by Sir J. F. W. Herschel, *&c. pp. xi, 488:*
2 maps, text illust. 8^0. *London*, 1849.
> Geology, by C. Darwin.
> Mineralogy, by Sir H. T. De La Beche.
> Zoology, by R. Owen.
> Botany, by Sir W. Hooker.
> Ethnology [including Anthropology], by J. C. Prichard.

———— Third edition, superintended by the Rev. R. Main, *&c. pp. xviii, 429: 2*
maps, text illust. 8^0. *London*, 1859.
> The section Mineralogy was revised for this edition by W. H. Miller; that on Ethnology by
> T. Wright and M. D'Avezac.

———— Fourth edition, superintended by the Rev. R. Main, *&c. pp. xvi, 392:*
2 maps, text illust. 8^0. *London*, 1871.
> In this edition the section Geology was revised by J. Phillips; that on Botany by J. D. Hooker;
> and that on Ethnology by E. B. Tylor.

———— Fifth edition. Edited by Sir R. S. Ball. *pp. xii, 450: 1 pl., 6 maps, text*
illust. 8^0. *London*, 1886.
> In this edition the section Geology was revised by A. Geikie; that on Mineralogy by W. J.
> Sollas. The remaining sections were rewritten: —
> Anthropology, by E. B. Tylor.
> Zoology, by H. N. Moseley.
> Botany, by J. D. Hooker.

Great Britain and Ireland. — ADMIRALTY.

Deep-sea Soundings in the North Atlantic Ocean between Ireland and New-foundland, made in H.M.S. Cyclops, Lieut. Commander J. Dayman . . . 1857. *pp. 73: 1 chart.* 8⁰. *London*, 1858.

Appendix A. Prof. Huxley's Report on the examination of specimens of bottom.

Great Britain and Ireland. — ADMIRALTY.

Deep-sea Soundings in the North Atlantic Ocean between Newfoundland, the Azores, and England, made in H.M.S. Gorgon, Commander J. Dayman . . . 1858. *pp. 30: 1 chart.* 8⁰. *London*, 1859.

Great Britain and Ireland. — ADMIRALTY.

Deep-sea Soundings in the Bay of Biscay and Mediterranean Sea, made in H.M.S. Firebrand, Commander J. Dayman . . . in . . . 1859. *pp. 47.*

8⁰. *London*, 1860.

U.S. ALBATROSS

BIBLIOGRAPHY OF WORKS RELATING TO DEEP-SEA STUDIES
RESULTING FROM THE U.S. ALBATROSS
(AFTER TOWNSEND, C. H., 1901)

1. 1884. Gill, Theodore. Diagnoses of new genera and species of deep-sea fish-like ver-tebrates. Proc. U.S. Nat. Mus., 1883, vol. 6, pp. 253–260.
2. 1884. Gill, Theodore and John A. Ryder. Diagnoses of new genera of Nemichthyoid eels. Proc. U.S. Nat. Mus. 1883, vol. 6, pp. 260–262.
3. 1884. Gill, Theodore. Deep-sea fishing fishes. Forest and Stream, vol. 21, Nov. 8, p. 284.
4. 1884. Gill, Theodore, and John A. Ryder. On the anatomy and relations of the Eury-pharyngidæ. Proc. U.S. Nat. Mus. 1883, vol. 6, pp. 262–273.
5. 1884. Tanner, Z. L., Lieut., U.S.N. Report on the work of the U.S. F.C. steamer FISH HAWK for the year ending Dec. 31, 1882, and on the construction of the steamer ALBATROSS. Rep. U.S.F.C. 1882, pp. 3–34, 3 pls.
6. 1884. Gill, Theodore. The ichthyological peculiarities of the Bassalian Fauna. Science, vol. 3, No. 68, pp. 620–622, 3 cuts.
7. 1884. Gill, Theodore. Three new families of fishes added to the deep-sea fauna in a year. Am. Nat., vol. 18, p. 433.
8. 1885. Gill, Theodore, and John A. Ryder. On the literature and systematic relations of the Saccopharyngoid fishes. Proc. U.S. Nat. Mus., 1884, vol. 7, pp. 48–65, 1 pl.
9. 1884. Smith, Sidney I. Report on the Decapod Crustacea of the ALBATROSS dredgings off the east coast of the United States in 1883. Rep. U.S.F.C. 1882, vol. 10, pp. 345–426, 10 pls.
10. 1884. Verrill, A. E. Second catalogue of Mollusca recently added to the fauna of the New England coast and adjacent parts of the Atlantic, consisting mostly of deep-sea species, with notes on others previously recorded. Trans. Conn. Acad. Arts and Sciences, vol. 6, pp. 139–294, 5 pls.
11. 1884. Verrill, A. E. List of deep-water and surface Mollusca taken off the east coast of the United States by the U.S.F.C. steamers FISH HAWK and ALBATROSS, 1880–1883. Ext. Conn. Acad. Sci. Transactions, New Haven. The society. July. vol. 6, pp. 263–290.
12. 1885. Schroeder, Seaton, Lieut., U.S.N. Hydrographic work of the ALBATROSS in 1884. Bull. U.S.F.C. 1885, vol. 5, pp. 269, 270.
13. 1885. Verill, A. E. Results of the explorations made by the steamer ALBATROSS off the northern coast of the United States in 1883. Rep. U.S.F.C. 1883, part 11, pp. 503–699, 44 pls.
14. 1885. Verrill, A. E. Notice of the remarkable marine fauna occupying the outer banks off the southern coast of New England. No. 11. (Brief contributions to zoology from the museum of Yale College. No. LVII.) Work of the ALBATROSS in 1884. Am. Jour. Sci. 1885, third series, vol. 29, No. 170, Feb., pp. 149–157.
15. 1885. Verrill, A. E. Third catalogue of Mollusca recently added to the fauna of the New England coast and the adjacent parts of the Atlantic, consisting mostly of deep-sea

species, with notes on others previously recorded. Trans. Conn. Acad. of Arts and Sciences 1885, vol. 6, pp. 395–452, 3 pls.

16.   1885.   Bush, Katherine J. Additions to the shallow-water Mollusca of Cape Hatteras, N. C., dredged by the U.S.F.C. steamer ALBATROSS in 1883 and 1884. Trans. Conn. Acad. of Arts and Sciences 1885, vol. 6, pp. 453–480, 1 pl.

17.   1885.   Smith, Sidney I. On some new or little-known Decapod Crustacea, from recent Fish Commission dredgings off the east coast of the United States: Proc. U.S. Nat. Mus. 1884, vol. 7, pp. 493–511.

18.   1885.   Nye, Jr., Willard. Notes upon octopus, flying fish, etc., taken during the ALBATROSS cruise in January, 1884. Bull. U.S.F.C. 1885, vol. 5, pp. 189–190.

19.   1886.   Bean, Tarleton H. Description of a new species of Plectromus (P. crassiceps) taken by the U.S. Fish Commission. Proc. U.S. Nat. Mus. 1885, vol. 8, pp. 73, 74.

20.   1886.   Goode, G. Brown, and Tarleton H. Bean. Description of Leptophidium cervinum and L. marmoratum, new fishes from deep water off the Atlantic and Gulf coasts. Proc. U.S. Nat. Mus. 1885, vol. 8, pp. 422–424.

21   1886.   Goode, G. Brown, and Tarleton H. Bean. Descriptions of new fishes obtained by the United States Fish Commission mainly from deep water off the Atlantic and Gulf coasts. Proc. U.S. Nat. Mus. 1885, vol. 8, pp. 589–605.

22.   1886.   Goode, G. Brown, and Tarleton H. Bean. Descriptions of thirteen species and two genera of fishes from the BLAKE Collection. Bull. Mus. Comp. Zoology, vol. 12, No. 5, pp. 153–170.

23.   1886.   Fewkes, J. Walter. Report on the Medusæ collected by the U.S.F.C. steamer ALBATROSS in the region of the Gulf Stream, in 1883–84, Rep. U.S.F.C. 1884, part 12, pp. 927–980, 10 pls.

24.   1886.   Fewkes J. Walter. On a collection of Medusæ made by the Steamer ALBATROSS in the Caribbean Sea and Gulf of Mexico. Proc. U.S. Nat. Mus. 1885. vol. 8, pp. 397–402.

25.   1886.   Rathbun, Richard. Report upon the Echini collected by the U.S.F.C. steamer ALBATROSS in the Caribbean Sea and Gulf of Mexico, January to May, 1884. Proc. U.S. Nat. Mus. 1885, vol. 8, pp. 83–89.

26.   1886.   Rathbun, Richard. Notice of a collection of stalked Crinoids made by the steamer ALBATROSS in the Gulf of Mexico and Caribbean Sea, 1884 and 1885. Proc. U.S. Nat. Mus. 1885, vol. 8, pp. 628–635.

27.   1886.   Rathbun, Richard. Report upon the Echini collected by the U.S.F.C. steamer ALBATROSS in the Gulf of Mexico from January to March, 1885. Proc. U.S. Nat. Mus. 1885, vol. 8, pp. 606–620.

28.   1886.   Tanner, Z. L. Report on the work of the U.S.F.C. steamer ALBATROSS for the year ending December 31, 1884. Rep. U.S.F.C. 1884, part 12, pp. 3–116, 3 pls.

29.   1886.   Smith, Sidney I. On some genera and species of Penæidæ, mostly from recent dredgings of the U.S. Fish Commission. Proc. U.S. Nat. Mus. 1885, vol. 8, pp. 170–190.

30.   1886.   Smith, Sidney I. Description of a new crustacean allied to Homarus and Nephrops. Proc. U.S. Nat. Mus. 1885, vol. 8, pp. 167–170.

31.   1886.   Verrill, A. E. Notice of recent additions to the Marine Invertebrata of the northeastern coast of America, with descriptions of new genera and species and critical remarks on others. Part V. Annelida, Echinodermata, Hydroida, Tunicata. Proc. U.S. Nat. Mus. 1885, vol. 8, pp. 424–448.

32.   1886.   Washburn, F. L. Deep-sea dredging on the U.S.S. ALBATROSS. Trans. Am. Fish. Soc., pp. 17–21.

33.   1887.   Benedict, James E. Descriptions of 10 species and a new genus of Annelids from the dredgings of the steamer ALBATROSS. Proc. U.S. Nat. Mus. 1886, vol. 9, pp. 547–553, 6 pls.

34.   1887.   Collins, Capt. J. W. Report on the discovery and investigation of fishing grounds made by the ALBATROSS during a cruise along the Atlantic coast and in the Gulf of Mexico, with notes on the Gulf fisheries. Rep. U.S.F.C. 1885, part 13, pp. 217–311, 10 pls.

35.   1887.   Tanner, Z. L. Report on the work of the U.S.F.C. steamer ALBATROSS for the year ending December 31, 1885. Rep. U.S.F.C. 1885, part 13, pp. 3–89, 5 pls., 9 figs.

36. 1887. Tanner, Z. L. Record of hydrographic soundings and dredging stations occupied by the steamer ALBATROSS in 1886. Bull. U.S.F.C. 1886, vol. 6, pp. 277–285.

37. 1887. Smith, Sidney I. Report on the Decapod Crustacea of the ALBATROSS dredgings off the east coast of the United States during the summer and autumn of 1884. Rep. U.S.F.C. 1885, part 13, pp. 605–705, 20 pls.

38. 1887. Collins, J. W. Notes on an investigation of the great fishing banks of the western Atlantic. Bull. U.S.F.C. 1886, vol. 6, pp. 369–381.

39. 1888. Fewkes, J. Walter. Are there deep-sea Medusæ? Amer. Jour. Sci., 1888, third series, vol. 35, No. 206, Feb., pp. 166–179.

40. 1889. Dall, William Healey. A preliminary catalogue of the Shellbearing Marine Mollusks and Brachiopods of the southeastern coast of the United States, with illustrations of many of the species. Bull. U.S. Nat. Mus., No. 37, 221 pp., 74 pls.

41. 1889. Fewkes, J. Walter. Report on the Medusæ collected by the U.S.F.C. steamer ALBATROSS in the region of the Gulf Stream in 1885–86. Rep. U.S.F.C. 1886, part 14, pp. 513–536, 1 pl.

42. 1889. Smith, Sanderson. Lists of the dredging stations of the U.S. Fish Commission, the U.S. Coast Survey, and the British steamer CHALLENGER, in North American waters, from 1867 to 1887, together with those of the principal European government expeditions in the Atlantic and Arctic oceans. Rep. U.S.F.C. 1886, part 14, pp. 871–1017, 5 chts.

43. 1890. Bean, Tarleton H. Notes on fishes collected at Cozumel, Yucatan, by the U.S. Fish Commission, with descriptions of new species. Bull. U.S.F.C. 1888, vol. 8, pp. 193–206, 2 pls.

44. 1890. Bean, Tarleton H. Scientific results of explorations by the U.S.F.C. steamer ALBATROSS, VIII. Description of a new cottoid fish from British Columbia. Proc. U.S. Nat. Mus. 1889, vol. 12, pp. 641, 642.

45. 1890. Dall, William Healey. Scientific results of explorations by the U.S.F.C. steamer ALBATROSS. VII. Preliminary report on the collection of Mollusca and Brachiopoda obtained in 1887–88. Proc. U.S. Nat. Mus. 1889, vol. 12, pp. 219–362.

46. 1890. Agassiz, Alexander. Notice of *Calamocrinus diomedae,* a new Stalked Crinoid from the Galapagos, dredged by the U.S.F.C. steamer ALBATROSS, Lieut. Commander Z. L. Tanner, U.S.N., commanding. Bull. Mus. Comp. Zool., vol. 20, pp. 165–167.

47. 1890. Jordan, David Starr. Scientific results of explorations by the U.S.F.C. steamer ALBATROSS IX. Catalogue of fishes collected at Port Castries, St. Lucia, by the steamer ALBATROSS, Nov., 1888. Proc. U.S. Nat. Mus. 1889, vol. 12, pp. 645–652.

48. 1890. Jordan, David Starr, and Charles Harvey Bollman. Scientific results of explorations of the U.S.F.C. steamer ALBATROSS. IV. Descriptions of new species of fishes collected at the Galapagos Islands and along the coast of the United States of Colombia, 1887–88. Proc. U.S. Nat. Mus. 1889, vol. 12, pp. 149–183.

49. 1890. Stearns, Robert E. C. Scientific results of explorations by the U.S.F.C. steamer ALBATROSS. XVII. Descriptions of new West American land, fresh-water, and marine shells, with notes and comments. Proc. U.S. Nat. Mus. 1890, vol. 13, pp. 205–225.

50. 1891. Gilbert, Charles H. Scientific results of explorations by the U.S.F.C. steamer ALBATROSS. XII. A preliminary report on fishes collected by the steamer ALBATROSS on the Pacific coast of North America during the year 1889, with descriptions of 12 new genera and 92 new species. Proc. U.S. Nat. Mus. 1890, vol. 13, pp. 49–126.

51. 1891. Gilbert, Charles H. Scientific results of explorations by the U.S.F.C. steamer ALBATROSS. XIX. A supplementary list of fishes collected at the Galapagos Islands and Panama, with descriptions of one new genus and three new species. Proc. U.S. Nat. Mus. 1890, vol. 13, pp. 449–455.

52. 1891. Agassiz, A. Three letters from Alexander Agassiz to Hon. Marshall McDonald, U.S. Commissioner of Fish and Fisheries, on the dredging operations off the west coast of Central America to the Galapagos, to the west coast of Mexico, and in the Gulf of California, in charge of Alexander Agassiz, carried on by the U.S.F.C. steamer ALBATROSS. Bull. Mus. Comp. Zool., vol. 21, pp. 186–200.

53. 1891. Bean, Tarleton H. Scientific results of explorations by the U.S.F.C. steamer AL-BATROSS. XI. New fishes collected off the coast of Alaska and the adjacent region southward. Proc. U.S. Nat. Mus. 1890, vol. 13, pp. 37–45.

54. 1891. Jordan, David Starr. Scientific results of explorations by the U.S.F.C. steamer AL-BATROSS. XVIII. List of fishes obtained in the harbor of Bahia, Brazil, and in adjacent waters. Proc. U.S. Nat. Mus. 1890, vol. 13, pp. 313–336.

55. 1891. Vasey, George, and J. N. Rose. Scientific results of explorations by the U.S.F.C. steamer ALBATROSS. XVI. Plants collected in 1889 at Socorro and Clarion Islands, Pacific Ocean. Proc. U.S. Nat. Mus. 1890, vol. 13, pp. 145–149.

56. 1891. White, Charles A. Scientific results of explorations by the U.S.F.C. steamer AL-BATROSS. X. On certain Mesozoic fossils from the islands of St. Pauls and St. Peters in the Straits of Magellan. Proc. U.S. Nat. Mus. 1890, vol. 13, pp. 13, 14, 2 pls.

57. 1891. Benedict, J. E., and Mary J. Rathbun. The genus *Panopeus*. Proc. U.S. Nat. Mus. 1891, vol. 14, pp. 355–385, pls. xix–xxiv.

58. 1891. Townsend, C. H. Report upon the pearl fishery of the Gulf of California. Bull. U.S. Fish Com. 1889, vol. 9, pp. 91–94, 3 pls.

59. 1892. Agassiz, Alexander. Reports of an exploration off the west coasts of Mexico, Central and South America, and off the Galapagos Islands, in charge of Alexander Agassiz, by U.S.F.C. steamer ALBATROSS, during 1891. I. *Calamocrinus diomedae* a new Stalked Crinoid, with notes on the apical system and the homologies of Echinoderms. Mem. Mus. Comp. Zoology, 1892, vol. 17, 96 pp., 32 pls.

60. 1892. Agassiz, Alexander. Reports on the dredging operations off the west coast of Central America to the Galapagos, to the West Coast of Mexico, and in the Gulf of California, in charge of Alexander Agassiz, carried on by the U.S.F.C. steamer AL-BATROSS. II. General sketch of the expedition of the ALBATROSS from Feb. to May, 1891. Bull. Mus. Comp. Zool, 1892, vol. 23, pp. 1–90, 22 pls.

61. 1892. Goode, G. Brown, and T. H. Bean. The present condition of the study of deep-sea fishes. Proc. Am. Ass. Adv. Sci., vol. 40, p. 324.

62. 1892. Gilbert, Charles H. Scientific results of explorations by the U.S.F.C. steamer AL-BATROSS. XXI. Descriptions of apodal fishes from the tropical Pacific. Proc. U.S. Nat. Mus. 1891, vol. 14, pp. 347–352.

63. 1892. Gilbert, Charles H. Scientific results of explorations by the U.S.F.C. steamer AL-BATROSS. XXII. Descriptions of thirty-four new species of fishes collected in 1888 and 1889, principally among the Santa Barbara Islands and in the Gulf of California. Proc. U.S. Nat. Mus. 1891, vol. 14, pp. 539–566.

64. 1892. Goës, A. Reports on the dredging operations off the west coast of Central America to the Galapagos, to the west coast of Mexico, and in the Gulf of California, in charge of Alexander Agassiz, carried on by the U.S.F.C. steamer ALBATROSS during 1891. III. On a peculiar type of Arenaceous Foraminifer from the American tropical Pacific, *Neusina agassizi*. Bull. Mus. Comp. Zool. 1892, vol. 23, pp. 195–198, 1 pl.

65. 1892. Dall, William H. Scientific results of explorations by the U.S.F.C. steamer AL-BATROSS. XX. On some new or interesting West American shells obtained from the dredgings of the U.S.F.C. steamer ALBATROSS in 1888, and from other sources. Proc. U.S. Nat. Mus. 1891, vol. 14, pp. 173–191, 3 pls.

66. 1892. Rathbun, Richard. The U.S. Fish Commission, some of its work. Century Mag. 1892, vol. 43, Mar., pp. 679–697; 20 cuts.

67. 1892. Verrill, A. E. The Marine Nemerteans of New England and adjacent waters. Trans. Conn. Acad. Arts and Sciences 1892, vol. 8, pp. 382–456; 7 pls, 9 figs.

68. 1892. Verrill, A. E. Marine Planarians of New England. Trans. Conn. Acad. Arts and Sciences 1892, vol. 8, pp. 459–520, 5 pls., 2 figs.

69. 1892. Cruise of the ALBATROSS. Bull. Am. Geog. Soc. 1892, vol. 24, No. 3, pp. 464–467.

70. 1893. Tanner, Z. L. Report upon the investigations of the U.S.F.C. steamer ALBATROSS from July 1, 1889, to June 30, 1891. Rep. U.S.F.C. 1889–1891, part 17, pp. 207–342, 1 pl.

71. 1893. Brooks, William K. The genus Salpa. Mems. Biol. Lab. Johns Hopk. Univ. 1893, 11, pp. 1–371, 57 pls.

72. 1893. Benedict, James E. Corystoid crabs of the genera *Telmessus* and *Erimacrus*. Proc. U.S. Nat. Mus. 1892, vol. 15, pp. 223–30, 3 pls.

73. 1893. Benedict, James E. Preliminary descriptions of 37 new species of Hermit Crabs of the genus *Eupagurus* in U.S. Nat. Museum. Proc. U.S. Nat. Mus. 1892, vol. 15, pp. 1–26.

74. 1893. Beard, J. Carter. The Abysmal depths of the sea. Cosmopolitan Magazine, Mar., pp. 532–538, 11 cuts.

75. 1893. Beecher, Charles E. The development of *Terebratalia obsoleta* Dall. Trans. Conn. Acad. Arts and Sciences 1893, vol. 9, pp. 392–399, 3 pls.

76. 1893. Faxon, Walter. Reports on dredging operations off the west coast of Central America, to the Galapagos, to the west Coast of Mexico, and in the Gulf of California, in charge of Alexander Agassiz, carried on by U.S.F.C. steamer ALBATROSS during 1891. VI. Preliminary descriptions of new species of Crustacea. Bull. Mus. Comp. Zool. 1893, vol. 24, pp. 149–220.

77. 1893. Rathbun, Mary J. Catalogue of the crabs of the family Periceridae in the U.S. National Museum. Proc. U.S. Nat. Mus. 1892, vol. 15, pp. 231–277, pls., XXVIII–XL.

78. 1893. Ludwig, Hubert. Reports on the dredging operations off the west coast of Central America, to the Galapagos, to the west coast of Mexico, and to the Gulf of California, in charge of Alexander Agassiz, carried on by the U.S.F.C. steamer ALBATROSS in 1897. IV. Vorläufiger Bericht über die erbeuteten Holothurien. Bull. Mus. Comp. Zool, 1893, vol. 24, pp. 105–114.

79. 1893. Scudder, Samuel H. Reports on the dredging operations off the west coast of Central America, to the Galapagos, to the west coast of Mexico, and in the Gulf of California, in charge of Alexander Agassiz, carried on by the U.S.F.C. steamer ALBATROSS during 1891. VII. The Orthoptera of the Galapagos Islands. Bull. Mus. Comp. Zool. 1893, vol. 25, pp. 1–26, 12 pls.

80. 1893. Schimkéwitsch, W. M. Reports on the dredging operations off the west coast of Central America, to the Galapagos, to the west coast of Mexico, and in the Gulf of California, in charge of Alexander Agassiz, carried on by the U.S.F.C. steamer ALBATROSS during 1891. VIII. Compte-Rendu sur les Pantopodes. Bull. Mus. Comp. Zool. 1893, vol. 25, pp. 27–44, 2 pls.

81. 1894. Tanner, Z. L. Report upon the investigations of the U.S.F.C. steamer ALBATROSS for the year ending June 30, 1892. Rep. U.S.F.C. 1892, part 18, pp. 1–64, 1 pl.

82. 1894. Stearns, Robert E. C. Scientific results of explorations by the U.S.F.C. steamer ALBATROSS. XXV. Report on the Mollusk fauna of the Galapagos Islands, with descriptions of new species. Proc. U.S. Nat. Mus. 1893, vol. 16, pp. 353–450, 1 pl., 1 map.

83. 1894. Stearns, Robert E. C. The shells of the Tres Marias and other localities along the shores of Lower California and the Gulf of California. Proc. U.S. Nat. Mus. 1894, vol. 17, pp. 139–204.

84. 1894. Rathbun, Richard. A summary of the fishery investigations conducted in the North Pacific Ocean and Bering Sea from July 1, 1888 to July 1, 1892, by the U.S.F.C. steamer ALBATROSS. Bull. U.S.F.C. 1892, Vol. 12, pp. 127–201, 5 chts.

85. 1894. Rathbun, Mary J. Scientific results of explorations by the U.S.F.C. steamer ALBATROSS. XXIV. Descriptions of new genera and species of crabs from the west coast of North America and the Sandwich Islands. Proc. U.S. Nat. Mus. 1893, vol. 16, pp. 223–260.

86. 1894. Rathbun, Mary J. Catalogue of the crabs of the family Maiidæ in the U.S. National Museum. Proc. U.S. Nat. Mus. 1893, vol. 16, pp. 63–103, pls. III–VIII.

87. 1894. McMurrich, J. Playfair. Scientific results of explorations by the U.S.F.C. steamer ALBATROSS. XXIII. Report on the Actiniæ collected by the ALBATROSS during the winter of 1887–88. Proc. U.S. Nat. Mus. 1893, vol. 16, pp. 119–216, 17 pls.

88. 1894. Studer, Théophile. Reports on the dredging operations off the west coast of Central America, to the Galapagos, to the west coast of Mexico, and in the Gulf of California in charge of Alexander Agassiz, carried on by the U.S.F.C. steamer ALBATROSS during 1891. X. Note preliminaire sur les Alcyonaires. Bull. Mus. Comp. Zool. 1894, vol. 25, pp. 55–70.

89. 1894. Clarke, Samuel F. Reports on the dredging operations off the west coast of Central

America to the Galapagos, to the west coast of Mexico, and in the Gulf of California, in charge of Alexander Agassiz, carried on by the U.S.F.C. steamer ALBATROSS during 1891. XI. The Hydroids. Bull. Mus. Comp. Zool. 1894, vol. 25, pp. 71–78, 5 pls.

90.   1894.   Woodworth, W. McM. Reports on dredging operations off the west coast of Central America, to the Galapagos, to the west coast of Mexico, and in the Gulf of California, in charge of Alexander Agassiz, carried on by the U.S.F.C. steamer ALBATROSS during 1891. IX. Report on the Turbellaria. Bull. Mus. Comp. Zool. 1894, vol. 25, pp. 49–52, 1 pl.

91.   1894.   Ludwig, Hubert. Reports on an exploration off the west coasts of Mexico, Central and South America, and off the Galapagos Islands, in charge of Alexander Agassiz, by the U.S.F.C. steamer ALBATROSS during 1891. XII. The Holothurioidae. Mem. Mus. Comp. Zool. 1894, vol. 17, No. 3, pp. 1–183, 19 pls.

92.   1894.   Bergh, Rudolph. Reports on the dredging operations off the west coast of Central America to the Galapagos to the west coast of Mexico, and in the Gulf of California, in charge of Alexander Agassiz, carried on by the U.S.F.C. steamer ALBATROSS during 1891. XIII. Die Opisthobranchien. Bull. Mus. Comp. Zool. 1894, vol. 25, pp. 125–233, 12 pls.

93.   1894.   McDonald, Marshall. The salmon fisheries of Alaska. Bull. U.S.F.C. 1892, vol. 12, pp. 1–20, 9 pls.

94.   1894.   Mann, Albert. List of Diatomacea from a deep-sea dredging in the Atlantic Ocean off Delaware Bay, by the ALBATROSS. Proc. U.S. Nat. Mus. 1893, vol. 16, pp. 303–312.

95.   1894.   Eigenmann, Carl I., and C. H. Beeson. A revision of fishes of the subfamily Sebastinæ of the Pacific coast of America. Proc. U.S. Nat. Mus. 1894, vol. 17, pp. 375–407.

96.   1894.   Ortmann, Arnold. Reports on the dredging operations off the west coast of Central America to the Galapagos, to the west coast of Mexico, and in the Gulf of California, in charge of Alexander Agassiz, carried on by the steamer ALBATROSS during 1891. XIV. The Pelagic Schizopoda. Bull. Mus. Comp. Zool. 1894, vol. 25, pp. 99–110, 1 pl.

97.   1894.   Hickson, Sydney J. The fauna of the deep sea. 12 mo. xvi + 169 pp. 23 ills. Appleton's N.Y. (Modern science series, edited by Sir John Lubbock.)

98.   1895.   Goode, G. Brown, and Tarleton H. Bean. Scientific results of explorations by the U.S.F.C. steamer ALBATROSS. XXVIII. On *Cetomimidae* and *Rondeletiidae*, two new families of bathybial fishes from the Northwestern Atlantic. Proc. U.S. Nat. Mus. 1894, vol. 17, pp. 451–454.

99.   1895.   Goode, G. Brown, and Tarleton H. Bean. Scientific results of explorations by the U.S.F.C. steamer ALBATROSS. XXIX. A revision of the order Heteromi, deep-sea fishes, with a description of the new generic types, *Macdonaldia* and *Lipogenys*. Proc. U.S. Nat. Mus. 1894, vol. 17, pp. 455–470.

100.   1895.   Goode, G. Brown, and Tarleton H. Bean. Scientific results of explorations by the U.S.F.C. steamer ALBATROSS. XXX. On *Harriotta*, a new type of Chimæroid fish from the deep waters of the Northwestern Atlantic. Proc. U.S. Nat. Mus. 1894, vol. 17, pp. 471–473, 1 pl.

101.   1895.   Goode, George Brown, and Tarleton H. Bean. Oceanic ichthyology, a treatise on the deep-sea and pelagic fishes of the world, based chiefly upon the collections made by the steamers BLAKE, ALBATROSS, and FISH HAWK, in the northwestern Atlantic, with an atlas containing 417 figures. Spec. Bull. U.S. Nat. Mus. xxxv + 553 pp. Atlas, xxiii + 26 pp, 123 pls.

102.   1895.   Goode, G. Brown, and Tarleton H. Bean. New deep-sea fishes. Am. Nat., vol. 29, pp. 281.

103.   1895.   Goode, G. Brown, and Tarleton H. Bean. More deep-sea fishes. Am. Nat., vol. 29, pp. 376, 3 pls.

104.   1895.   Gilbert, Charles H. The ichthyological collections of the steamer ALBATROSS during the years 1890 and 1891. Rep. U.S.F.C. 1893, part 19, pp. 393–476, 16 pls.

105.   1895.   Dall, William Healey. Scientific results of explorations by U.S.F.C. steamer ALBATROSS. XXXIV. Report on Mollusca and Brachiopoda dredged in deep water,

chiefly near the Hawaiian Islands, with illustrations of hitherto unfigured species from Northwest America. Proc. U.S. Nat. Mus. 1894, vol. 17, pp. 675–733, 10 pls.

106.   1895.   Dall, W. H. Synopsis of a review of the genera of recent and Tertiary Mactridae and Mesodesmatidae. Proc. Malacological Soc. (Lond.), vol. 1, pt. 5, Mar., pp. 203–213.

107.   1895.   Benedict, James E. Scientific results of explorations by the steamer ALBATROSS XXXI. Descriptions of new genera and species of crabs of the family Lithodidae, with notes on the young of *Lithodes camtschaticus* and *Lithodes brevipes*. Proc. U.S. Nat. Mus. 1894, vol. 17, pp. 478–488.

108.   1895.   Bigelow, Robert Payne. Scientific results of explorations by the U.S.F.C. steamer ALBATROSS. XXXII. Report on the Crustacea of the order Stomatopoda collected by the steamer ALBATROSS between 1885 and 1891, and on other specimens in the U.S. National Museum. Proc. U.S. Nat. Mus. 1894, vol. 17, pp. 489–550, 3 pls.

109.   1895.   Giesbrecht, Wilhelm. Reports on the dredging operations off the west coast of Central America to the Galapagos, to the west coast of Mexico, and in the Gulf of California, carried on by ALBATROSS, during 1891. XVI. Die Pelagischen Copepoden. Bull. Mus. Comp Zool., 1895, vol. 25, pp. 243–263, 4 pls.

110.   1895.   Faxon, Walter. Reports on an exploration off the west coasts of Mexico, Central and South America, and off the Galapagos Islands, by the steamer ALBATROSS, during 1891. XV. The Stalk-eyed Crustacea. Mem. Mus. Comp. Zool., 1895, vol. 18, pp. 1–292, 67 pls.

111.   1895.   Muller, G. W. Reports on the dredging operations off the west coast of Central America, to the Galapagos, to west coast of Mexico, and in the Gulf of California, carried on by U.S.F.C. steamer ALBATROSS, during 1891. XIX. Die Ostracoden. Bull. Mus. Comp. Zool. 1895, vol. 27, pp. 153–170, 3 pls.

112.   1895.   Hartlaub, C. Reports on the dredging operations off the west coast of Central America, to the Galapagos, to the west coast of Mexico, and in the Gulf of California, carried on by the steamer ALBATROSS, during 1891. XVIII. Die Comatuliden. Bull. Mus. Comp. Zool. 1895, vol. 27, pp. 137–152, 4 pls.

113.   1895.   Rathbun, Mary J. Descriptions of a new genus and four new species of crabs from the Antillean region. Proc. U.S. Nat. Mus. 1894, vol. 17, pp. 83–86.

114.   1895.   Rathbun, Mary J. Notes on the crabs of the family Inachidae in the U.S. National Museum. Proc. U.S. Nat. Mus. 1894, vol. 17, pp. 43–75, 1 pl.

115.   1895.   Rathbun, Mary J. The genus Callinectes. Proc. U.S. Nat. Mus. 1895, vol. 18, pp. 349–375.

116.   1895.   Verrill, A. E. Descriptions of new species of starfishes and Ophiurans, with a revision of certain species formerly described; mostly from collections made by the U.S. Commission of Fish and Fisheries. Proc. U.S. Nat. Mus. 1894, vol. 17, pp. 245–297.

117.   1895.   Verrill, A. E. Distribution of the Echinoderms of Northeastern America. (Brief contributions to zoology from museum of Yale College, No. LVIII.) Am. Jour. Sci. 1895, Third Series, vol. 49, No. 290, Feb., pp. 127–141.

118.   1895.   Verrill, A. E. Supplement to the Marine Nemerteans and Planarians of New England. Trans. Conn. Acad. of Arts and Sciences 1895, vol. 9, pp. 523–534.

119.   1896.   Tanner, Z. L., and F. J. Drake. Report upon the operations of the U.S.F.C. steamer ALBATROSS for the year ending June 30, 1894. Rep. U.S.F.C. 1894, part 20, pp. 197–278, 2 pls., cht.

120.   1896.   Drake, F. J., Lieut. Commander U.S.N. Report upon the investigations of the steamer ALBATROSS for the year ending June 30, 1895. (Abstract.) Rep. U.S.F.C. 1895, part 21, pp. 125–168.

121.   1896–   Jordan, David Starr, and Barton Warren Evermann. Fishes of North and Middle
        1900   America. A descriptive catalogue of the species of fish-like vertebrates found in the waters of North America north of the Isthmus of Panama. Bull. 47, U.S. Nat. Mus., Parts I–IV, lviii–3313 pp., 392 pls.

122.   1896.   Verrill, A. E. The Opisthoteuthidae, a remarkable new family of deep-sea Cephalopoda, with remarks on some points in molluscan morphology. Am. Jour. Sci. 1896, fourth series, vol. 2, No. 7–July, pp. 74–80, 7 figs.

123.   1896.   Dall, W. H. Diagnoses of new species of Mollusks from the west coast of America. Proc. U.S. Nat. Mus, 1895. vol. 18, pp. 7–20.

124.   1896.   Goes, Axel. Reports on the dredging operations off the west coast of Central America to the Galapagos, to the west coast of Mexico, and in the Gulf of California, carried on by the U.S.F.C. steamer ALBATROSS, during 1891. XX. The Foraminifera. Bull. Mus. Comp. Zool. 1896, vol. 29, pp. 1–103, 9 pls.

125.   1897.   Tanner, Z. L. Commander, U.S. Navy. Deep-sea exploration: A general description of the steamer ALBATROSS, her appliances and methods. Bull. U.S.F.C. 1896, vol. 16, pp. 257–424, 40 pls., 76 figs.

126.   1897.   Gilbert, C. H., and Frank Cramer. Report on the fishes dredged in deep water near the Hawaiian Islands, with descriptions and figures of 23 new species. Proc. U.S. Nat. Mus. 1896, vol. 19, pp. 403–435.

127.   1897.   Gilbert, Charles Henry. Descriptions of 22 new species of fishes collected by the steamer ALBATROSS. Proc. U.S. Nat. Mus. 1896, vol. 19, pp. 437–457.

128.   1897.   Benedict, James E. A revision of the genus Synidotea. Proc. Acad. Nat. Sci. Phil. 1897, pp. 389–404, 13 cuts.

129.   1897.   Richardson, Harriet. Description of a new genus and species of Sphaeromidae from Alaskan waters. Proc. Biol. Soc. Wash. 1897, vol. 11, pp. 181–183.

130.   1897.   Dall, W. H. Notice of some new or interesting species of shells from British Columbia and the adjacent region. Nat. Hist. Soc. B. C., Bull. No. 2, pp. 1–18, pl. 1–2.

131.   1897.   Gill, Theo., and C. H. Townsend. Diagnoses of new species of fishes found in Bering Sea. Proc. Biol. Soc. Wash. 1897, vol. 11, pp. 231–234.

132.   1897.   Verrill, A. E., and Katharine J. Bush. Revision of the genera of Ledidae and Nuculidae of the Atlantic coast of the United States. (Brief contributions to zoology from the museum of Yale University. No. L.) Am. Jour. Sci. 1891, 4th series, vol. 3, No. 13, Jan., pp. 51–63, 21 figs.

133.   1897.   Maas, Otto. Reports on an exploration off the west coasts of Mexico, Central and South America, and off the Galapagos Islands, by the ALBATROSS, in 1891. XXI. Die Medusen. Mem. Mus. Comp. Zool. 1897, vol. 32, pp. 7–92, 14 pls., 1 map.

134.   1897.   Hansen, H. J. Reports on the dredging operations off the west coast of Central America to the Galapagos, to the west coast of Mexico, and in the Gulf of California, carried on by the U.S.F.C. steamer ALBATROSS, during 1891. XXII. The Isopoda. Bull. Mus. Comp. Zool. 1897, vol. 31, pp. 93–130, 6 pls., chart.

135.   1897.   Rathbun, Mary J. Synopsis of the American species of Ethusa, with description of a new species. Proc. Biol. Soc. Wash. 1897 vol. 11, pp. 109–110.

136.   1897.   Rathbun, Mary J. Synopsis of the American species of Palicus Philippi (=Cymopolia roux), with descriptions of six new species. Proc. Biol. Soc. Wash. 1897, vol. 11, pp. 93–99.

137.   1898.   Agassiz, A. Reports on dredging operations off the west coast of Central America to the Galapagos, to the west coast of Mexico, and in the Gulf of California, carried on by the steamer ALBATROSS during 1891. XXIII. Preliminary report on the Echini. Bull. Mus. Comp. Zool. 1898, vol. 32, pp. 69–86, 13 pls., chart.

138.   1898.   Benedict, James E. The Arcturidae in the U.S. Nat. Mus. Proc. Biol. Soc. Wash., vol. 12, pp. 41–51.

139.   1898.   Drake, F. J. Records of observations made on board the U.S.F.C. steamer ALBATROSS during the year ending June 30, 1896. Rep. U.S.F.C. 1896, part 22, pp. 357–386.

140.   1898.   Rathbun, Mary J. The Brachyura of the biological expedition to the Florida Keys and the Bahamas in 1893. Bull. Lab. Nat. Hist. Univ. of Iowa, vol. 4, pp. 250–294, pls. 1–9.

141.   1898.   Verrill, Addison E., and Katharine J. Bush. Revision of the deep-water Mollusca of the Atlantic coast of North America, with descriptions of new genera and species. Part I. Bivalvia. Proc. U.S. Nat. Mus., vol. 20, pp. 775–901.

142.   1898.   Mark, E. L. Reports on the dredging operations off the west coast of Central America to the Galapagos, to the west coast of Mexico, and in the Gulf of California, carried on by the U.S.F.C. steamer ALBATROSS during 1891.

XXIV.—Preliminary report on *Branchiocerianthus urceolus,* a new type of Actinian. Bull. Mus. Comp. Zool. 1890, vol. 32, pp. 147–154, 3 pls.

143.　1898.　Richardson, Harriet. Description of a new parasitic Isopod of the genus Aega, from the southern coast of the United States. Proc. Biol. Soc. Wash. 1898, vol. 12, pp. 39–40.

144.　1899.　Flint, James M. Recent Foraminifera. A descriptive catalogue of specimens dredged by the U.S.F.C. steamer ALBATROSS. Ann. Rep. Smith. Institution 1897; Rep. U.S. Nat. Mus., Part I, pp. 249–350, 80 pls.

145.　1899.　Dall, W. H. Synopsis of the American species of the family Diplodontidae. Jour. of Conch. (Brit.), Oct., pp. 244–246.

146.　1899.　Dall, William I. Synopsis of the recent and Tertiary Leptonacea of North America and the West Indies. Proc. U.S. Nat. Mus., vol. 21, pp. 874–897, 2 pls.

147.　1899.　Bush, Katherine J. Revision of the marine Gastropods referred to Cyclostrema, Adeorbis, Vitrinella, and related genera, with descriptions of some new genera and species belonging to the Atlantic fauna of America. Trans. Conn. Acad. Arts and Sciences 1899, vol. 10, pp. 97–143.

148.　1899.　Lütken, C. F., and Th. Mortensen. Reports of an exploration off the west coasts of Mexico, Central and South America, and off the Galapagos Islands, in charge of Alexander Agassiz, by the steamer ALBATROSS, during 1891. XXV.—The Ophiuridae. Mems. Mus. Comp. Zool. 1899, vol. 23, pp. 93–208, 22 pls., chart.

149.　1899.　Garman, S. Reports of an exploration off the west coasts of Mexico, Central and South America, and off the Galapagos Islands, in charge of Alexander Agassiz, by the ALBATROSS during 1891. XXVI. The Fishes. Mems. Mus. Comp. Zool. 1899, vol. 24, 431, pp., 97 pls., chart.

150.　1899.　Bean, Barton A. Notes on the capture of rare fishes. Proc. U.S. Nat. Mus., vol. 21, pp. 639, 640.

151.　1899.　Rathbun, Mary J. The Brachyura collected by the U.S.F.C. steamer ALBATROSS on the voyage from Norfolk, Va., to San Francisco, Cal., 1887–88. Proc. U.S. Nat. Mus., vol. 21, pp. 567–616.

152.　1899.　Richardson, Harriet. Key to the Isopods of the Pacific coast of North America, with descriptions of 22 new species. Proc. U.S. Nat. Mus., vol. 21, pp. 815–869.

153.　1899.　Gilbert, Charles H. Report on fishes obtained by the steamer ALBATROSS in the vicinity of Santa Catalina Island and Monterey Bay. Rep. U.S.F.C. 1898, part 24, pp. 25–29, 2 pls.

154.　1899.　Gilbert, Charles Henry. On the occurrence of *Caulolepis longidens* Gill, on the coast of California. Proc. U.S. Nat. Mus., vol. 21, pp. 565, 566.

155.　1899.　Woodworth, W. McM. Reports on the dredging operations off the west coast of Central America, to the Galapagos, to west coast of Mexico, and in the Gulf of California, carried on by the steamer ALBATROSS during 1891. XXVII. Preliminary account of *Planktonemertes agassizii,* a new pelagic Nemertean. Bull. Mus. Comp. Zool. 1899, vol. 35, pp. 1–4, 1 pl.

156.　1899.　Moser, Commander Jefferson F. The salmon and salmon fisheries of Alaska. Report of the operations of the ALBATROSS for the year ending June 30, 1898. Bull. U.S.F.C. 1898, part 18, pp. 1–178, 63 pls., 26 figs., cht.

157.　1899.　Smith, Hugh M. Exploring expedition to the mid-Pacific Ocean. Science (U.S.), June 9, pp. 796–798.

158.　1899.　Smith, Hugh M. The deep-sea exploring expedition of the steamer ALBATROSS. Nat. Geog. Mag., vol. 10, No. 8, pp. 290–296, 3 ills.

159.　1899.　Verrill, A. E. Descriptions of imperfectly known and new Actinians, with critical notes on other species, III. (Brief Contributions to Zoology from the Museum of Yale College, no. LX.) Am. Jour. Sci., fourth series, vol. 7, 1899, pp. 143–146, 20 figs.

160.　1899.　Verrill, A. E. Revision of certain genera and species of starfishes, with descriptions of new forms. Trans. Conn. Acad. Arts and Sciences 1899, vol. 10, pp. 145–234, 8 pls.

161.　1899.　Verrill, A. E. North American Ophiuroidea. I. Revision of certain families and genera of West Indian Ophiurans. II. A faunal catalogue of the known species of

West Indian Ophiurans. Trans. Conn. Acad. Arts and Sciences 1899, vol. 10, pp. 301–386, 2 pls.

162.   1899–   Agassiz, A. Explorations of the ALBATROSS in the Pacific Ocean. Letters to U.S.
       1900   Commissioner of Fisheries. Science, Dec., 1899; Jan. and April, 1900.

163.   1900.   Agassiz, A. Explorations of the ALBATROSS in the Pacific Ocean. (Extract from a letter to Hon. George M. Bowers, U.S. Commissioner of Fish and Fisheries, dated Papeete Harbor, Tahiti Island, Sept. 30, 1899, on the trip of the ALBATROSS from San Francisco to Papeete.) Am. Journ. Sci. 1900, fourth series, vol. 9, No. 49, Jan., pp. 33–43.

164.   1900.   Moore, H. F. The ALBATROSS South Sea Expedition. Rep. U.S.F.C. 1900, part 26, pp. 137–161.

165.   1900.   Baker, Ray Stannard. The Bottom of the Sea. McClure's Mag., Dec., pp. 160–170, 8 cuts.

166.   1900.   Dall, William H. Synopsis of the Solenidae of North America and the Antilles. Proc. U.S. Nat. Mus., vol. 22, pp. 107, 112.

167.   1900.   Rathbun, Mary J. Synopsis of North American Invertebrates. VII. The cyclometopous or cancroid crabs of North America. Am. Nat., vol. 34, Feb., pp. 131–143.

168.   1900.   Nutting, Charles Cleveland. American Hydroids. Part I. The Plumularidae. U.S. Nat. Mus. Special Bulletin, 285 pp., 34 pls.

169.   1901.   Benedict, James E. The hermit crabs of the *Pagurus bernhardus* type. Proc. U.S. Nat. Mus., vol. 23, pp. 451–466.

170.   1901.   Dall, William H. Synopsis of the family Tellinidae and of the North American species. Proc. U.S. Nat. Mus., vol. 23, pp. 285–326.

171.   1901.   Dall, William H. Synopsis of the family Cardiidae and of the North American species. Proc. U.S. Nat. Mus., vol. 23, pp. 381–392.

172.   1901.   Richardson, Harriet. Key to the Isopods of the Atlantic coast of North America, with descriptions of new and little-known species. Proc. U.S. Nat. Mus., vol. 23, pp. 493–579.

173.   1901.   Jordan, David Starr, and John Otterbein Snyder. A list of fishes collected in Japan by Keinosuke Otaki and by the U.S.F.C. steamer ALBATROSS, with descriptions of 14 new species. Proc. U.S. Nat. Mus., vol. 23, pp. 335–380, 12 pls.

174.   1901.   Jordan, David Starr, and John Otterbein Snyder. A review of the lancelets, hagfishes, and lampreys of Japan, with a description of two new species. Proc. U.S. Nat. Mus., vol. 23, pp. 725–734, 1 pl.

175.   1901.   Benedict, James E. Four new symmetrical hermit crabs (Pagurids) from the West India region. Proc. U.S. Nat. Mus., vol. 23, pp. 771–776.

176.   1901.   Dall, William Healey. Synopsis of the Lucinacea and of the American species. Proc. U.S. Nat. Mus., vol. 23, pp. 779–833, 4 pls.

177.   1901.   Jordan, David Starr, and John Otterbein Snyder. A review of the apodal fishes or eels of Japan, with descriptions of 19 new species. Proc. U.S. Nat. Mus., vol. 23, pp. 837–890, 22 figs.

178.   1901.   Gill, Theodore, and John A. Ryder. Note on Eurypharynx and an allied new genus. Zool. Anz. 1884, 7, pp. 119–123.

179.   1901.   Gill, Theodore. What are the Saccopharyngoid fishes? Nature, 1884, vol. 29, Jan. 10, p. 236.

180.   1901.   Schulze, Frz. Eilhard. Amerikanische Hexactinelliden nach dem Materiale der Albatross-Expedition. Herausgegeben mit Unterstützung d. kgl. preuss. Akademie der Wissenschaften, 1899. Jena, Gust. Fischer. 4°, 126 pp. Atlas von 19 Taf.

181.   1901.   Murray, Sir John. Address to the geographical section of the British association. Scottish Geog. Mag., 1899, vol. 15, Oct., pp. 505–522, map.

DUTCH SIBOGA EXPEDITION 1899–1900*

* 1901–1961; titles as listed by E. J. Brill.

Siboga-Expeditie

UITKOMSTEN

OP

ZOOLOGISCH, BOTANISCH,

OCEANOGRAPHISCH EN

GEOLOGISCH GEBIED

VERZAMELD IN

NEDERLANDSCH OOST-INDIË 1899–1900

AAN BOORD H. M. SIBOGA ONDER COMMANDO VAN

Luitenant ter zee 1^e kl. G. F. TYDEMAN

UITGEGEVEN DOOR

Dr. MAX WEBER

Em. Prof. in Amsterdam, Leider der Expeditie

(met medewerking van de Maatschappij ter bevordering van het Natuurkundig
Onderzoek der Nederlandsche Koloniën)

N. V. BOEKHANDEL EN DRUKKERIJ
VOORHEEN
E. J. BRILL
LEIDEN

A SURVEY OF THE FASCICULES

Fascicules containing an index (or at least an exhaustive table of contents) are indicated by "+ I".

As long as in stock, some fascicules, marked *"n.s.a."* (i.e. not separately available), are reserved for subscribers to full sets (i.e. exclusive of the out of print fascicules).

Livr. 1 (mon. XLIV). C. Ph. Sluiter. Die Holothurien der Siboga-Expedition. Okt. 1901. 4to. (IV, 142 S., 10 (3 farb.) Taf., + I)

Livr. 2 (mon. IX). E. S. Barton [=E. S. Gepp-Barton]. The genus *Halimeda*. Dec. 1901. 4to. (iv, 32 p., 4 pl.)

Livr. 3 (mon. I). M. Weber. Introduction et description de l'expédition du Siboga. Janvier 1902. In-4°. (iv, 160 p., liste des stations (16 p.), 86 ill., 2 diagr., 3 croquis cartogr., 2 cartes dépl. en coul. [La route de l'exp. *et* Détails de la route], + I)

Livr. 4 (mon. II). G. F. Tydeman. Description of the ship and appliances used for scientific exploration. Avril 1902. 4to. (iv, 32 p., 11 fig., 3 pl.)

Livr. 5 (mon. XLVII). H. F. Nierstrasz. The Solenogastres of the Siboga-Expedition, [I]. June 1902. 4to. (iv, 48 p., 6 (ptly col.) pl.)

Livr. 6 (mon. XIII). Die Gorgoniden der Siboga-Expedition, I. J. Versluys. Die Chrysogorgiidae. Juli 1902. 4to (IV, 120 S., 170 Fig., + I)

Livr. 7 (mon. XVIa). [The Madreporaria of the Siboga-Expedition, I]. A. Alcock. Report on the deep-sea Madreporaria. Aug. 1902. 4to. (iv, 52 p., 5 pl., + I)

Livr. 8 (mon. XXV). C. Ph. Sluiter. [Die Gephyrea der Siboga-Expedition]. Sipunculiden und Echiuriden, nebst Zusammenstellung der überdies aus dem indischen Archipel bekannten Arten. Okt. 1902. 4to. (IV, 54 S., 3 Fig., 4 Taf., + I)

Livr. 9 (mon VIa). The Porifera of the Siboga Expedition, I. G. C. J. Vosmaer and J. H. Vernhout. The genus *Placospongia*. Nov. 1902. 4to. (iv, 18 p., 5 (2 col.) pl.)

Livr. 10 (mon. XI). O. Maas. Die Scyphomedusen der Siboga-Expedition. März 1903. 4to. (VIII, 92 S., 12 (11 farb.) Taf.)

Livr. 11 (mon. XII). F. Moser. Die Ctenophoren der Siboga-Expedition. April 1903. 4to. (IV, 34 S., 4 fab. Taf.)

Livr. 12 (mon. XXXIV). [Die Amphipoda der Siboga-Expedition, III]. P. Mayer. Die Caprellidae. Juni 1903. 4to (IV, 160 S., 10 (8 gefalt.) Taf., + I)

Aus dem Inhalt: Literatur seit 1890. *Systematik* [totale Revision der Gruppe] (Definition der Species, Nomenclatur, Genuscharaktere, Tabellarische Übersicht der Gattungen, Schlüssel zu den Gattungen). *Faunistik* [totale Revision] (Regionen; Listen: Mittelmeer, Schwarzes Meer, Atlantischer Ocean bis zu 60° N. B., Nordische Gewässer von 60° N. B. ab, Westküste von Amerika, Grosser Ocean, Indischer Ocean). Morphologie, Biologie und Phylogenie. Die Funde der Sib.-Exp. Alphabetisches Verzeichnis der Gattungen, Arten und Varietäten im systematischen Teil.

Livr. 13 (mon. III). G. F. Tydeman. Hydrographic results of the Siboga-Expedition. July 1903. 4to. (iv, 96 p., 18 fig., 24 charts and plans, 3 col. fold. charts of depths, + I)
Contents: General observations and results of surveying. Hydrographic notices. Heights (Table) of islands and mountains in the Indian Archipelago. Depths of the seas in the Indian Archipelago. Geographical positions (Table) in the Eastern part of the Indian Archipelago. Soundings (Table) by H.N.M.S.S. "Siboga" on her scientific cruise. List (Table) of deep sea soundings by H.N.M. surveying S.S. "Bali" in the Eastern part of the East Indian Archipelago. Data concerning the voyages of the S.S. "Siboga".

Livr. 14 (mon. XLIII). J. C. H. de Meijere. Die Echinoidea der Siboga-Expedition. Jan. 1904. 4to (IV, 252 S., 23 Taf.)

Livr. 15 (mon. XLVa). R. Koehler. Ophiures de l'expédition du Siboga, I. Ophiures de mer profonde. Févr. 1904. In-4°. (iv, 176 p., 36 pl., + I)

Livr. 16 (mon. LII). [The Opisthobranchia of the Siboga-Expedition, II]. J. J. Tesch. The Thecosomata and Gymnosomata. April 1904. 4to (iv, 92 p., 6 pl.)

Livr. 17 (mon. LVIa). Die Tunicaten der Siboga-Expedition, I. C. PH. SLUITER. Die socialen und holosomen Ascidien. Mai 1904. 4to. (IV, 126 S., 15 (2 farb.) Taf., + I)
Die Nummer "mon. LVIa" ist irrtumlich zweimal verwendet worden, nämlich auch zur Bezeichnung des Inhalts von *Livr.* 24.

Livr. 18 (mon. LXI). A. WEBER-VAN BOSSE and M. FOSLIE. The Corallinaceae of the Siboga-Expedition. Aug. 1904. 4to. (iv, 110 p. 34 fig., 16 pl.)

Livr. 19 (mon. VIII). S. J. HICKSON and H. M. ENGLAND. The Stylasterina of the Siboga-Expedition. Jan. 1905. 4to. (iv, 26 p., 3 pl.)

Livr. 20 (mon. XLVIII). H. F. NIERSTRASZ. Die Chitonen [Polyplacophora] der Siboga-Expedition. Jan. 1905. 4to (IV, 114 S., 8 (3 farb.) Taf.)

Livr. 21 (mon. XLVb). R. KOEHLER. Ophiures de l'expédition du Siboga, II. Ophiures littorales. Mars 1905. In-4°. (iv, 142 p., 18 pl., + I)

Livr. 22 (mon. XXVIbis). S. F. HARMER. The Pterobranchia of the Siboga-Expedition. With an account of other species. July 1905. 4to (vi, 132 p., 2 fig., 14 (ptly col.) pl.)
Regarding Pterobranchia see also *Livr.* 105.

Livr. 23 (mon. XXXVI). W. T. CALMAN. The Cumacea of the Siboga-Expedition. Aug. 1905. 4to. (iv, 24 p., 4 fig., 2 fold. pl.)

Livr. 24 (mon. LVIa). Die Tunicaten der Siboga-Expedition, *Supplement zur I. Abt.* [*Livr. 17*]. C. PH. SLUITER. Die socialen und holosomen Ascidien. Sept. 1905. 4to (IV S., S. 129–140, 1 gefalt. Taf.)
Seite 129 der Lfg. 24 folgt gleich nach Seite 126 der Lfg. 17; die Seiten 127 und 128 existieren nicht. Die Nummer "mon. LVIa" ist irrtümlich zweimal verwendet worden, nämlich auch zur Bezeichnung des Inhalts von Lfg. 17.

Livr. 25 (mon. L). [Die Opisthobranchia der Siboga-Expedition, III]. R. BERGH. DIE OPISTHOBRANCHIATA. Okt. 1905. 4to. (IV, 248 S., 20 (5 farb.) Taf., + I)

Livr. 26 (mon. X). O. MAAS. [Hydromedusae]. Die craspedoten Medusen der Siboga-Expedition. Okt. 1905. 4to. (IV, 84 S., 14 farb. Taf.)

Livr. 27 (mon. XIIIa). Die Gorgoniden der Siboga-Expedition, II. J. VERSLUYS. Die Primnoidae. Jan. 1906. 4to. (IV, 188 S., 179 Fig., 10 Taf., 1 gefalt. Karte [Weltkarte zur Erl. d. horizontalen Verbreitung der Primnoidae], + I)

Livr. 28 (mon. XXI). G. FOWLER. The Chaetognatha of the Siboga-Expedition. With a discussion of the synonymy and distribution of the group. April 1906. 4to. (viii, 86 p., 3 pl., 6 fold. maps [regarding the distribution of the Chaetognatha over the world], + I)

Livr. 29 (mon. LI). [Die Prosobranchier der Siboga-Expedition, VIII]. J. J. TESCH. Die Heteropoden. [Mit einer Ubersicht der bis jetzt beschriebenen Arten und Gattungen]. Aug. 1906. 4to. (VI, 112 S., 14 (1 farb.) Taf., + I)

Livr. 30 (mon. XXX). G. W. MULLER. Die Ostracoden der Siboga-Expedition. Sept. 1906. 4to. (VI, 40 S., 9 Taf.)

Livr. 31 (mon. IVbis). F. E. SCHULZE. Die Xenophyophoren der Siboga-Expedition. Okt. 1906. 4to (IV, 18 S., 3 farb. Taf.)

Livr. 32 (mon. LIV). M. BOISSEVAIN. The Scaphopoda of the Siboga-Expedition, treated together with the known Indo-Pacific Scaphopoda. Dec. 1906. 4to. (iv, 76 p., 39 fig., 6 pl.)

Livr. 33 (mon. XXVI). J. W. SPENGEL. Studien über die Enteropneusten der Siboga-Expedition, nebst Beobachtungen an verwandten Arten. Mai 1907. 4to (IV, 128 S., 20 Abb., 17 farb. (14 gefalt.) Taf.)

Livr. 34 (mon. XX). H. F. NIERSTRASZ. Die Nematomorpha der Siboga-Expedition. Aug. 1907. 4to. (IV. 22 S., 3 farb. Taf.)

Livr. 35 (mon. XIIIc). Die Alcyoniden der Siboga-Expedition. I. S. J. HICKSON, Coralliidae—II. J. VERSLUYS, *Pseudocladochonus hicksoni* nov. gen. nov. spec. Ein neuer Telestide der Siboga-Expedition. Aug. 1907. 4to. (IV, 8 S., 1 farb. Taf.; 32 S., 16 Fig., 2 Taf.)

Livr. 36 (mon. XXXIa). The Cirripedia of the Siboga-Expedition, A. P. P. C. HOEK. Cirripedia pedunculata. Oct. 1907. 4to (iv, 128 p., 10 pl.) [Index is contained in Livr. 67]

Livr. 37 (mon. XLIIa). Die Crinoiden der Siboga-Expedition, I. L. DÖDERLEIN. Die gestielten Crinoiden. Nov. 1907. 4to. (VI, 54 S., 7 Fig., 23 Taf.)

Livr. 38 (mon. IX). A. D. LENS and TH. VAN RIEMSDIJK. The Siphonophora of the Siboga-Expedition. May 1908. 4to. (iv, 130 p., 52 fig., 24 (1 col. fold.) pl., + I)

Livr. 39 (mon. XLIX la). The Prosobranchia of the Siboga-Expedition, I. M. M. SCHEPMAN. Rhipidoglossa and Docoglossa. With an appendix by R. BERGH. [Pectinobranchiata]. July 1908. 4to (iv, 108 p., 3 fig., 9 pl.) [Index is contained in Livr. 66]
Livr. 39 contains an Appendix to Livraisons 39 and 58.

Livr. 40 (mon. XL). J. C. C. LOMAN. Die Pantopoden der Siboga-Expedition. Mit Berücksichtigung der Arten Australiens und des tropischen Indik. Sept. 1908. 4to (VI, 90 S., 4 Fig., 1 gefalt. Tab., 15 (1 gefalt.) Taf., + I)

Livr. 41 (mon. LVIc). [Die Tunicaten der Siboga-Expedition, III]. J. E. W. IHLE. Die Appendicularien. Nebst Beiträgen zur Kenntnis der Antomie dieser Gruppe. Dez. 1908. 4to. (VI, 127 S., 10 Fig., 4 Taf.)

Livr. 42 (mon. XLIX 2). [Die Prosobranchier der Siboga-Expedition, VII]. M. M. SCHEPMAN and H. F. NIERSTRASZ. Parasitische Prosobranchier. Sept. 1909. 4to. (IV, 28 S., 2 Taf.)

Livr. 43 (mon. XLIX 1b). The Prosobranchia of the Siboga-Expedition, II. M. M. SCHEPMAN. Taenioglossa and Ptenoglossa. Sept. 1909. 4to. (iv p., p. 109–232, pl. XXVI) [Index is contained in Livr. 66]

Livr. 44 (mon. XXIXa). The Copepoda of the Siboga-Expedition, I.A. SCOTT. Freeswimming, littoral and semi-parasitic Copepoda. Sept. 1909. 4to. (vi, 324 p., 69 pl., + I)

Livr. 45 (mon. LVIb). Die Tunicaten der Siboga-Expedition, II., C. PH. SLUITER. Die merosomen Ascidien (Krikobranchia excl. Clavelinidae). Dez. 1909. 4to. (IV, 112 S., 2 Fig., 8 (6 farb.) Taf., + I)

Livr. 46 (mon. XLIX 1c). The Prosobranchia of the Siboga-Expedition, III. M. M. SCHEPMAN. Gymnoglossa. Dec. 1909. 4to. (iv p., p. 233–246, 1 pl.) [Index is contained in Livr. 66]

Livr. 47 (mon. XIIIb). The Gorgonacea of the Siboga-Expedition, III. C. C. NUTTING. The Muriceidae. Febr. 1910. 4to. (iv, 108 p., 22 pl., + I)

Livr. 48 (mon. XIII b1). The Gorgonacea of the Siboga-Expedition, IV. C. C. NUTTING. The Plexauridae. May 1910. 4to. (iv, 20 p., 4 pl.)
A revision of the Plexauridae is contained in *Livr. 124.*

Livr. 49 (mon. LVId). [Die Tunicaten der Siboga-Expedition, IV]. J. E. W. IHLE. Die Thaliaceen (einschliesslich Pyrosomen). Juli 1910. 4to. (IV, 56 S., 6 Fig., 1 Taf.)

Livr. 50 (mon. XIII b2). The Gorgonacea of the Siboga-Expedition, V. C. C. NUTTING. The Isidae. June 1910. 4to. [iv, 24 p., 6 pl.]

Livr. 51 (mon. XXXVII). H. J. HANSEN. The Schizopoda of the Siboga-Expedition. July 1910. 4to (iv, 124 p., 3 fig., 16 pl.)

Livr. 52 (mon. XIII b3). The Gorgonacea of the Siboga-Expedition, VI. C. C. NUTTING. The Gorgonellidae. Sept. 1910. 4to. (iv, 40 p., 11 pl.)

Livr. 53 (mon. XVa). The Actiniaria of the Siboga-Expedition, I. J. PLAYFAIR MCMURRICH. Ceriantharia. Oct. 1910. 4to. (iv, 48 p., 14 fig., 1 pl.)

Livr. 54 (mon. XIII b4). The Gorgonacea of the Siboga-Expedition, VII. C. C. NUTTING. The Gorgonidae. Oct. 1910. 1910. 4to. (iv, 12 p., 3 pl.)

Livr. 55 (mon. XXXIXa). The Decapoda of the Siboga-Expedition, I [b]. J. G. DE MAN. Family Penaeidae [containing a list of all the species known (Sept. 1910)]. Febr. 1911. 4to. (vi, 132 p., + I)
The number "mon. XXXIXa" has erroneously been used twice, viz. also to indicate the contents of Livr. 69. Livr. 69 contains the plates regarding the Family Penaeidae and some Errata to Livr. 55.
See also *Livr. 93.*

Livr. 56 (monogr. LXII). A. and E. S. GEPP. The Codiaceae of the Siboga-Expedition, including a monograph of Flabellarieae and Udoteae. Febr. 1911. 4to. (iv, 150 p., 22 pl., +I)

Livr. 57 (mon. XIII b5). The Gorgonacea of the Siboga-Expedition, VIII. C. C. NUTTING. The Scleraxonia. June 1911. 4to. (iv, 62 p., 12 pl.)
A revision of Scleraxonia is contained in *Livr. 130.*

Livr. 58 (mon. XLIX 1d). The Prosobranchia of the Siboga-Expedition, IV. M. M. SCHEPMAN. Rachiglossa. Sept. 1911. 4to. (iv p., p. 247–364, pl. XVIII–XXIV (1 col)) [Index is contained in Livr. 66]
An *Appendix* to *Livr. 58* is contained in *Livr. 39.*

Livr. 59 (mon. VI al). The Porifera of the Siboga-Expedition, II. G. C. J. VOSMAER. The genus *Spirastrella*. Nov. 1911. 4to. (iv, 70 p., 1 graph, 14 pl.)

Livr. 60 (mon. XXXIX al). The Decapoda of the Siboga-Expédition, II. J. G. DE MAN. Family Alpheidae. Dec. 1911. 4to. (vi p., p. 133–466, + I)
The *plates* regarding the Family Alpheidae are contained in *Livr*. 74. See also *Livr*. 93.

Livr. 61 (mon. LIIIa). Les Lamellibranches de l'expédition du Siboga, [I]. P. PELSENEER. Partie anatomique. Déc. 1911. In-4°. (vi, 126 p., 26 pl., + I)

Livr. 62 (mon. XXIV 1a). Polychaeta errantia of the Siboga-Expedition, I. R. HORST. Amphinomidae. July 1912. 4to. (iv, 44 p., 10 pl.) *n.s.a.*

Livr. 63 (mon. LIIIb). Les Lamellibranches de l'expedition du Siboga, [II]. PH. DAUTZENBERG et A. BAVAY. Partie systematique. 1 Pectinides. Nov. 1912. In-4°, (iv p., p. 127–168, pl. (en coul.) XXVII, XXVIII, + I) *n.s.a.*

Livr. 64 (mon. XLIX 1e). The Prosobranchia of the Siboga-Expedition, V. M. M. SCHEPMAN. Toxoglossa. April 1913. 4to. (iv p., p. 365–452, 1 fig., pl. XXV–XXX) [Index is contained in Livr. 66] *n.s.a.*

Livr. 65 (mon. LVII). M. WEBER. Die Fische der Siboga-Expedition. Mai 1913. 4to. (XVI, 710 S., 123 Fig. 12 (1 farb., 1 gefaltete) Taf., + I) *n.s.a.*

Livr. 66 (mon. XLIX 1f). The Prosobranchia of the Siboga-Expedition, VI, M. M. SCHEPMAN. Pulmonata, and Opisthobranchia [I]: Tectibranchiata, tribe Bullomorpha. May 1913. 4to. (iv p., p. 453–494, pl. XXXI, XXXII) [Contains a combined index to Livr. 39, 43, 46, 58, 64, and 66] *n.s.a.* Fascicule (*livraison*) 66 has erroneously been titled Prosobranchia VI; it only contains Pulmonata and Opisthobranchia I.

Livr. 67 (mon XXXIb). The Cirripedia of the Siboga-Expedition, B. P. P. C. HOEK. Cirripedia sessilia. May 1913. 4to. (iv p., p. 129–275, 2 fig., pl. XI–XXVII) [Contains a combined index to Livr. 36 and 67] *n.s.a.*

Livr. 68 (mon. LIXa). A. WEBER-VAN BOSSE. Liste des algues du Siboga [*en francais*], 1. Myxophyceae, Chlorophyceae, Phaeophyceae. Avec le concours de TH. REINBOLD [*in deutscher Sprache*]. Sept. 1913. In-4°. (iv, 186 p., 52 fig., 5 pl. (1 en coul.)) [L'index se trouve dans la livr. 108] *out of print* Voir aussi *Livr*. 108.

Livr. 69 (mon. XXXIXa). The Decapoda of the Siboga-Expedition, [Supplement to Ib]. J. G. DE MAN. *Plates* of Part 1 (Family Penaeidae). Aug. 1913. 4to (vi p., 10 pl.)
The number "mon. XXXIXa" has erroneously been used twice, viz. also to indicate the contents of *Livr*. 55 (*Livr*. 55 contains *the text* regarding the Family Penaeidae). Besides the *plates* regarding the Family Peneaidae *Livr*. 69 contains some *Errata* to the *text* of *Livr*. 55. See also *Livr*. 93.

Livr. 70 (mon. VIIa). A. BILLARD. Les Hydröides de l'expedition du Siboga, I. Plumulariidae. Oct. 1913. In-4°. (iv, 116 p., 96 fig., 6 pl., + I) *n.s.a.*

Livr. 71 (mon. XXXIXb). Die Decapoda brachyura der Siboga-Expedition, I. J. E. W. IHLE. Dromiacea. Nov. 1913. 4to. (IV, 96 S., 38 Fig., 4 Taf.) *n.s.a.*
The title "Decapoda brachyura, I" has erroneously been used twice, viz. also to indicate the contents of *Livr*. 82. However, *Livr*. 82 contains Decapoda brachyura, IV.

Livr. 72 (mon. XXXIIa). H. F. NIERSTRASZ. Die Isopoden der Siboga-Expedition, I. Isopoda chelifera. Dex. 1913. 4to. (IV, 56 z, 3 Taf.) *n.s.a.*

Livr. 73 (mon. XVII). A. J. VAN PESCH. The Antipatharia of the Siboga-Expedition. Febr. 1914. 4to. (viii, 260 p., 262 fig., 8 pl., 1 fold. list, + I) *n.s.a.*

Livr. 74 (*supplement to* mon. XXXIX al). The Decapoda of the Siboga-Expedition, Supplement to II. J. G. DE MAN. *Plates* of Part II: Family Alpheidae. Jan. 1915. 4to. (vi p., 23 (1 fold.) pl., 2 p. of addenda) *n.s.a.*
The *text* regarding the Family Alpheidae is contained in *Livr*. 60. See also *Livr*. 93.

Livr. 75 (mon. XXVIIIa). S. F. HARMER. The Polyzoa of the Siboga-Expedition, I. Entoprocta, Ctenostomata and Cyclostomata. Oct. 1915. 4to. (vi, 180 p., 12 pl., + I) *n.s.a.*
Some additions to *Livr*. 75 are contained in *Livr*. 105.

Livr. 76 (mon. XXXIX a2). The Decapoda of the Siboga-Expedition, III. J. G. DE MAN. Families Eryonidae, Palinuridae, Scyllaridae and Nephropsidae. [With lists of all the species known, 1915]. April 1916. 4to. (vi, 122 p., 4 pl., + I) *n.s.a.*
Regarding Scyllaridae and Nephropsidae: see also *Livr*. 93.

Livr. 77 (mon. XIV). S. J. HICKSON. The Pennatulacea of the Siboga-Expedition. With a general survey of the order. Sept. 1916. 4to. (x, 266 p., 45 fig., 10 (1 col. fold.) pl., 1 col. fold. map [Track of the Sib.-Exp. with Distribution of the Pennatulacea]. + I)

Livr. 78 (mon. XXXIX b1). Die Decapoda brachyura der Siboga-Expedition, II. J. E. W. Ihle. Oxystomata: Dorippidae. Oct. 1916. 4to. (IV S., S. 97–158, Fig. 39–77) *n.s.a.*

Der Titel "Decapoda brachyura, II" ist irrtümlich zweimal verwendet worden, nämlich auch zur Bezeichnung des Inhalts von *Livr.* 84. *Livr.* 84 sol aber heissen: Decapoda brachyura, V.

Livr. 79 (mon. LXV). O. B. Böggild. Meeresgrundproben der Siboga-Expedition. Dez. 1916. 4to. (IV, 50 S., 1 Taf., 1 farb., gefaltete Karte (Die Beschaffenheit des Meeresgrundes))

Livr. 80 (mon. XXIV 1b). Polychaeta errantia of the Siboga-Expedition, II. R. Horst. Aphroditidae and Chrysopetalidae. July 1917. 4to. (iv p., p. 45–144, 5 fig., pl. XI–XXIX) *n.s.a.*

Livr. 81 (mon. XLVIa). Die Asteriden der Siboga-Expedition, I. L. Döderlein. Die Gattung *Astropecten* und ihre Stammegeschicte [mit einer systematischen Übersicht der Arten und Varietaten von Astropecten]. Juli 1917. 4to. (VI, 192 S., 8 Fig., 12 Abb., 17 Taf., + I) *n.s.a.*

Der Titel "Asteriden I" ist irrtumlich zweimal verwendet worden, nämlich auch zur Bezeichnung des Inhalts von *Livr.* 91. *Livr.* 91 soll aber heissen: Asteriden IV.

Livr. 82 (mon. XXXIXc). The Decapoda brachyura of the Siboga-Expedition, IV [erroneously numbered I]. J. J. Tesch. Hymenosomidae, Retroplumidae, Ocypodidae, Grapsidae and Gecarcinidae. Febr. 1918. 4tl. (iv, 148 p., 6 pl., + I) *n.s.a.*

As to "Decapoda brachyura. I" see *Livr.* 71.

Livr. 83 (mon. XLIIb). [The Crinoids of the Siboga-Expedition, II]. A. H. Clark. The unstalked Crinoids. March 1918. 4to. (x, 300 p., 17 fig., 28 pl., + I) *n.s.a.*

The original of plate X of this *Livr.* was lost in transit from Washington to Holland. Unfortunately, Mr. Austin H. Clark having deceased meanwhile, substitution is not possible.

Livr. 84 (mon. XXXIX c1). The Decapoda brachyura of the Siboga-Expedition, V [erroneously numbered II]. J. J. Tesch. Goneplacidae and Pinnotheridae. Aug. 1918. 4to. (iv p., p. 149–296, pl. VII–XVIII, + I) *n.s.a.*

As to "Decapoda brachyura, II" see *Livr.* 78.

Livr. 85 (mon. XXXIX b2). Die Decapoda brachyura der Siboga-Expedition, III. J. E. W. Ihle. Oxystomata: Calappidae, Leucosiidae, Raninidae. Aug. 1918. 4to. (IV S., S. 159–322, Fig. 78–148) *n.s.a.*

Livr. 86 (mon. XXXVIII). [The Decapoda of the Siboga-Expedition, Ia]. H. J. Hansen. The Sergestidae. June 1919. rto. (iv, 66 p., 14 fig., 5 pl) *n.s.a.*

Livr. 87 (mon. XXXIX a3). The Decapoda of the Siboga-Expedition, IV. J. G. de Man. Families Pasiphaeidae, Stylodactylidae, Hoplophoridae, Nematocarcinidae, Thalassocaridae, Pandalidae, Psalidopodidae, Gnathophyllidae, Processidae, Glyphocrangonidae and Crangonidae. Jan. 1920. 4to. (iv, 318 p., 25 pl., + I) *n.s.a.*

Regarding Processidae and Crangonidae: see also *Livr.* 93.

Livr. 88 (mon. XLVIb). Die Asteriden der Siboga-Expedition, II. L. Döderlein. Die Gattung *Luidia* und ihre Stammesgeschichte. Oct. 1920. 4to. (VI S., S. 193–294, 5 Fig., Taf. XVIII–XX, + I) *n.s.a.*

Der Titel "Asteriden II" ist irrtümlich zweimal verwendet worden, nämlich auch zur Bezeichnung des Inhalts von *Livr.* 98. *Livr.* 98 soll aber heissen Asteriden v.

Livr. 89 (mon. LIXb). A. Weber-van Bosse. Liste des algues du Siboga, II. Rhodophyceae. *1e partie: Protoflorideae, Nemalionales, Cryptonemiales. Avril 1921. In-4°.* (vi p., p. 187–310, fig. 53–109, pl. VI–VIII) [l'Index se trouve dans la livr. 108] *n.s.a.*

Livr. 90 (mon. XVIb). The Madreporaria of the Siboga-Expedition, II. C. J. van der Horst. Madreporaria fungida. April 1921. 4to. (vi p., p. 53–98, 6 pl., + I) *n.s.a.*

Livr. 91 (mon. XLVI 1). Die Asteriden der Siboga-Expedition, IV [irrtumlicherweise genannt I]. L. Döderlein. Porcellanasteridae, Astropectinidae, Benthopectinidae. Dez. 1921. 4to. (IV, 48 S., 7 Fig., 13 Taf.) *n.s.a.*

Betreffs Asteriden I sieh *Livr.* 81.

Livr. 92 (mon. XVIc). The Madreporaria of the Siboga-Expedition, III. C. J. van der Horst. Eupsammidae. April 1922. 4to. (vi p., p. 99–128, 9 fig., pl. VII, VIII)

Livr. 93 (mon. XXXIX a4). The Decapoda of the Siboga-Expedition, V. J. G. de Man. On a collection of macrurous decapod Crustacea of the Siboga-Expedition, chiefly Penaeidae and Alpheidae [containing also some species of Processidae, Crangonidae, Scyllaridae and Nephropsidae]. Sept. 1922. 4to. (vi, 52 p., 4 pl.) *n.s.a.*

Regarding Penaeidae: see also *Livrs. 55 and 69.*

Regarding Alpheidae: see also *Livrs. 60 and 74.*

Regarding Processidae and Crangonidae: see also *Livr.* 87.
Regarding Scyllaridae and Nephropsidae: see also *Livr.* 76.

Livr. 94 (mon. LIXc). A. WEBER-VAN BOSSE. Liste des algues du Siboga, III. Rhodophyceae, 2e *partie: Ceramiales.* Janvier 1923. In-4°. (iv p., p. 311–392, fig. 110–142, pl. IX, X) [l'Index se trouve dans la livr. 108] *n.s.a.*

Livr. 95 (mon. XXXIIb). H. F. NIERSTRASZ und G. A. BRENDER À BRANDIS. Die Isopoden der Siboga-Expedition, II. Isopoda genuina. 1. Epicaridea. Febr. 1923. 4to. (VII S., S. 57–122, Taf. IV–IX, + I) *n.s.a.*

Livr. 96 (mon. XVId). The Madreporaria of the Siboga-Expedition, IV. H. BOSCHMA. *Fungia patella.* Nov. 1923. 4to. (vi p., p. 129–148, pl. IX, X)

Livr. 97 (mon. LVIII). M. WEBER. Die Cetaceen der Siboga-Expedition Vorkommen und Fang der Cetaceen im Indo-Australischen Archipel. Dez. 1923. 4to. (VI, 38 S., 5 Abb., 3 Taf.)

Livr. 98 (mon. XLVI 2). Die Asteriden der Siboga-Expedition, V [irrtümlicherweise genannt II]. L. DÖDERLEIN. Pentagonasteridae. Mai 1924. 4to (IV S., S. 49–70, Taf. XIV–XIX)
Betreffs Asteriden II sieh *Livr.* 88.

Livr. 99 (mon. XXIV 1c). Polychaeta errantia of the Siboga-Expedition, III. R. HORST. Nereidae and Hesionidae. May 1924. 4to. (iv p., p. 145–198, pl. XXX–XXXVI)

Livr. 100 (mon. LXVI). A. WICHMANN. Geologische Ergebnisse der Siboga-Expedition. März 1925. 4to. (VIII, 164 S., 33 Fig. und Abb., + I) *n.s.a.*

Livr. 101 (mon. XXXI bis). [The Cirripedia of the Siboga-Expedition, C]. P. N. VAN KAMPEN und H. BOSCHMA. Die Rhizocephalen. Juni 1925. 4to. (IV, 62 S., 45 Fig., 3 Taf.)
Die Nummer "mon. XXXIbis" ist irrtümlich zweimal verwendet worden, nämlich auch zur Bezeichnung des Inhalts von *Livr.* 116.

Livr. 102 (mon. XXXIX a5). The Decapoda of the Siboga-Expedition, VI. J. G. DE MAN. Axiidae [with a list of all the species of Axiidae known, July 1925]. Dec. 1925. 4to. (viii, 128 p., 10 pl., + I) *n.s.a.*

Livr. 103 (mon. VIIb). A. BILLARD. Les Hydroides de l'expedition du Siboga, II. Synthecidae et Sertularidae. Dec. 1925. In-4°. (iv p., p. 117–232, 58 fig., pl. VII–IX, + I)

Livr. 104 (mon. XXXV). H. J. HANSEN. The Stomatopoda of the Siboga-Expedition. July 1926. 4to. (iv, 48 p., 2 pl.)

Livr. 105 (mon. XXVIIIb). S. F. HARMER. The Polyzoa of the Siboga-Expedition, II. Cheilostomata Anasca. With additions to previous reports [viz. A few additions to Part I (Livr. 75) and to the Report on the Pterobranchia (Livr. 22)]. Sept. 1926. 4to. (viii p., p. 181–502, 23 fig., pl. XIII–XXXIV, + I)
Additions to *Livr.* 105 are contained in *Livr.* 145.

Livr. 106 (mon. VI). [The Porifera of the Siboga-Expedition, IV]. I. IJIMA. The Hexactinellida of the Siboga-Expedition. [with a list of recognizably known recent Hexactinellida, arranged systematically]. Jan. 1927. 4to. (viii, 384 p., 36 fig., 26 pl., + I)

Livr. 107 (mon. IV). J. HOFKER. The Foraminifera of the Siboga-Expedition, I. Families Tinoporidae, Rotaliidae, Nummulitidae, Amphisteginidae. Nov. 1927. 4to. (iv, 78 p., 11 fig., 38 (1 col.) pl.) *out of print.*
Regarding Rotaliidae and Amphisteginidae: see also *Livr.* 142.

Livr. 108 (mon. LIXd). A. WEBER-VAN BOSSE. Liste des algues du Siboga. IV. Rhodophyceae, 3e *partie:* Gigartinales, Rhodymeniales et tableau de la distribution des Chlorophycées, Phaeophycées et Rhodophycées de l'Archipel malaisien. April 1928. In-4°. (vi p., p. 393–534, fig. 143–213, pl. XI–XVI) [contient un index collectif pour les livraisons 68, 89, 94 et 108] *n.s.a.*
Dans quelques exemplaires de la livraison 108 il se trouve la page de titre de l'ancien volume XXXIV. Comme la matière initialement destinée à constituer le volume XXXIV a été remplacée par d'autres études, cet ancien titre n'est plus valable. Voir aussi *Livr.* 68.

Livr. 109 (mon. XXXIX a6). The Decapoda of the Siboga-Expedition, VII. J. G. DE MAN. The Thalassinidae and Callianassidae collected by the Siboga-Expedition, with some remarks on the Laomediidae. Dec. 1928. 4to. (vi, 188 p., 20 pl., + I) *n.s.a.*

Livr. 110 (mon. IVa). J. HOFKER. The Foraminifera of the Siboga-Expedition, II. Families Astrorhizidae, Rhizamminidae, Reophacidae, Anomalinidae, Peneroplidae. With an introduction on the life-cycle of the Foraminifera. Jan. 1930. 4to (iv p., p. 79–170, fig. 12–33, pl. XXXIX–LXIV) *out of print*

The number "mon. IVa" has erroneously been used twice, viz. also to indicate the contents of *Livr.* 142.

Livr. 111 (mon. VI a2). The Porifera of the Siboga-Expedition, III. M. BURTON. Calcarea. April 1930. 4to. (iv, 18 p., 8 fig.)

Livr. 112 (mon. XXXIX c2). Die Decapoda brachyura der Siboga-Expedition, VI. H. J. FLIPSE. Oxyrrhyncha: Parthenopidae. [Mit einer tabellarischen Übersicht über die Verbreitung der Genera im west-amerikanischen, atlantischen und indo-pazifischen Gebiet und einer Liste aller bekannten Parthenopidae mit Angabe von Synonymen]. Juni 1930. 4to. (IV, 96 S., 44 Fig.)

Livr. 113 (mon. XXXIIIa). Les Amphipodes de l'expédition du Siboga, I. J. M. PIRLOT. Les Amphipodes Hypérides (à l'exception des Thaumatopsidae [Cystisomatidae] et des Oxycephalidae). Sept. 1930. In-4°. (iv, 56 p., 11 fig.)

Livr. 114 (mon. XXXIIc). H. F. NIERSTRASZ. Die Isopoden der Siboga-Expedition, III. Isopoda genuina. 2. Flabellifera. März 1931. 4to. (IV S., S. 123–234, 129 Fig., Taf. X, XI) *n.s.a.*

Livr. 115 (mon. XIIId). A. J. THOMSON and L. M. I. DEAN. The Alcyonacea of the Siboga-Expedition, [III]. With an addentum to the Gorgonacea. [With the assistance of W. R. SHERRIFFS regarding Dendronephthya, D. CHALMERS regarding Siphonogorgia, and J. J. SIMPSON regarding Stereonephthya and some related genera]. Sept. 1931. 4to. (iv, 228 p., 1 fig., 28 (16 col.) pl., + I)
This fascicule includes besides Alcyonacea a few representatives of the orders Stolonifera and Telestacea.

Livr. 116 (mon. XXXIbis). [Die Cirripedien der Siboga-Expedition, D]. H. BOSCHMA. Die Rhizocephalen. *Supplement.* Dez. 1931. 4to. (IV, 68 S., 41 Fig.)
Die Nummer "mon. XXXIbis ist irrtümlich zweimal verwendet worden, nämlich auch zur Bezeichnung des Inhalts von *Livr.* 101.

Livr. 117 (mon. XXXIIIb). Les Amphipodes de l'expédition du Siboga, II. J. M. PIRLOT. Les Amphipodes Gammarides. 1. Les Amphipodes fouisseurs: Phoxocephalidae, Oedicerotidae. Fevr. 1932. In-4°. (vi p., p. 57–114, fig. 12–34) *n.s.a.*

Livr. 118 (mon. LIIIc). The Lamellibranchia of the Siboga-Expedition, [III]. B. PRASHAD. Systematic part 2. Pelecypoda (exlusive of the Pectinidae). July 1932. 4to. (iv, 354 p., 9 pl., fold. sketch-map [Track of the Siboga-Expedition], + I) *n.s.a.*

Livr. 119 (mon. XXXIX a7). Die Decapoda der Siboga-Expedition, VIII. A. J. VAN DAM. Galatheidea: Chirostylidae. [Mit einer Liste aller bekannten Chirostylidae]. Mai 1933. 4to. (IV, 46 S., 50 Fig.)

Livr. 120 (mon. XXXIIIc). Les Amphipodes de l'expédition du Siboga, II. J. M. PIRLOT. Les Amphipodes Gammarides. 2. Les Amphipodes de la mer profonde. i: Lysianassidae, Stegocephalidae, Stenothoidae, Pleustidae, Lepechinellidae. Oct. 1933. 4to. (vi p., p. 115–167, fig. 35–60, + I)

Livr. 121 (mon. XXVIIIc). S. F. HARMER. The Polyzoa of the Siboga-Expedition, III. Cheilostomata Ascophora. 1. Family Reteporidae. Febr. 1934. 4to. (viii p., p. 503–640, fig. 24–48, pl. XXXV–XLI, + I)

Livr. 122 (mon. XXXIIId). Les Amphipodes de l'expédition du Siboga, II. J. M. PIRLOT. Les Amphipodes Gammarides. 2. Les Amphipodes de la mer profonde. ii: Hyperiopsidae, Pardaliscidae, Astyridae nov. fam., Tironidae, Calliopiidae, Paramphithoidae, Amathillopsidae nov. fam., Eusiridae, Gammaridae, Aoridae, Photidae, Ampithoidae, Jassidae. Oct. 1934. In-4°. (viii p., 167–236, fig. 61–100, + I)

Livr. 123 (mon. XXIXb). The Copepoda of the Siboga-Expedition, II. W. H. LEIGH-SHARPE. Commensal and parasitic Copepoda. Dec. 1934. 4to. (viii, 44 p., 39 fig., + I) *n.s.a.*
Some copies of fascicule 123 contain a title-page for volume XIV. As the material originally meant for volume XIV has been substituted by other studies, this old title-page is no more valid.

Livr. 124 (mon. XIII b7). Die Gorgonacea der Siboga-Expedition, *Supplement I.* G. STIASNY. Revision der Plexauridae. Juni 1935. 4to. (VI, 106 S., 27 Abb., 2 Tab., 7 Taf., + I)
The main study of Plexauridae is contained in *Livr.* 48.

Livr. 125 (mon. XLVI 3). Die Asteriden der Siboga-Expedition, VI [irrtümlicherweise genannt III]. L. DÖDERLEIN. Oreasteridae. Okt. 1935. 4to. (IV S., S. 71–110, Taf. XX–XXVII)
Betreffs Asteriden III sieh *Livr.* 126.

Livr. 126 (mon. XLVIc). Die Asteriden der Siboga-Expedition, III. L. DÖDERLEIN. Die Unterfamilie Oreasterinae. März 1936. 4to. (VI S., S. 295–370, Taf. XXI–XXXII, + I)
Der Titel "Asteriden III' ist irrtümlich zewimal verwendet worden, nämlich auch zur Bezeichnung des Inhalts von *Livr.* 125. soll aber heissen: Asteriden VI.

Livr. 127 (mon. XXXIIIe). Les Amphipodes de l'expédition du Siboga, II. J. M. PIRLOT. Les Amphipodes Gammarides.
2. Les Amphipodes de la mer profonde. iii: Addendum et partie générale.
3. Les Amphipodes littoraux. i: Lysianassidae, Ampeliscidae, Leucothoidae, Stenothoidae, Phliantidae, Colomastigidae, Ochlesidae, Liljeborgiidae, Oedicerotidae, Synopiidae, Eusiridae, Gammaridae.
Nov. 1936. In-4°. (viii p., p. 237–328, fig. 101–146)

Livr. 128 (mon. XXII). G. STIASNY-WIJNHOFF. [Die Nemertini der Siboga-Expedition:] Polystilifera. Dez. 1936. 4to. (XII, 214 S., 90 Fig., 16 Taf., + I)

Livr. 129 (mon. XXVII). J. W. JACKSON and G. STIASNY. The Brachiopoda of the Siboga-Expedition. April 1937. 4to. (iv, 20 p., 2 pl.)

Livr. 130 (mon. XIII b8). Die Gorgonacea der Siboga-Expedition, *Supplement II.* G. STIANSNY. Revision der Scleraxonia mit Ausschluss der Militodidae und Coralliidae. Juni 1937. 4to. (VI, 138 S., 40 Fig., 8 Taf., + I)
The main study of Scleraxonia is contained in *Livr.* 57.

Livr. 131 (mon. XXXIX c3). The Decapoda brachyura of the Siboga-Expedition, VII. J. E. LEENE. Brachygnatha: Portunidae. [With a tabular view of the geographical distribution of the genera mentioned in this part]. June 1938. 4to. (iv, 156 p., 87 fig., +I) *n.s.a.*

Livr. 132 (mon. XXXIIIf). Les Amphipodes de l'expedition du Siboga.
II. J. M. PIRLOT. Les Amphipodes Gammarides.
3. Les Amphipodes littoraux. ii: Familles des Dexaminidae, Talitridae, Aoridae, Photidae, Ampithoidae, Corophiidae, Jassidae, Cheluridae et Podoceridae.
I. J. M. PIRLOT. Les Amphipodes Hypérides. *Addendum:* Familles des Lanceolidae. Cystisomatidae et Oxycephalidae; La sexualité chez *Cystisoma* Guérin Méneville.
Oct. 1938. In-4°. (viii p., p. 329–388, fig. 147–163), + I)

Livr. 133 (mon. XXIV 2). Polychètes sédentaires de l'expédition du Siboga, I. F. MESNIL et P. FAUVEL. Maldanidae, Cirratulidae, Capitellidae, Sabellidae et Serpulidae. Mai 1939. In-4°. (iv, 42 p., 12 fig.)

Livr. 134 (mon. LVa). W. ADAM. Les Cephalopoda de l'expédition du Siboga, I. Le genre *Sepiotheutis* Blainville 1824. Mai 1939. In-4°. (iv, 34 p., 3 fig., 3 tables dépl., 1 pl., + I)

Livr. 135 (mon. LVb). W. ADAM. Les Cephalopoda de l'expédition du Siboga, II. Révision des espèces indo-malaises du genre *Sepia* Linné 1758. — III. Révision du genre *Sepiella* (Gray) Steenstrup 1880. Nov. 1939. In-4°. (iv p., p. 35–122, 17 fig., 7 tabl. dépl., 4 pl. + I)

Livr. 136 (mon. XXXIId). H. F. NIERSTRASZ. Die Isopoden der Siboga-Expedition, IV. Isopoda genuina. 3. Gnathiidea, Anthuridea, Valvifera, Asellota, Phreatocoidea. Sept. 1941. 4to. (VI S., S. 235–308, 66 Fig.)

Livr. 137 (mon. XLVIIb). H. A. STORK. Solenogastres der Siboga-Expedition, II. Nov. 1941. 4to. (IV S., S. 49–71, 41 Fig., 2 Taf.)

Livr. 138 (mon. XXVbis). [Pogonophora de l'expédition du Siboga, I]. M. CAULLERY. *Siboglinum* Caullery 1914. Type nouveau d'invertébres, d'affinités à préciser. Avril 1944. In-4°. (iv, 26 p., 89 fig.)

Livr. 139 (mon. XXIV 2bis). Polychètes sedentaires de l'expédition du Siboga, [II]. M. CAULLERY. Ariciidae, Spionidae, Chaetopteridae, Chlorhaemidae, Opheliidae, Oweniidae, Sabellariidae, Sternaspidae, Amphictenidae, Ampharetidae, Terebellidae. Sept. 1944. In-4°. (vi, 204 p., 157 fig.)

Livr. 140 (mon. XXXIX a8). [The Decapoda of the Siboga-Expedition, IX]. L. B. HOLTHUIS. The Hippolytidae and Rhynchocinetidae collected by the Siboga and Snellius expeditions. With remarks on other species. March 1947. 4to. (iv, 100 p., 15 fig., + I)

Livr. 141 (mon. XXXIX a9). The Decapoda of the Siboga-Expedition, X. L. B. HOLTHUIS. The Palaemonidae collected by the Siboga and Snellius expeditions. With remarks on other species. 1. Subfamily Palaemoninae. July 1950. 4to. (iv, 268 p., 52 fig., +I) (A bibliography to *livr.* 141 is contained in *Livr.* 143)

Livr. 142 (mon. IVa). J. HOFKER.The Foraminifera of the Siboga-Expedition, III. Ordo Dentata, subordines Protonoraminata, Biforaminata, Deuteroforaminata. (Families Valvulinidae, Bolivinidae, Buliminellidae, Buliminidae, Uvigerinidae, Virgulinidae, Cassidulinidae, Cerabuliminidae, Eponidae, Cibicidae, Epistominidae, Robertinidae, Alabaminidae, Laticarinidae, Conorbidae, Amphisteginidae, Pulvinulinidae, Cymbaloporettidae, Valvulineridae. Rotaliidae). June 1951. 4to. (x, 514 p., 348 fig., + I)

The number "mon. IVa" has erroneously been used twice, viz. also to indicate the contents of *Livr.* 110. Regarding Amphisteginidae and Rotaliidae: see also *Livr.* 107.

Livr. 143 (mon. XXXIX a10). The Decapoda of the Siboga-Expedition, XI. L. B. HOLTHUIS. The Palaemonidae collected by the Siboga and Snellius expeditions. With remarks on other species. 2. Subfamily Pontoniinae. [With a list of all species of Pontoniinae known up to 1941 and a combined bibliography to Livr. 141 and 143]. June 1952. 4to. (iv, 254 p., 110 fig., 1 fold. table, + I)

Livr. 144 (mon. LVc). W. ADAM. Les Cephalopoda de l'expédition du Siboga, IV. Cephalopodes à l'exclusion des genres *Sepia, Sepiella* et *Sepiotheutis.* Nov. 1954. In-4°. (vi. p., p. 123–198, 40 fig., 4 tabl., 4 pl., + I)

Livr. 145 (mon. XXVIIId). S. F. HARMER. The Polyzoa of the Siboga-Expedition, IV. Cheilostomata Ascophora. 2. Ascophora, except Reteporidae. With additions to part II (Anasca) [Livr. 105]. Oct. 1957. 4to. (xvi p., p. 641–1148, fig. 49–118, 1 ptr., pl. XLII–LXXIV, + I)

Livr. 146 (mon. XXV 3). E. C. SOUTHWARD. Pogonophora of the Siboga-Expedition [II] Oct. 1961. 4to. (iv, 22 p., 12 fig.)

Livr. 147 (mon. XXIV 1a) M. H. PETTIBONE. Polychaeta erranta. pt. IV (1970)

Price of a complete set of the SIBOGA-EXPEDITION *on application.*

DANISH INGOLF EXPEDITION 1898–1899*

THE DANISH

INGOLF-EXPEDITION

PUBLISHED AT THE COST OF THE GOVERNMENT

BY

THE DIRECTION OF THE ZOOLOGICAL MUSEUM OF THE UNIVERSITY.

COPENHAGEN.

H. HAGERUP.

PRINTED BY BIANCO LUNO A. S.

1902–1946.

429

THE DANISH INGOLF-EXPEDITION

HITHERTO PUBLISHED

SWEDISH DEEP–SEA EXPEDITION 1947–1948*

* 1966 as listed in "*Reports of*" Vols. I–X.

GÖTEBORGS KUNGL. VETENSKAPS- OCH VITTERHETS-SAMHÄLLE

Suenska Djupharsexpeditionen, 1947–1948

REPORTS OF THE SWEDISH DEEP–SEA EXPEDITION 1947–1948

EDITED BY
HANS PETTERSSON
SCIENTIFIC LEADER OF THE EXPEDITION
NILS JERLOV
BÖRJE KULLENBERG

———————————

VOLUME I–X

———————————

Sponsored and distributed by the
SWEDISH NATURAL SCIENCE RESEARCH COUNCIL
STOCKHOLM 23–SWEDEN

REPORTS OF THE SWEDISH DEEP-SEA EXPEDITION VOL. I–X

CONTENTS

VOLUME I (1966)

THE SHIP, ITS EQUIPMENT, AND THE VOYAGE

VOLUME II (1966)

ZOOLOGY

VOLUME III (1966)

PHYSICS AND CHEMISTRY

VOLUME IV (1966)

BOTTOM INVESTIGATIONS

VOLUME V (1966)

SEDIMENT CORES FROM THE EAST PACIFIC

VOLUME VI (1966)

SEDIMENT CORES FROM THE WEST PACIFIC

VOLUME VII (1966)

SEDIMENT CORES FROM THE NORTH ATLANTIC OCEAN

VOLUME VIII (1966)

SEDIMENT CORES FROM THE MEDITERRANEAN SEA AND THE RED SEA

VOLUME IX (1966)

SEDIMENT CORES FROM THE INDIAN OCEAN

VOLUME X (1966)

SPECIAL INVESTIGATIONS

DANISH GALATHEA EXPEDITION 1950–1952*

* From 1956–1970 as listed by Danish Science Press.

438

GALATHEA REPORT

Scientific Results
of The Danish Deep-Sea Expedition
Round the World 1950–52

ISSUED BY THE GALATHEA COMMITTEE
PRESIDENT: H. R. H. PRINCE AXEL OF DENMARK

EDITORIAL COMMITTEE:

Anton F. Bruun, D. Sc., LL. D.
Leader of the Expedition

Sv. Greve, Captain, R. D. N.
Commander of the Galathea

R. Spärck, D. Sc.
Professor of Zoology, Vice-President of the Committee

EDITOR:

Torben Wolff, Ph. D.
Deputy Leader of the Expedition

DANISH SCIENCE PRESS, LTD.
COPENHAGEN
1961

439

VOLUME 1 (1959)

BRUUN, ANTON F.: General Introduction to the Reports and List of Deep-Sea Stations, 42 pp., 11 text-figs., 4 pls.

STEEMANN NIELSEN, E. & E. AABYE JENSEN: Primary Oceanic Production. The Autotrophic Production of Organic Matter in the Oceans. 88 pp., 41 text-figs.

BRAARUD, TRYGVE: A Red Water Organism from Walvis Bay, 2 pp., 2 text-figs.

ZOBELL, CLAUDE E. & RICHARD Y. MORITA: Deep-Sea Bacteria. 16 pp., 4 text-figs.

KIILERICH, A.: Bathymetric Features of the Philippine Trench. 17 pp., 12 text-figs., 4 pls.

KRAMP. P. L.: *Stephanoscyphus* (Scyphozoa). 15 pp., 12 text-figs., 1 pl.

MILLAR, R. H.: Ascidiacea, 21 pp., 20 text-figs., 1 pl.

DAHL, ERIK: Amphipoda from Depths Exceeding 6000 Meters. 31 pp., 20 text-figs.

PICKFORD, GRACE E.: Vampyromorpha, 11 pp., 1 text-fig.

ROFEN, ROBERT R.: The Whale-Fishes: Families Cetomimidae, Barbourisiidae and Rondeletiidae (Order Cetunculi). 6 pp., 2 pls.

VOLUME 2 (1956)

CARLGREN, OSCAR: Actiniaria from Depths Exceeding 6000 Metres. 7 pp., 7 text-figs., 1 pl.

KRAMP, P. L.: Hydroids from Depths Exceeding 6000 meters. 4 pp., 6 text-figs.

MADSEN, F. JENSENIUS: *Primnoella krampi* n. sp. A New Deep-Sea Octocoral. 2 pp., 1 text-fig.

MADSEN, F. JENSENIUS: Echinoidea, Asteroidea, and Ophiuroidea from Depths Exceeding 6000 Metres. 11 pp., 5 text-figs., 1 pl.

HANSEN, BENT: Holothurioidea from Depths Exceeding 6000 Metres. 22 pp., 25 text-figs.

HANSEN, BENT & F. JENSENIUS MADSEN: On two Bathypelagic Holothurians from the South China Sea, *Galatheathuria* n. g. *aspera* (Théel) and *Enypniastes globosa* n. sp. 5 pp., 2 text.figs., 1 pl.

GISLÉN, TORSTEN: Crinoids from Depths Exceeding 6000 Metres. 2 pp., 1 pl.

KIRKEGAARD, J. B.: Benthic Polychaeta from Depths Exceeding 6000 Metres. 16 pp., 13 text-figs.

KIRKEGAARD, J. B.: Pogonophora. *Galathealinum bruuni* n. gen. n. sp., a New Representative of the Class. 5 pp., 2 text-figs.

WOLFF, TORBEN: Isopoda from Depths Exceeding 6000 Metres. 73 pp., 56 text-figs.

FAGE, LOUIS: Les Pycnogonides du genre *Nymphon*. 7 pp., 10 text-figs.

FAGE, LOUIS: Les Pycnogonides (excl. le genre *Nymphon*). 16 pp., 22 text-figs.

KIRKEGAARD, J. B.: Pogonophora. First Records from the Eastern Pacific. 4 pp., 4 text-figs.

WOLFF, TORBEN: Crustacea Tanaidacea from Depths Exceeding 6000 Metres. 55 pp., 54 text-figs.

WIESER, WOLFGANG: Some Free-Living Marine Nematodes. 11 pp., 4 text-figs.

VOLUME 3 (1959)

LEMCHE, HENNING and KARL GEORG WINGSTRAND: The Anatomy of *Neopilina galatheae* Lemche, 1957 (Mollusca Tryblidiacea). 63 pp., 1 text-fig., 56 pls.

SCHMIDT, W. J.: Bemerkungen zur Schalenstruktur von *Neopilina galatheae*. 5 pp., 2 pls.

MUNK, O.: The Eyes of *Ipnops murrayi* Günther, 1878. 9 pp., 3 text-figs., 2 pls.

VOLUME 4 (1961)

KIRKEGAARD, J. B.: Pogonophora III. The Genus *Lamellisabella*. 4 pp., 3 text-figs.

WOLFF, TORBEN: Description of a Remarkable Deep-Sea Hermit Crab, with Notes on the Evolution of the Paguridea. 22 pp., 11 text-figs.

MADSEN, F. JENSENIUS: The Porcellanasteridae. A Monographic Revision of an Abyssal Group of Sea-Stars. 142 pp., 43 text-figs., 13 pls.

MADSEN, F. JENSENIUS: On the Zoogeography and Origin of the Abyssal Fauna, in View of the Knowledge of the Porcellanasteridae, 42 pp., 2 text-figs.

NIELSEN, JØRGEN: Heterosomata (Pisces). 8 pp., 3 text-figs., 1 pl.

VOLUME 5 (1961)

AKESSON, BERTIL: Some Observations on *Pelagosphaera* Larvae (Sipunculoidea). 11 pp., 1 text-fig., 6 pls.

COHEN, DANEL M.: On the Identity of the Species of the Fish Genus *Argentina* in the Indian Ocean. 3 pp., 1 text-fig.

BARNARD, J. LAURENS: Gammaridean Amphipoda from Depths of 400 to 6000 Metres, 106 pp., 83 text-figs.

WOLFF, TORBEN: Animal Life from a Single Abyssal Trawling. 34 pp., 26 text-figs., 4 pls.

KNUDSEN, JØRGEN: The Bathyal and Abyssal *Xylophaga* (Pholadidae, Bivalvia). 47 pp., 41 text-figs.

VOLUME 6 (1962)

WOLFF, TORBEN: The Systematics and Biology of Bathyal and Abyssal Isopoda Asellota. 320 pp., 184 text-figs., 19 pls.

VOLUME 7 (1964)

KIILERICH, A.: Hydrographical Data. 22 pp., 1 text-fig.

CASTLE, P. H. J.: Deep-Sea Eels: Family Synaphobranchidae, 14 pp., 2 text-figs.

TIXIER-DURIVAULT, A: Stolonifera et Alcyonacea. 16 pp., 30 text-figs.

MILLAR, R. H.: Ascidiacea: Additional material. 4 pp., 4 text-figs., 1 pl.

LÉVI, CLAUDE: Spongiaires des zones bathyale, abyssale et hadale. 50 pp., 63 text-figs., 10 pls.

NIELSEN, JØRGEN G.: Fishes from Depths Exceeding 6000 Metres. 12 pp., 8 text-figs.

KNUDSEN, JØRGEN: Scaphopoda and Gastropoda from Depths Exceeding 6000 Metres. 12 pp., 10 text-figs.

MUNK, OLE: The Eyes of Three Benthic Deep-Sea Fishes Caught at Great Depths. 13 pp., 1 text-fig., 3 pls.

VOLUME 8 (1966)

THEISEN, BIRGIT: On the Cranial Morphology of *Ipnops murrayi* Günther, 1878, with Special Reference to the Relations between the Eyes and the Skull. 12 pp., 3 text-figs., 2 pls.

MUNK, OLE: Ocular Degeneration in Deep-Sea Fishes. 11 pp., 9 pls.

NIELSEN, JØRGEN G.: On the Genera *Acanthonus* and *Typhlonus* (Pisces, Brotulidae). 15 pp., 10 text-figs., 3 pls.

NIELSEN, JØRGEN G.: Synopsis of the Ipnopidae (Pisces, Iniomi) with Description of Two New Abyssal Species. 27 pp., 15 text-figs., 3 pls.

ANDERSEN, KNUD P.: Classification of Ipnopidae by Means of Principal Components and Discriminant Functions. 14 pp., 16 text-figs.

McCAIN, JOHN C.: *Abyssicaprella galatheae*, a New Genus and Species of Abyssal Caprellid (Amphipoda: Caprellidae). 5 pp., 3 text-figs.

VERVOORT, W.: Bathyal and Abyssal Hydroids. 78 pp., 66 text-figs.

ZENKEVITCH, L. A.: The Systematics and Distribution of Abyssal and Hadal (Ultra-abyssal) Echiuroidea. 9 pp., 10 text-figs., 1 pl.

VOLUME 9 (1968)

LARSEN, BIRGER: Sediments from the Central Philippine Trench. 15 pp., 11 text-figs.

LANG, KARL: Deep-Sea Tanaidacea. 187 pp., 128 text-figs., 10 pls.

MUNK, OLE: On the Eye of the so-called Preorbital Light Organ of the Isospondylous Deep-Sea Fish *Bathylact. nigricans* Goode & Bean, 1896, 8 pp., 6 text-figs., 2 pls.

NIELSEN, JØRGEN G. AND VERNER LARSEN: Synopsis of the Bathylaconidae (Pisces, Isospondyli) with a New Eastern Pacific Species. 18 pp., 10 text-figs., 3 pls.

NIELSEN, JØRGEN G., ASE JESPERSEN and OLE MUNK. Spermatophores in Ophidioidea (Pisces, Percomorphi). 16 pp., 6 text-figs., 9 pls.

List of Papers Resulting in Whole or in Part from the Galathea Expedition. 2 pp.

VOLUME 10 (1969)

NIELSEN, JØRGEN G.: Systematics and Biology of the Aphyonidae (Pisces, Ophidioidea). 82 pp., 57 text-figs., 4 pls.

MILLAR, R. H.: Ascidiacea: Some Further Specimens. 8 pp., 8 text-figs.

JONES, N. S.: The Systematics and Distribution of Cumacea from Depths Exceeding 200 Metres. 82 pp., 35 text-figs.

VOLUME 11 (1970)

LIST OF PAPERS RESULTING IN WHOLE OR IN PART FROM

THE GALATHEA DEEP–SEA EXPEDITION ROUND THE WORLD 1950–52

(Supplementary to the lists given in vol. 1, 1959, pp. 18–19 and vol. 9, 1968, pp. 255–256.)

CARTON, Y., 1970: Le genre *Paranicothoe,* un nouveau representant de la famille des Nicothoidae. – J. Parasitology **56,** 4: 47–48.

CASTLE, P. H. J., 1969: The eel genera *Congrina* and *Coloconger* off southern Mozambique and their larval forms. – J. L. B. Smith Inst. Ichthyologi Spec. Publ. No. 6: 1–9, 2 text-figs.

COLLETTE, BRUCE B. & N. V. PARIN. 1970: Needlefishes (Belonidae) of the Eastern Atlantic Ocean. – Atlantide Rep. No. 11:7–60, 13 text-figs.

COOK, PATRICIA L., 1967: Polyzoa (Bryozoa) from West Africa. The Pseudostega, the Cribrimorpha and some Ascophora Imperfecta. – Bull. Br. Mus. nat. Hist. (Zool.) **15,** 7: 321–351, 14 text-figs., 2 pls.

–1968: Polyzoa from West Africa. The Malacostega. Part I. – Ibid. **16,** 3: 113–160, 20 text-figs., 3 pls.

–1968: Bryozoa (Polyzoa) from the Coasts of Tropical West Africa. – Atlantide Rep. No. 10: 115–262, 2 text-figs., 4 pls.

ERBEN, H. K., G. FLAJS & A. SIEHL, 1968: Über die Schalenstruktur von Monoplacophoren. – Abh. math.-naturw. Kl. Akad. Wiss. Mainz Nr. **1:** 1–24, 3 text-figs., 17 pls.

GOSLINE, W. A., 1969: The Morphology and Systematic Position of the Alepocephaloid Fishes. – Bull. Br. Mus. (Nat. Hist.) Zool. 18, 6: 185–218, 14 text-figs.

HANSEN, BENT, 1968: Brood-protection in a Deep-sea Holothurian, *Oneirophanta mutabilis* Théel. – Nature **217:** 1062–1063, 1 text-fig.

–1968: The Taxonomy and Zoogeography of the Deep-Sea Holothurians in Their Evolutionary Aspects. – Stud. trop. Oceanogr. Miami **5:** 480–501, 13 text-figs.

HERRING, J. L., 1961: The Genus *Halobates* (Hemiptera: Gerridae). – Pacific Insects **3** (2–3): 223–305, 117 text-figs.

JOHNSEN, P., 1970: Notes on African Acridoidea in Danish Museums (*Orthoptera*). – Nat. Jutlandica **15:** 121–162, 10 pls.

KEVAN, D. KEITH McE., 1968: Orthoptera-Caelifera from Rennell and Bellona Islands (Solomon Islands). – Nat. Hist. Rennell Isl., Br. Solomon Isls. **5:** 75–77.

KIRKEGAARD, J. B., 1954: The zoogeography of the abyssal Polychaetes. – On the Distribution of the Deep Sea Bottom Fauna. Int. Union biol. Sci., Ser. B, No. 16: 40–42.

KNUDSEN, J., 1969: Remarks on the biology of abyssal bivalves. – Proc. Third Europ. Malac. Congr., Malacologia **9,** 1: 271–272.

KRAMP, P. L., 1968: The Hydromedusae of the Pacific and Indian Oceans, Sections II and III. – Dana Rep. No. 72: 1–200, 367 text-figs.

–1968: The Scyphomedusae collected by the Galathea Expedition 1950–52. – Vidensk. Meddr dansk naturh. Foren. **131:** 67–98, 2 text-figs.

KREFFT, G., 1968: Knorpelfische (Chondrichthyes) aus dem tropischen Ostatlantik. – Atlantide Rep. No. 10: 33–76, 3 text-figs., 4 pls.

KUSTANOWICH, S., 1963: Distribution of Planktonic Foraminifera in Surface Sediments of the South-West Pacific Ocean. – N. Z. J. Geol. Geophys. **6:** 534–564, 12 text-figs., 3 pls.

MADSEN, F. JENSENIUS, 1954: Some general remarks on the distribution of the echinoderm fauna of the deep-sea. — On the Distribution of the Deep Sea Bottom Fauna. Int. Union biol. Sci., Ser. B. No. 16: 31–36.

— 1970: West African Ophiuroids. — Atlantide Rep. No. 11: 151–230, 49 text-figs.

LOUWERENS, C. J., 1970: The Carabidae (Col.) of Rennell and Bellona Islands, with a few records from Guadalcanal. — Nat. Hist. Rennell Isl., Br. Solomon Isls. 6: 87–92, 1 text-fig.

MEINEL. W., 1962: Über den Suspensionsmodus des Mandibularbogens und den splanchnischen Apparent bei *Stylephorus chordatus* Shaw 1791 (Lampridiformes, Stylephoridae). — Naturwiss. Hft. 19: 453–454, 2 text-figs.

MEINEL, W., 1968: Über die Ableitung der Suspensionsmodi des ersten und zweiten splanchnischen Bogens am Neurocranium der Fische. — Ber. oberhess. Ges. Nat.-u. Heilk., Neue Folge, Naturw. Abt. 36: 59–79, 2 text-figs.

MIYAKE, SADAYOSHI & KEIJI BABA, 1970: The Crustacea Galatheidae from the tropical-subtropical region of West Africa, with a list of the known species. — Atlantide Rep. No. 11: 61–97, 9 text-figs.

MUNK, OLE, 1966: Ocular Anatomy of Some Deep-Sea Teleosts. — Dana Rep. No. 70: 1–63, 28 text-figs., 16 pls.

PARIN, N. V., 1968: Scomberesocidae (Pisces, Synentognathi) of the Eastern Atlantic Ocean. — Atlantide Rep. No. 10: 275–290, 5 text-figs.

PETERSEN, BØRGE, 1968: New Records of Rhopalocera (Lep.) from Rennell. — Nat. Hist. Rennell Isl., Br. Solomon Isls. 5: 105–110, 2 pls.

PIERROT-BULTS, A. C., 1970: Variability in *Sagitta planctonis* Steinhaus, 1896 (Chaetognatha) from West-African waters in comparison to North Atlantic samples. — Atlantide Rep. No. 11: 141–149, 4 text-figs.

POULSEN, ERIK M., 1969: Ostracoda-Myodocopa. Part III A. Halocypriformes-Thaumatocypridae and Halocypridae. — Dana Rep. No. 75: 1–100, 40 text-figs.

RICHARDS, W. J., 1968: Eastern Atlantic Triglidae (Pisces, Scorpaeniformes). — Atlantide Rep. No. 10: 77–114, 13 text-figs., 1 pl.

SAINT-LAURENT, M., 1967: Révision des genres *Catapaguroides* et *Cestopagurus* et description de quatre genres nouveaux. I. *Catapaguroides* A. Milne-Edwards et Bouvier et *Decaphyllus* nov. gen. (Crustacés Décapodes Paguridae). — Bull. Mus. natn. Hist. nat., 2. sér., 39, 5–6: 923–954, 1100–1119, 57 text-figs.

SHIH, CHANG-TAI, 1969: The Systematics and Biology of the Family Phronimidae (Crustacea: Amphipoda). — Dana Rep. No. 74: 1–100, 22 text-figs.

SPOEL, S. VAN DER, 1970: The pelagic Mollusca from the "Atlantide" and "Galathea" Expeditions collected in the East Atlantic. — Atlantide Rep. No. 11: 99–139, 24 text-figs.

STOCK, J. H., 1968: Pycnogonida collected by the Galathea and Anton Bruun in the Indian and Pacific Oceans. — Vidensk. Meddr dansk naturh. Foren. 131: 7–65, 22 text-figs.

UTINOMI, HUZIO, 1968: Pelagic, shelf and shallow-water Cirripedia from the Indo-West-pacific. — Ibid. 131: 161–186, 8 text-figs.

WEITZMAN, STANLEY H., 1967: The Osteology and Relationships of the Astronesthidae, a Family of Oceanic Fishes. — Dana Rep. Nol 71: 1–54, 31 text-figs.

VANDEL, A., 1970: Les isopodes terrestres des iles Rennell et Bellona. — Nat. Hist. Rennell Isl., Br. Solomon Isls. 6: 139–153, 13 text-figs.

WOLFF, TORBEN, 1969: The Fauna of Rennell and Bellona, Solomon Islands. — Phil. Trans. Roy. Soc. B 255: 321–343, 14 text-figs.

— 1970: Lake Tegano on Rennell Island, the Former Lagoon of a Raised Atoll. — Nat. Hist. Rennell Isl., Br. Solmon Isls. 6: 7–29, 6 text-figs., 8 pls.

— 1970: The Concept of the Hadal or Ultra-abyssal Fauna. — Deep-Sea Res. 17. 983–1003, 4 text figs.

WOOD, E. J. FERGUSON: Diatoms in the Ocean Deeps. — Pacif. Sci. 10 377–481, 4 text-figs.

YONGE, C. M., 1957: *Neopilina*: Survival from the Paleozoic. — Discovery 17, 6: 255–256, 1 text-fig.

— 1957: Reflexions on the monoplacophoran, *Neopilina galatheae* Lemche. — Nature 179: 672–673.

ZENKEVITCH, L. A., 1966: Shallow-Water Echiuroidea from the Galathea Expedition. — Vidensk. Meddr dansk naturh. Foren. 129: 275–277, 1 text-fig.

Soviet Marine Research Institutes

MONOGRAPHS AND JOURNALS IN THE FIELD OF OCEANOGRAPHY*

I

INSTITUTIONS CARRYING ON RESEARCH ON THE MARINE FLORA AND FAUNA OF THE USSR

Department and Name of Institution	Place	Date of Foundation	Main Expedition Ships
(A) Academy of Sciences of the USSR			
1. Zoological Institute	Leningrad		
2. Botanical Institute	Leningrad		
3. Murman Marine Biological Institute	Dal'naya, Zelenetskaya Guba, Murmansk	1936	*Professor Derjugin*
4. Institute of Oceanology	Moscow	1941	*Vitiaz*
5. Black Sea Station of the Institute of Oceanology	Gelendzhik		*Academician S. Vavilov*
6. Acoustic Institute	Moscow	1951	*P. Lebedev S. Vavilov Lomonosov*
7. Institute of Marine Hydrophysics	Moscow		
8. Black Sea Hydrophysical Station of the Institute of Marine Hydrophysics	Katsiveli, Crimea	1929	
(B) Academy of Sciences of the Ukrainian SSR			
9. Sevastopol Biological Station	Sevastopol	1871–1872	Alexander Kovalevsky
10. Odessa Biological Station	Odessa	1954	

* From L. A. Zenkevitch, 1963.

Department and Name of Institution	Place	Date of Foundation	Main Expedition Ships
11. Laboratory of the Odessa Biological Station	Vilkovo, Odessa Province	1954	
12. Karadag Biological Station	Karadag, Crimea	1914	

(C)   Karelo–Finnish Branch of the Academy of Sciences of the USSR

13. White Sea Biological Station	Cape Kartesh, Chupa Guba, White Sea	1949	

(D)   University Marine Stations

14. Novorossiysk Biological Station of Rostov University	Novorossiysk	1921	
15. White Sea Biological Station of Moscow University	Velikaya Salma, Kandalaksha Gulf, White Sea	1938	
16. Peterhof Biological Institute of Leningrad University	Petrodvorets	1920	

(E)   Institutes of Fisheries

17. All-Union Institute of Fisheries and Oceanography (VNIRO)	Moscow	1933 (1921*)	
18. Pacific Ocean Institute of Fisheries and Oceanography	Vladivostok	1929 (1925)	*Zhemchug, Almaz, Isumrud, Ogon, Ozlik,* and others
19. Kamchatka Branch of the Pacific Ocean Institute of Fisheries	Petropavlovsk on Kamchatka	1932	
20. Sakhalin Branch of the Pacific Institute of Fisheries	Antonovo, Chekhov District, Sakhalin	1932	
21. Amur Branch of the Pacific Ocean Institute of Fisheries	Khabarovsk	1945	
22. Polar Institute of Fisheries and Oceanography	Murmansk	1933 (1929)	*Sevastopol, Knipovich, Academician, Berg Persey II, Professor Masyatzev, Alazan*

* Emerged in 1933 when the Central Institute of Fisheries was united with the State Oceanographic Institute I.

Department and Name of Institution	Place	Date of Foundation	Main Expedition Ships
23. Baltic Institute of Fisheries and Oceanography	Kaliningrad	1945	
24. Azov–Black Seas Institute of Fishery and Oceanography	Kerch	1921	*Grot, Donetz,* and others
25. Azov Institute of Marine Fishery	Rostov-on-Don	1955	*Professor Vasnetzov*
26. Latvian Institute of Marine Fishery	Riga	1945	
27. Latvian Laboratory of Commercial Ichthyology	Riga	1945	
28. Estonian Laboratory of Commercial Ichthyology	Tallin	1944	
29. Caspian Institute of Fisheries and Oceanography*	Astrakhan	1897	
30. Azerbaijan Institute of Fishery*	Baku	1912	
31. Georgian Scientific Experimental Laboratory	Batumi	1932	
32. Aral Institute of Fisheries and Oceanography	Aralsk	1929	
33. Scientific Research Laboratory for Seaweeds	Arkhangel	1930	
34. Kura Experimental Sturgeon Hatchery	Baku	—	
35. Institute of Lake and River Fisheries (VNIORKH)	Leningrad	1914	
36. Ob'–Tazov Branch of the Institute of Lake and River Fisheries	Tobolsk	1932	
37. Siberian Branch of the Institute of Lake and River Fisheries	Krasnoyarsk	1908	
38. Scientific Research Institute of Marine Fisheries of the Ukrainian SSR	Odessa	1932	
39. State Oceanographic Institute of the Hydrometeorological Administration* (GOI)	Moscow	1942	*Schokalsky, Veojkov*
40. All-Union Arctic and Antarctic Institute of the Ministry of the Merchant Marine	Leningrad 1959	1919	
41. Kandalaksha State Nature Reserve	Kandalaksha (White Sea)	1939	

* Up to 1917 the Astrakhan Ichthyological Laboratory.
* Up to 1917 the Baku Ichthyological Laboratory.
* Was founded in 1942 separately from the State Oceanographic Institute (GOIN), which had been reorganized in 1933 into the All-Union Institute of Fisheries and Oceanography.

Department and Name of Institution	Place	Date of Foundation	Main Expedition Ships
42. Astrakhan Nature Reserve	Astrakhan	1919	
43. "Gassan–Kuli" Nature Reserve	Krasnovodsk	1933	
44. "Kzil–Agach" Nature Reserve	Lenkoran'	1929	

II

MAJOR RUSSIAN MONOGRAPHS IN THE FIELD OF OCEANOGRAPHY

Andriashev, A. P. Essay on the Animal Geography and Origin of the Fish of the Bering Sea and Adjacent Waters. 1933.

Andriashev, A. P. The Fish of the Northern Seas of the USSR. 1954.

Arkhangelsky, A. D. and Strahov, N. M. Geological Structure and History of the Development of the Black Sea. 1958.

Berezkin, V. A. The Dynamics of the Sea. 1938.

Berg, L. S. The Aral Sea. 1908.

Berg, L. S. Fresh-water Fish of the USSR. 1948–1949.

Blinov, L. K. Hydrochemistry of the Aral Sea. 1956.

Brodsky, K. A. Copepods. 1950.

Brujevitch, S. B. Hydrochemistry of the Central and Southern Caspian. 1937.

Datzke, V. G. Organic Substances in the Waters of the South Seas of the USSR. 1959.

Derjavin, A. N. The Caspian Mysids. 1939.

Derjavin, A. N. A Survey of the History of the Caspian Fauna and of the Bodies of Fresh Water of Azerbaijan and the Caspian Aquatic Fauna, from the Symposium "Azerbaijan Animal World." 1951.

Derjugin, K. M. The Fauna of the Kola Guba and Its Environment. 1915.

Derjugin, K. M. The Fauna of the White Sea and Its Environment. 1929.

Derjugin, K. M. The Mogil'noye Relict Lake. 1926.

Djakonov, A. M. The Echinoderms of the Barents, Kara and White Seas. Proceedings of the Leningrad Society of Naturalists. 1926, **56,** (2).

Djakonov, A. M. Brittle Stars (Ophiuroidea) of the Seas of the USSR. Classification Keys to USSR Fauna, No. 55, 1954. Zoological Institute of the Academy of Sciences of the USSR.

Esipov, V. K. The Fish of the Kara Sea. 1952.

Filatova, Z. A. Zoogeographical Zonation of the Northern Seas of the USSR according to the Distribution of the Bivalves. 1957.

Gaevskaya, N. S. (Editor). Classification Keys to the Fauna and Flora of the Northern Seas of the USSR. 1937.

Grimm, O. A. The Caspian Sea and Its Fauna. (Works of the Aral–Caspian Expedition 1876–1877.)

Gurjanova, E. F., Zachs, I. G., and Uschakow, P. V. Das Litoral des Kola-Fjords. 1928–1930.

Gurjanova, E. F. Gammaridae of the Seas of the USSR and Adjacent Waters. 1951.

Gurjanova, E. F. The Gammaridae of the Northern Part of the Pacific Ocean. 1962.

Issatchenko, B. L. Research on Arctic Ocean Micro-organisms. 1914.

Ivanov, A. V. Commercial Water Invertebrates. 1955.

Ivanov, A. V. The Pogonphora. 1959, 1960.

Jashnov, V. A. Plankton Productivity of the Northern Seas of the USSR. 1940.

Jouse, A. P. Stratigraphic and Geographical Investigations in the Northwestern Part of the Pacific Ocean. 1962.

Klenova, M. V. The Geology of the Sea. 1948.

Klenova, M. V. (1960). Geology of the Barents Sea. Ac. Sci. USSR (R).

Kluge, G. A. (1962). Bryozoa of the Seas of the USSR. Ac. Sci. USSR (R).

Knipovitch, N. M. The Basis of the Hydrology of the European Arctic Ocean. 1906.

Knipovitch, N. M. The Hydrology of the Sea of Azov. 1927.

Knipovitch, N. M. Hydrological Research in the Sea of Azov. 1932.

Knipovitch, N. M. Hydrological Research in the Black Sea. 1932.

Knipovitch, N. M. The Hydrology of Seas and Brackish Waters. 1938.

Lindberg, G. U. The Quaternary Period in the Light of the Biogeographical Data. 1955.

Markovsky, J. M. Invertebrate Fauna in the Lower Stream of the Rivers in the Ukraine, its Environmental Conditions and its Utilization. 1953–1955.

Maslov, N. A. Bottom-living Fish in the Fishery Industry in the Barents Sea. Proceedings of the Polar Institute of Fisheries and Oceanography, no. 8. 1944.

Meisner, V. I. Fisheries. 1933. (Ed. "Snabtechisdat" L.)

Milashevitch, K. O. The Molluscs of the Black and Azov Seas. 1916.

Moiseev, P. A. Cod and Dab of the Far Eastern Seas. 1933.

Mordukhai-Boltovskoy, F. D. The Caspian Fauna in the Azov–Black Sea Basin. 1960.

Morosowa-Wodjanitzkaja, N. V. Phytoplankton of the Black Sea. 1940–1957.

Naumov, D. V. (1960). Hydroids and Hydromedusa in Seawater, Brackish Water and Fresh-water Basins of the USSR. Ac. Sci. USSR (R).

Nikitin, B. N. Vertical Distribution of Plankton in the Black Sea. 1926–1929 and 1938–1945.

Nikolsky, G. V. Fish of the Aral Sea. 1940.

Saidova, Kh. M. (1962). The Ecology of the Foraminifera and Paleogeography of the Far East Seas of USSR and Northwestern part of the Pacific Ocean. Ac. Sci. USSR (R).

Samoilov, N. V. River Mouths. 1952.

Schimkevitch, V. M. Pantopoda. USSR Fauna, Parts 1 and 2, 1929, 1930.

Schmidt, P. J. *Pisces marium orientateium Imperii Rossici.* 1904.

Schmidt, P. J. Fish of the Pacific Ocean. 1948.

Schmidt, P. J. The Migration of Fish. 1947.

Schmidt, P. J. Fish of the Sea of Okhotsk. 1950.

Schokalsky, J. M. Oceanography. 1917.

Schokalsky, J. M. Physical Oceanography. 1933.

Schorygin, A. A. Nutrition and Nutrient Correlations of Caspian Sea Fish. 1952.

Schuleikin, V. V. The Physics of the Sea. 1932, 1937, 1941.

Sinova, E. S. The Algae of the Murman. 1912–14.

Sinova, E. S. The Algae of the White, Black, Japan, Chukotsk Seas. 1928–54.

Snezhinsky, V. A. Practical Oceanography. 1954.

Soldatov, V. K. and Lindberg, G. U. A Survey of the Fish of Far Eastern Seas. 1930.

Soldatov, V. K. Commercial Ichthyology. Vol. I, 1934; Vol. II, 1938.

Sovinsky, V. K. An Introduction to the Study of the Fauna of Ponto–Caspian–Aral Sea Basin. Notes of the Kiev Society of Naturalists. 1904, 18.

Suvorov, E. K. The Foundations of Ichthyology. 1948.

Svetovidov, A. N. Gadiforms, Fauna of the USSR. Fishes, 1948, **9**, 4.

Svetovidov, A. N. Clupeidae, USSR Fauna. Fishes, 1952, **11**, 1.

Uschakov, P. V. (Editor). The Fauna and Flora of the Chukotsk Sea. 1952.

Uschakov, P. V. Okhotsk Sea Fauna and Its Environment. 1953.

Uschakov, P. V. Polychaete Worms of the Far Eastern Seas of the USSR. 1955.

Vinogradov, A. P. Chemical Composition of Marine Organisms. (Works of the Biochemistry and Geochemistry Laboratory of the Academy of Sciences, USSR 3—1935, 4—1936, 6—1944.)

Vinogradov, A. P. The Chemical Composition of Marine Organisms. The Sears Foundaton for Marine Research, New Haven, 1953.

Vize, V. Yu. The Seas of the Soviet Arctic. 1948.

Vorobieff, V. P. The Benthos of the Azov Sea. 1945.

Zenkevitch, L. A. Fauna and the Biological Productivity of the Sea. Vol. I, 1947; Vol. II, 1951.

Zenkevitch, L. A. The Seas of the USSR, Their Fauna and Flora. 1951 and 1955.

Zernov, S. A. Textbook on Hydrobiology. 1934 and 1949.

Zernov, S. A. The Problem of the Study of Life in the Black Sea. 1913.

Zinova, A. D. Classification Key for Brown Algae. 1953.

Zinova, A. D. Classification Key for Red Algae of the Northern Seas. 1955.

Zubov, N. N. Oceanographic Tables. 1931 and 1940.

Zubov, N. N. Sea Waters and Ice. 1938.

Zubov, N. N. Arctic Ice. 1945.

Zubov, N. N. Dynamic Oceanography. 1947.

Zubov, N. N. The Bases of the Study of the World-Ocean Straits. 1950.

Zubov, N. N. In the Centre of the Arctic. 1948.

III

THE MAIN RUSSIAN SERIALS AND PROCEEDINGS OF SCIENTIFIC INSTITUTES CONTAINING THE RESULTS OF RESEARCH DONE IN THE FIELD OF MARINE BIOLOGY

Contemporary	Year	Number of Volumes or Parts	Publications to Which the Present Series are Successors
Transactions of the Institute of Oceanology of the Academy of Sciences of the USSR	1946–1962	1–53	
Transactions of the Institute of Marine Hydrophysics of the Academy of Sciences of the USSR	1948–1958	1–24	
Transactions of the All-Union Institute of Marine Fisheries and Oceanography	1935–1962	1–44	Transactions of the Scientific Institute of Fisheries, Vols. 1–4, 1924–1930.
			Transactions of the Central Institute of Fisheries, Vols. 1–4, 1931–1932.
			Transactions of the All-Union Institute of Fisheries, Vols. 1–3, 1933–1934.

Contemporary	Year	Number of Volumes or Parts	Publications to Which the Present Series are Successors
			Transactions of the Floating Marine Scientific Institute, Vols. 1–2, 1926–1927. Transactions of the Marine Scientific Institute (Berichte des wissenschaftlichen Meeresinstituts), Vols. 3–4, 1928–1930. Transactions of the State Oceanographical Institute (GOIN), Vols. 1–3, 1932–1933.
Transactions of the Sevastopol Biological Station of the Academy of Sciences of the USSR	1936–1962	1–14	
Transactions of the Murman Marine Biological Institute of the Academy of Sciences of the USSR	1948–1962	1–5	Works of the Murman Biological Station of the Academy of Sciences, USSR, Vols. 1–3, 1925–1929. (Travaux de la Station Biologique de Murman)
Transactions of the Karadag Biological Station of the Academy of Sciences of the Ukrainian SSR (Travaux de la Station Biologique de Karadag de l'Académie des Sciences de l'URSS)	1930–1957	1–14	
Transactions of the Novorossiysk Biological Station	1937–1938	1–3	
Transactions of the Aral Branch of the All-Union Institute of Marine Fisheries and Oceanography	1933–1935	1–5	
Transactions of the Azov–Black Sea Institute of Fisheries and Oceanography	1940–1962	1–19	Transactions of Kerch Ichthyological Laboratory, Vol. 1, 1926–1927. Transactions of the Azov–Black Sea Scientific Fishery Station, Vols. 1–9, 1927–1939. Transactions of the Azov–Black Sea Scientific and

Contemporary	Year	Number of Volumes or Parts	Publications to Which the Present Series are Successors
Transactions of the Caspian Institute of Fisheries and Oceanography	1957	13–16	Commercial Expedition, Vols. 1–16, 1926–1955. (Bulletin of the Pacific Scientific Institute of Fisheries and Oceanography) Transactions of the Ichthyological Laboratory attached to the Administration of the Caspian–Volga Fish and Seal Industries, Vol. 1, 1909.
Transactions of the "N. M. Knipovitch" Polar Institute of Sea Fisheries and Oceanography	1938–1962	1–13	
Transactions of the Pacific Ocean Institute of Fisheries and Oceanography (Abhandlungen der wissenschaftlichen Fischerei-Expedition im Asowschen und Schwarzen Meer)	1930–1962	5–47	Transactions of the Pacific Ocean Scientific–Commercial Station, Vols. 1–4, 1928–1929.
Transactions of the State Oceanographical Institute	1947–1962	1–65	
Fauna of the USSR (published by the Zoological Institute of the Academy of Sciences of the USSR)	1917–1962		Fauna of Russia and Adjacent Countries, Vols. 1–26, 1911–1917.
Research on the Seas of the USSR (published by the Zoological Institute of the Academy of Sciences of the USSR)	1925–1937	1–25	
Research on the Far Eastern Seas of the USSR (published by the Zoological Institute of the Academy of Sciences of the USSR)	1927	1–7	
Key to the Classification of the Fauna of the USSR (published by the Zoological Institute of the Academy of Sciences of the USSR)	1933–1962		

Contemporary	Year	Number of Volumes or Parts	Publications to Which the Present Series are Successors
Tableaux analytiques de la Fauna de l'URSS (publiés par l'Institut Zoologique de l'Académie de Sciences de l'URSS)			
Transactions of the Zoological Institute of the Academy of Sciences of the USSR		1–28	
Transactions of the All-Union Hydrobiological Society	1949–1962	1–12	
Russian Hydrobiological Journal (published by the Volga Biological Station, Saratov)	1921–1928	1–7	
Zoological Journal (published by the Academy of Sciences of the USSR)	1916–1962	1–40	
Oceanology (published by the Academy of Sciences of the USSR).	1961	1–2	
Problems of Ichthyology (published by the Academy of Sciences of the USSR)	1961	1–2	
Transactions of the Arctic and Antarctic Institute	1959–1962	226–256	
Transactions of the Arctic Institute	1933–1959	1–225	

Author Index

Index of Genera and Species

Subject Index

Numbers in *italics* indicate pages where illustrations appear.

Abra profundorum, gross anatomy, *278*
Abyssal, benthos, vertical distribution, *232*
  biomass, 254
  bivalve species, 277
  fauna, origin and antiquity, 357, 360
  faunal province (AFP), *102,* 117, 142, 245, 248,
    251, 315, 317
    Antarctica, 315
    Arctic, 192, 197
      zonation in, 197
    asellote isopods, 251
    Milne Edwards Deep, 241
    Southeastern Pacific, 152
    tropical submergence, 251
  faunal zones, 146
  genera, eyeless, 313
    polar emergence, 313
  level-bottom community change with sediment
    change, 341
  molluscs, 278, 359
  nomenclature, 73
  Plain, off Carolinas, 237
  reproductive peak, *293*
  zone of depth, 73, 129
  zoogeographic regions, *323*
    provinces, *323*
    areas, *323*
  zoogeography of Vinogradova, *321*
Actiniaria, 45, 62, 65, 68, 180, 235
Aegean Sea, 1, 4
  station, Forbes', animals captured, 5
Agassiz, Alexander, vi, 2, 17, 18, *19,* 21, 22, 316
  Concepts, 20
  publications of, 388-393
Agassiz, Louis, 2, 15, 17
Agonidae, 290
Aguja Point, 132
Alaska orocline, Tectonic framework, *183*
Albatross, R/V Swedish, 2, 17, *18,* 22, 38
  cruise tracks, *25*
  expedition reports, 432-437
  Stations, *25*
Albatross, U. S., 2, 18, *20,* 21, 22, 229, 407

  publications resulting from, 407-416
Albert I, Prince of Monaco, 2, 22
Alcyonaria, 45, 62, 65, 68, 180
Alepocephyalids, 291
Aleutians, 172
Aleutian Trench, 230, 235
Allen, J. A., *26,* 27
Alpha Rise, 167
Aluminaut, 71
Amphipod(a) (s), 45, 62, 65, 68, 174, 180, *182,*
  235, 236
Analysis sheet, benthic biomass, 68
Anarrhicharchidae, 290
Anchor dredge, *53*
  ice, 213, 214
Ancient forms, 359
  survival of, 359
Ancient seas, regression of, 360
  sea levels, 362
Anomalies biomass, 258, 259
Antarctic(a), 246, 253, 289, 307, 314, 351
  abyss, *205*
  area, *322*
  bottom water, 124, 198, 220
  circumpolar province, 326
  continent, 198, *199*
  convergence, 198, *199, 209,* 217, 218, 219
  deep-water region, 323, 326
  divergence, 198
  eurybathial stenothermal stenographs, 305
  fossil marine fauna, 206
  glaciation, 352
  High, 205, 212
  High Arctic, 210
    Ophiuroids, 210
    Ross Sea, 210
    fauna (in the past), 206
    increase in diversity, 208
    Miocene glaciation, 208
  intermediate water, 135, 198
  isopod genera, 208, *217, 218, 219*
  Low, 205
  Jurassic, 206

471